Size-Strain V

Diffraction Analysis of the Microstructure of Materials

Proceedings of the
5th Size-Strain Conference

held
October 7-9, 2007
in
Garmisch-Partenkirchen, Germany

Editors

Eric J. Mittemeijer[1], Paolo Scardi[2],
Andreas Leineweber[1] and Udo Welzel[1]

[1] Max Planck Institute for Metals Research, Stuttgart, Germany
[2] University of Trento, Italy

Supplement No. 27 of Zeitschrift für Kristallographie
Oldenbourg Verlag

The Size-Strain V Conference has been sponsored by:

- International Centre for Diffraction Data
- International Union of Crystallography
- Anton Paar GmbH
- Bruker AXS GmbH
- PANalytical B.V.
- Max Planck Society

Z. Kristallogr. Suppl. **27** (2008) vii-viii / **DOI** 10.1524/zksu.2008.0001

PREFACE

The registration of the results of scientific work is of utmost importance, not only to produce a fair chronicle of the historic development, but above all to facilitate communication between contemporaries. The printing of scientific works and especially the electronic publication of the present day have been the cardinal instruments to continuously accelerate the speed of scientific progress (seen as an average over centuries; long periods of stagnation occur at irregular spaced moments of time in every field of science). Exchange of knowledge and ideas is even more essential in a scientific field of strongly multidisciplinary nature. Therefore, the publication of the **Proceedings of Size-Strain V**, the fifth representative of a conference series specifically devoted to the "Diffraction Analysis of the Microstructure of Materials", is imperative. In view of the above remarks, this is done as a "hard copy" but, in particular, at the same time in electronic form in "open access" format, on the world wide web, as a Supplement (No. 27, 2008) of the Zeitschrift für Kristallographie.

The behaviour of solid materials depends, in the language of today, largely on the arrangement of the constituent atoms in the solid. We have learned that many crucial properties of materials cannot be understood on the basis of the idealized ordering of the atoms as could be indicated for crystalline solids. The deviations in the atomic arrangement, away from the hypothetical, ideal ordering, i.e. the crystal imperfections, to a very large extent determine the properties of materials. This non-ideality is comprised by the notion "microstructure" in materials science. It encompasses in particular the concentrations and distributions of defects as vacancies, dislocations, stacking and twin faults, the (associated) lattice distortions due to strains/stresses, the crystallite size and its distribution, and also the compositional inhomogeneity. The diffraction-line broadening methods devoted to this type of analysis have been summarized under the heading "size-strain analysis" and this explains the nickname of the conference series.

The roots of "size-strain analysis" can be traced back to shortly after the discovery of the diffraction of X-rays by crystals. In 1918 Paul Scherrer, in Göttingen, presented what is now called the Scherrer equation for the determination of crystallite size from the line width. The next great step forward was the recognition that not only the finite size of the crystallites gives rise to diffraction-line broadening, but that in particular lattice distortions ("Verhakungen" was the name used in the original paper; see what follows), especially in engineering materials, can simultaneously induce pronounced diffraction-line broadening as well. Then the major question is how size- and strain-broadening effects can be separated in the observed line broadening. The original work that can be considered as the "birth" of "size-strain analysis" was performed by Ulrich Dehlinger und Albert Kochendörfer, in Stuttgart. They published their results as a paper entitled (translated into English) "Line broadening by deformed metals", which was published in Zeitschrift für Kristallographie, 101 (1939) 134. In the introduction of this paper the following sentence occurs (translated into English): *"It is shown in section 2 that the line breadth can be separated unambiguously in contributions due to crystallite size and strain, provided that the diffraction-angle dependence of the line breadth is known".* It is fitting that the present proceedings, almost 80 years after Dehlinger and Kochendörfer, also appear in Zeitschrift für Kristallographie, with three of the Editors of the Proceedings having Stuttgart as basis. What is more: these proceedings demonstrate that even today "size-strain separation" is controversially discussed and subject of intensive research, which is partly due to the current availability of, as compared to the "old days", huge computer power, new radiation sources, as offered by synchrotrons, and new instruments.

The "Size-Strain" conferences are held once per three years (previous conferences: Liptovsky-Mikulas (Slovakia, 1995); Freiberg (Germany, 1998); Trento (Italy, 2001); Prague (Czech Republic, 2004)). The dominant themes of the conferences have been:
- lattice defects,
- residual stresses,
- texture in thin films and at surfaces,
- line-broadening analysis and line-profile fitting, and
- diffraction/microstructure modeling.

For "Size-Strain V" in Garmisch-Partenkirchen (Germany, 2007) this has not been essentially different, but, as can be recognized by reading the current proceedings, special emphasis was paid to "nanocrystalline materials, "plastic deformation", "planar faulting" and "instrumental techniques", in view of the recent developments in the field.

There is not much literature focusing on the "Diffraction Analysis of the Microstructure of Materials" in a concentrated way; most of the relevant work is published in a highly scattered and not always easily retrievable way. Therefore we end with a brief listing of publications which do provide such in-depth overviews:

- R.L. Snyder, J. Fiala and H.J. Bunge (Eds.), *Defect and Microstructure Analysis by Diffraction*, Oxford University Press, New York, 1999.

- E.J. Mittemeijer and P. Scardi (Eds.), *Diffraction Analysis of the Microstructure of Materials*, Springer, Berlin, 2004.

- E.J. Mittemeijer, U. Welzel and R. Kuzel (Eds.), *"State of the Art of Powder Diffraction"*, Zeitschrift für Kristallographie, 222 (2007) issue 3-4 (pp. 105 – 209).

- and then, of course, the current proceedings................

Eric J. Mittemeijer
Andreas Leineweber
Udo Welzel

Stuttgart, August 2008

Paolo Scardi

Trento, August 2008

Table of Contents

LINE BROADENING ANALYSIS; DETERMINATION OF CRYSTALLITE SIZE AND MICROSTRAIN

MATERIALS MICROSTRUCTURE

STRESSES AND STRAINS

THIN FILMS

PLANAR FAULTING

Diffraction analysis of layer disorder

M. Leoni[1,*]

[1] Department of Materials Engineering and Industrial Technologies, University of Trento, 38100 via Mesiano, 77, Trento, Italy
[*] Matteo.Leoni@unitn.it

Keywords: planar faulting

Abstract. Defects in the regular sequence of atomic planes can be frequently met in crystalline materials. Not rare is the case where type, quantity and sequence of this kind of defects influence the properties and behaviour of the material. As a matter of fact, planar defects modify the distribution of intensity in reciprocal space, introducing peculiar features in the observed diffraction patterns. Throughout the years, methods have been independently proposed in several scientific fields for extracting the stacking information from the diffraction patterns. An historical review of those studies will be given, starting with simple probabilistic approaches and passing through the matrix method, simulations and Monte Carlo analysis, to end with the OD theory and with the extraction and refinement of stacking probabilities directly from a powder pattern. The most relevant finding of the past and possible perspective for the future will be shown and commented.

Full paper published in Z. Kristallogr. **223** (2008) 561-568 / **DOI** 10.1524/zkri.2008.1214 (open access).

Peculiarities of the X-ray diffraction of oxygen-deficient perovskite-related materials with partial vacancy ordering

S. V. Cherepanova[1,*], S. V. Tsybulya[1]

[1] Boreskov Institute of Catalysis, pr. Lavrentieva 5, Novosibirsk 630090, Russia
[*] svch@catalysis.ru

Keywords: oxygen-deficient perovskite, vacancy ordering, 1D disordering, X-ray diffraction, simulation

Abstract. X-ray diffraction patterns have been simulated for perovskite-Grenier phase and perovskite-brownmillerite coherent heterogeneous systems. Such systems occur as a result of partial ordering in oxygen-deficient perovskite-related materials. Calculations have been carried out on the basis of the model of 1D-disordered crystal. Simulation shows that main diffraction effect characteristic for both systems is appearance of broad peak of diffuse scattering between 110 diffraction peak of cubic perovskite and 113 diffraction peak of Grenier phase or 114 diffraction peak of brownmillerite.

1. Introduction

The oxygen-deficient perovskite compounds ABO_{3-y} are widely investigated. Vacancies can be randomly distributed or ordered. In its turn ordering can be full or partial. For the case of y between 0-0.5, full ordering leads to the formation of series of perovskite-related compounds which can be represented by a homologous series formula $A_nB_nO_{3n-1}$ [1]. Partial ordering leads to intergrowth of the mentioned structures with the perovskite-type structure. In this paper we consider intergrowth in one direction which results in the formation of 1D nanostructured materials. These materials look as alternation of coherently connected lamellar blocks of nanosized thickness. X-ray diffraction patterns have been simulated for perovskite-Grenier phase and perovskite-brownmillerite 1D nanostructures. Analysis of the influence of these phenomena on the X-ray diffraction pattern has been carried out.

2. Simulation of X-ray diffraction patterns

Software developed [2] has been used for the calculation of X-ray diffraction patterns. The simulations are performed with use of the model of a one dimensionally (1D) disordered crystal. The model represents a statistical sequence of 2D periodic layers of different types.
Two types of layers have been chosen for description of the 1D nanostructure consisting of alternating lamellar domains with perovskite and Grenier phase structures. They are

octahedral layer (*O*-type layer) for description of perovskite structure (figure1a) and layer consisting of two octahedral and one tetrahedral layer (*OTO*-type layer) for description of Grenier phase structure (figure 1b). Perovskite-Grenier phase system has been considered on the example of $LaFeO_3$-$LaCa_2FeO_8$ one.
A model of a perovskite-brownmillerite system also has been described with use of two types of layers. As in previous case the perovskite structure has been described by sequence of octahedral layers (figure 1a). The layer for description of brownmillerite structure (figure 1c) consists of four layers: two octahedral and two tetrahedral (*OTOT′*-type layer). The perovskite-brownmillerite system has been considered by the example of $LaFeO_3$-$Ca_2Fe_2O_5$ one.
X-ray diffraction patterns have been simulated for different ratios of probabilities W_O and $W_{OTO}(W_{OTOT'})$ of appearance of *O* and *OTO* (*OTOT′*) layers. This ratio determines the ratio between average thicknesses of domains with perovskite and Grenier phase $\langle L_O\rangle/\langle L_{OTO}\rangle$ (brownmillerite $\langle L_O\rangle/\langle L_{OTOT'}\rangle$) structures. Also we varied the conditional probabilities of appearance of each type layer after the same one. This parameter influences on the average thickness of lamellar domains.
Calculations have been carried out for the average crystallite size of about 400 Å. Thicknesses have been defined as 3.93 Å for octahedral layer, 11.27 Å for *OTO*-type layer and 14.77 Å for *OTOT′*-type layer. It is proposed that perovskite is cubic (a_c=3.93Å) and domains of different structures are coherently connected with each other. So for all type layers the lattice parameters *a* and *b* have been defined equal to 5.56 Å($a = b = \sqrt{2}a_c$). On all figures diffraction peaks 200 and 020 for Grenier phase and brownmillerite structures coincide with each other due to equality of *a* and *b*.

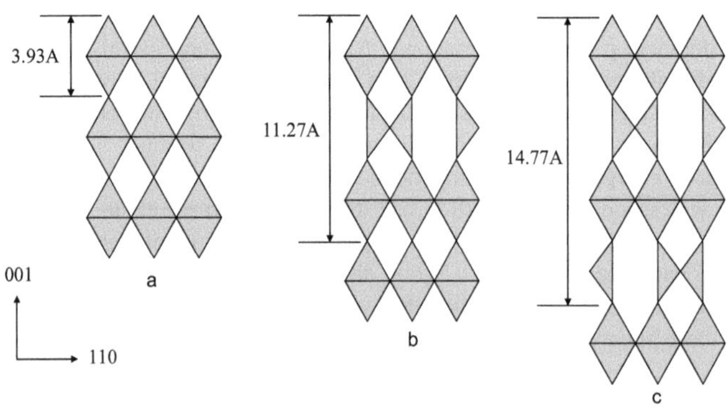

Figure 1. Vacancy ordering in fully ordered oxygen deficient ABO_{3-y} oxides: a) perovskite structure y=0; b) Grenier phase structure (y=1/3); c) brownmillerite structure (y=0.5).

3. Coherent connection of lamellar domains of perovskite and Grenier phase structure

Figure 2 shows X-ray diffraction pattern simulated for the $W_O : W_{OTO} = 0.85 : 0.15$. This ratio gives the ratio of average thicknesses of lamellar domains having perovskite and Grenier phase structures $\langle L_O \rangle / \langle L_{OTO} \rangle = 2$.

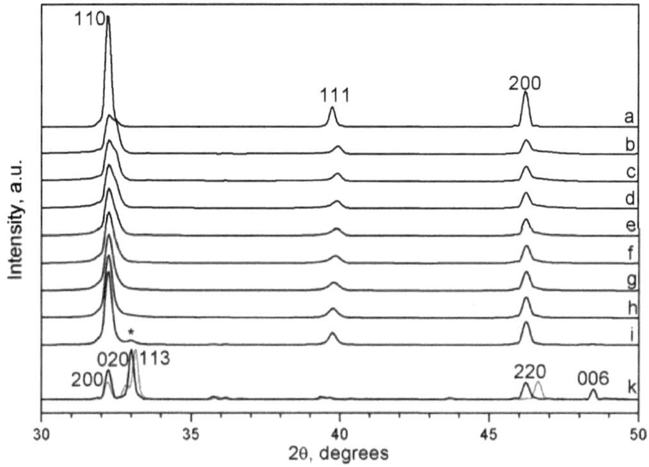

Figure 2. Simulated X-ray diffraction patterns for a) perovskite structure; perovskite-Grenier phase system, average thicknesses of domains with perovskite and Grenier phase structure in Å b) 26 and 13; c) 32 and 16; d) 38 and 19; e) 46 and 23; f) 56 and 28; g) 76 and 38; h) 112 and 56; i) 226 and 113;k) Grenier phase structure (black line for a=b, grey line for a>b).

Figure 3 shows X-ray diffraction pattern simulated for the ratio $W_O : W_{OTO} = 0.74 : 0.26$. This ratio determines average thickness of lamellar domains having perovskite structure equal to thickness of domains with Grenier phase structure, i.e. $\langle L_O \rangle / \langle L_{OTO} \rangle = 1$.

Figure 4 shows X-ray diffraction pattern simulated for the ratio $W_O : W_{OTO} = 0.59 : 0.41$. This ratio provides $\langle L_O \rangle / \langle L_{OTO} \rangle = 0.5$.

On the figures 2, 3 and 4 one can see that broad peak of diffuse scattering appears in the region between 110 diffraction peak of perovskite and 113 peak of Grenier phase.

When average thicknesses $\langle L_{OTO} \rangle$ of domains having Grenier phase structure are relatively large position of this peak is very close to the position of 113 diffraction peak of Grenier phase (figures 2i, 3h, 4g; the peak is marked by *).

Decrease in average thickness of domains leads to shifting this peak to smaller angles and to its broadening (figure 4f).

Further decrease in average thickness of domains leads to further shifting. But starting from some value of average thickness its position is almost not changed. Position of this peak depends on the ratio of thicknesses of lamellar domains with perovskite and Grenier phase structures. The larger $\langle L_O \rangle / \langle L_{OTO} \rangle$ ratio the closer position of this peak to the 110 diffraction peak of perovskite. After achieving almost unchangeable position the peak width begins to reduce. The narrower domains the narrower peak of diffuse scattering.

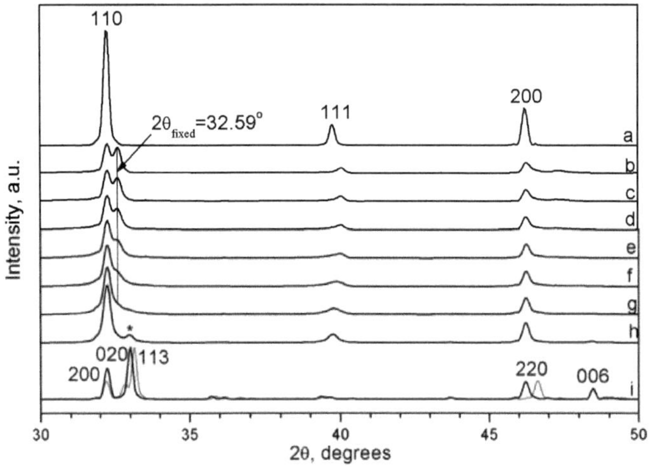

Figure 3. Simulated X-ray diffraction patterns for a) perovskite structure; perovskite-Grenier phase system, average thicknesses of domains with perovskite and Grenier phase structure in Å b) 16; c) 19; d) 23; e) 28; f) 38; g) 56; h) 113; i) Grenier phase structure (black line for a=b, grey line for a>b).

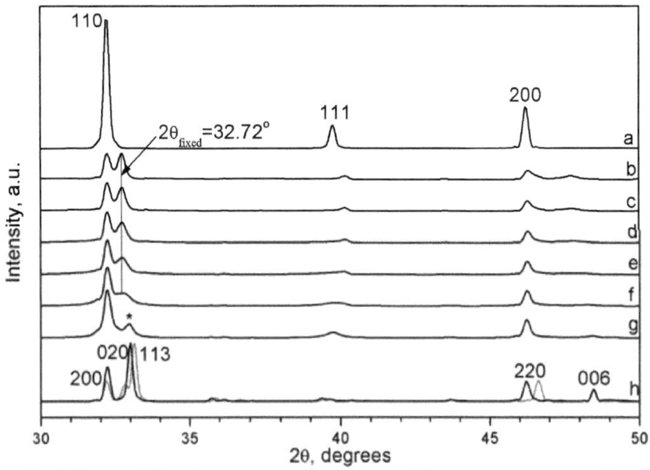

Figure 4. Simulated X-ray diffraction patterns for a) perovskite structure; perovskite-Grenier phase system, average thicknesses of domains with perovskite and Grenier phase structure in Å b) 10 and 19; c) 12 and 23; d) 14 and 28; e) 19 and 38; f) 28 and 56; g) 57 and 113; h) Grenier phase structure (black line for a=b, grey line for a>b).

4. Coherent connection of lamellar domains of perovskite and brownmillerite structure

Figure 5 shows an X-ray diffraction pattern simulated for the ratio $W_O : W_{OTOT'} = 0.88 : 0.12$. This ratio gives the ratio of average thicknesses of lamellar domains having perovskite and brownmillerite structures $\langle L_O \rangle / \langle L_{OTOT'} \rangle = 2$.

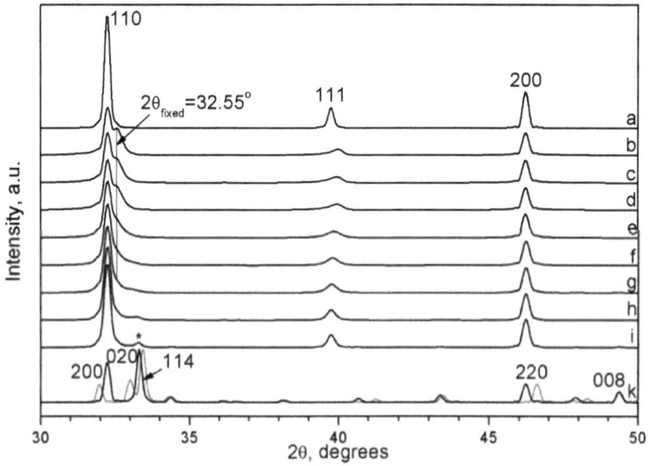

Figure 5. Simulated X-ray diffraction patterns for a) perovskite structure; perovskite-brownmillerite phase system, average thicknesses of domains with perovskite and brownmillerite structure in Å b) 36 and 18; c) 42 and 21; d) 50 and 25; e) 60 and 30; f) 74 and 37; g) 98 and 49; h) 248 and 74; i) 296 and 148; k) brownmillerite structure (black line for a=b, grey line for a>b).

Figure 6 shows X-ray diffraction pattern simulated for the ratio $W_O : W_{OTOT'} = 0.79 : 0.21$. This ratio determines average thickness of lamellar domains having perovskite structure equal to thickness of domains with brownmillerite structure, i.e. $\langle L_O \rangle / \langle L_{OTOT'} \rangle = 1$.

Figure 7 shows X-ray diffraction pattern simulated for the ratio $W_O : W_{OTOT'} = 0.65 : 0.35$. This ratio provides $\langle L_O \rangle / \langle L_{OTOT'} \rangle = 0.5$.

As one can see on the figures 5, 6 and 7 that the X-ray diffraction effects observed for the perovskite-brownmillerite system are similar to effects observed for the perovskite-Grenier phase system. The only difference is position of the peak of diffuse scattering. One can see comparing figures 2 and 5, 3 ($2\theta_{fixed}$=32.59°) and 6 ($2\theta_{fixed}$=32.72°), 4 ($2\theta_{fixed}$=32.72°) and 7 ($2\theta_{fixed}$=32.95°) that the peak for the perovskite-brownmillerite system is situated at the higher angle than peak for perovskite-Grenier phase system. It occurs due to difference in positions of the 113 diffraction peak of Grenier phase structure and 114 diffraction peak of brownmillerite structure (for compositionally-related structures). Grenier phase unit cell is represented by sequence of two octahedral layers and one tetrahedral layer (*OTO*) and brownmillerite unit cell consists of two octahedral and two tetrahedral layers (*OTOT'*) along *c*. Tetrahedral layers are thinner than octahedral so 1/3 of *OTO* thickness higher than 1/4 of *OTOT'* one.

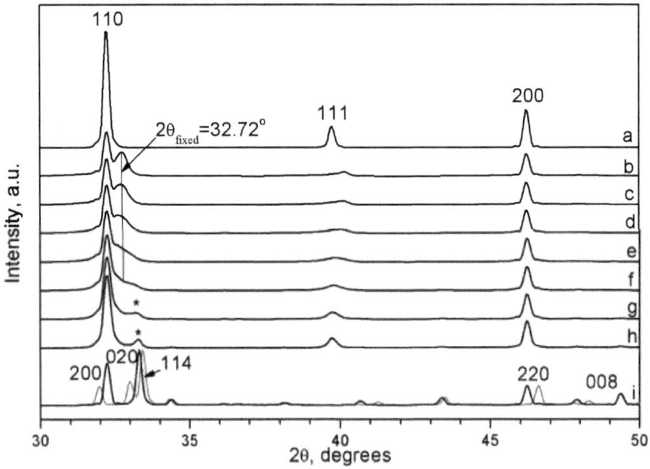

Figure 6. Simulated X-ray diffraction patterns for a) perovskite structure; perovskite-brownmillerite system; average thicknesses of domains with perovskite and brownmillerite structure in Å b) 21; c) 25; d) 30; e) 37; f) 49; g) 74; h) 148; i) brownmillerite structure (black line for a=b, grey line for a>b).

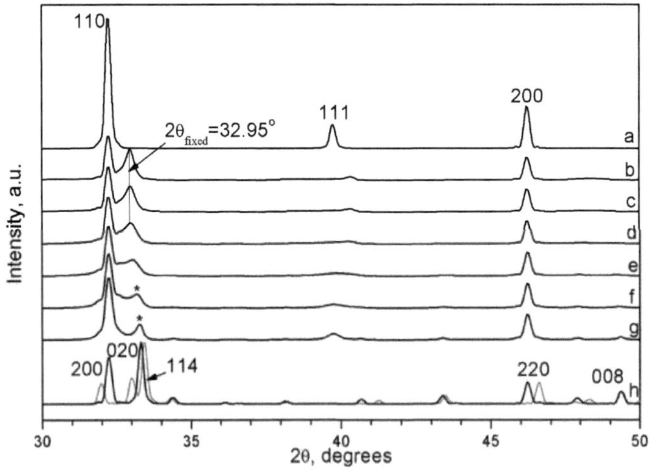

Figure 7. Simulated X-ray diffraction patterns for a) perovskite structure; perovskite-brownmillerite system, average thicknesses of domains with perovskite and brownmillerite structure in Å b) 13 and 25; c) 15 and 30; d) 19 and 37; e) 25 and 49; f) 37 and 74; g) 74 and 148; h) brownmillerite structure (black line for a=b, grey line for a>b).

To understand how average size of crystallites influence on the peak of diffuse scattering we have also simulated diffraction patterns for the average particle size of about 600 Å (for

$\langle L_O \rangle / \langle L_{OTOT'} \rangle = 0.5$) and compared them with diffraction patterns simulated above for the average particle size of about 400 Å with the same thickness ratio (figure 8). One can see that width of the peak of diffuse scattering is slightly depends on particle size. The width is less changes when peak is more broadened. Width of 110 and 200 peaks is fully depends on average particle size.

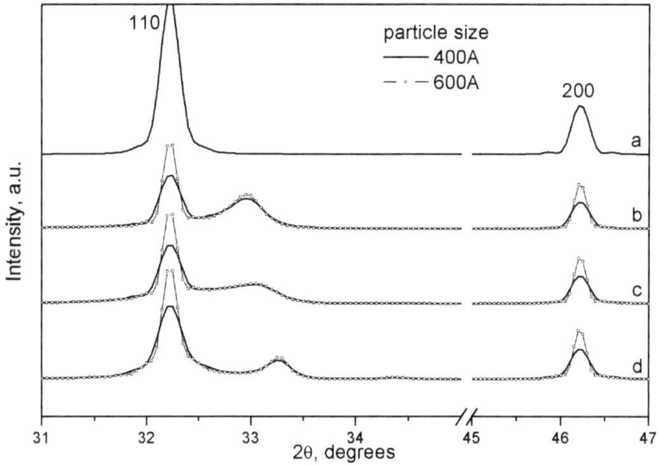

Figure 8. Simulated X-ray diffraction patterns for a) perovskite structure; perovskite-brownmillerite system, average thicknesses of domains with perovskite and brownmillerite structure in Å b) 15 and 30; c) 25 and 49; d) 74 and 148.

5. Some experimental facts

When peak of diffuse scattering is broad enough and its position is close to 110 diffraction peak of perovskite it is not visible separately and 110 diffraction peak looks asymmetric (figures 2d, 2e, 2f, 3g, 5e, 5f, 6f). Such effect was observed on the X-ray diffraction patterns of $La_{1-x}Ca_xFeO_{3-y}$ decomposed solid solution at x=0.4 and 0.6. HREM has confirmed that particles consist of lamellar domains with $LaFeO_3$ and $LaCa_2Fe_3O_8$ structures [3].
Sometimes asymmetry of the 110 peak is so slight that it is not visible (figures 2g, 2h, 5g). Thus the simulations show that at some values of microdomain thicknesses the X-ray diffraction patterns could be indexed to a cubic perovskite structure. Such contradiction between apparent (cubic perovskite) structure observed by XRD and real (microdomain) structure observed by HREM was revealed [4, 5]. The HREM study of several $Ca_xLa_{1-x}FeO_{3-y}$ samples showed the existence of a three-dimensional (3D) microdomain texture in which brownmillerite-type microdomains with a volume of $\sim 10^3$ nm were intergrowth within a Ca_2LaFeO_8 substructure[4]. This situation was not detected by conventional powder X-ray methods, where a simple cubic perovskite unit cell was observed. The structure of $CaTi_{0.4}Fe_{0.6}O_{3-\delta}$ was also investigated [5]. The powder XRD pattern was dominated by a simple cubic perovskite unit cell. TEM analysis revealed a 3D-microdomain structure consisting of disordered intergrowths of $CaTiO_3$- and $Ca_3TiFe_2O_8$-like phases.

6. Summary

Simulations show that diffraction effects are similar for perovskite-Grenier phase and perovskite-brownmillerite microdomain structures. The X-ray diffraction patterns are characterised by the presence of the peak of diffuse scattering in the region between the 110 diffraction peak of perovskite and 113 peak of Grenier phase or 114 peak of brownmillerite. Position and width of this peak is determined by average thickness of domains of Grenier phase structure $\langle L_{OTO} \rangle$ or brownmillerite one $\langle L_{OTOT'} \rangle$ and by the ratio of average thickness of domains of perovskite structure $\langle L_O \rangle$ to $\langle L_{OTO} \rangle$ or $\langle L_{OTOT'} \rangle$.

When average thicknesses of domains of Grenier phase or brownmillerite structure are more than 100Å the peak position is close to 113 diffraction peak of Grenier phase structure or 114 one of brownmillerite structure. Decrease in average thickness $\langle L_{OTO} \rangle$ or $\langle L_{OTOT'} \rangle$ leads to broadening and shifting the peak to smaller angles. At $\langle L_{OTO} \rangle < 50$ Å or $\langle L_{OTOT'} \rangle < 50$ Å the peak achieves an almost fixed position and after that it begins to become sharp. The peak position depends on $\langle L_O \rangle / \langle L_{OTO} \rangle$ or $\langle L_O \rangle / \langle L_{OTOT'} \rangle$ ratio. The larger the ratio, the closer the peak is to the 110 diffraction peak of cubic perovskite.

At some values of microdomain thicknesses the peak of diffuse scattering is so broad and close to 110 peak of perovskite that it is not visible separately and 110 peak looks asymmetric. When asymmetry is slight the X-ray diffraction patterns don't contain any features indicating the microdomain structure. In this case the X-ray diffraction patterns can be mistakenly ascribed to a cubic perovskite structure.

References

1. Grenier, J.-C., Pouchard, M. & Hagenmuller, P., 1981, in *Structure and Bonding. Ferrites, Transition Elements, Luminescence* (Berlin: Springer), pp. 2-25.
2. Cherepanova, S.V. & Tsybulya, S.V., 2004, *EPDIC8, Mater. Sci. Forum*, **443-4**, p. 87.
3. Isupova, L.A., Tsybulya, S.V., Kryukova, G.N., Rogov, V.A., Yakovleva, I.S. & Sadykov, V.A., 2003, in *Mixed Ionic Electronic System for advanced energy system*, edited by Orlovskaya & Browning (Boston: Kluwer Academic Publishers), pp. 137-156.
4. Alario-Franco, M.A., Gonzalez-Calbet, J.M., Vallet-Regi, M. & Grenier J.C., 1983, *J. Solid State Chem.*, **49**, 219.
5. Canalez-Vazquez, J., Figueiredo, F.M., Waerenborgh, J.C., Zhou, W., Frade, J.R. & Irvine, J.T.S., 2004, *J. Solid State Chem.*, **177**, 3105.

Acknowledgements. This work was supported by the RFBR under grant No 08-03-00964-a.

NANOCRYSTALLINE MATERIALS

Interference phenomena in nanocrystalline materials and their application in the microstructure analysis

D. Rafaja[1,*], V. Klemm[1], Ch. Wüstefeld[1], M. Motylenko[1], M. Dopita[1,2], M. Schwarz[3], T. Barsukova[3], E. Kroke[3]

[1] Institute of Materials Science, TU Bergakademie Freiberg, D-09599 Freiberg, Germany
[2] Department of Condensed Matter Physics, Charles University Prague, CZ-121 16 Prague, Czech Republic
[3] Institute of Inorganic Chemistry, TU Bergakademie Freiberg, D-09599 Freiberg, Germany
[*] Rafaja@ww.tu-freiberg.de

Keywords: nanocrystalline materials, partial coherence of crystallites, XRD, TEM

Abstract. Microstructural analytical methods with a high spatial resolution are needed for microstructure studies on nanocrystalline materials and nanocomposites in order to be able to explain the relationship between the microstructure and properties on the nanoscale. Theoretical considerations and experimental results presented in this contribution illustrate the capabilities of the X-ray diffraction that takes into account the phenomenon of the partial coherence of nanocrystallites for the microstructure analysis of nanostructured materials. The results obtained from X-ray diffraction were verified and complemented by the transmission electron microscopy with high resolution. Particularly, the use of the transmission electron microscopy in order to check the reliability of this X-ray diffraction approach is discussed. As experimental examples, nanocrystalline (Cr, Al) N coatings and boron nitride nanocomposites are presented in this contribution.

1. Introduction

Detailed analysis of the X-ray diffraction (XRD) line broadening is a well-established approach in the microstructure analysis of solids that is still being developed by a number of research groups. An overview of the development in this field before 2003 can be found in [1,2]. The main goal of the XRD line profile analysis is to identify and to quantify different microstructural features from the width and shape of the XRD lines, basically the crystallite size and the kind and density of lattice defects. As lattice defects, dislocations and stacking faults are usually considered, see, e.g., [3-29]. Basically, all well-established methods are based on the kinematical diffraction theory, which assumes that the irradiated volume of a sample under study consists of individual coherently diffracting domains (crystallites), which are non-coherent to each other. With regard to the XRD line profile analysis, the results of the kinematical diffraction theory are a constant XRD line broadening (independent of the size of the diffraction vector) that results from a small crystallite size, the XRD line broadening that increases with the size of the diffraction vector, which results from the presence of one-dimensional lattice defects (e.g. dislocations) and the XRD line broadening that varies with the diffraction indices, which results from the presence of the planar lattice defects.

In nanocrystalline materials, a departure of the XRD line broadening from its expected dependence on the diffraction vector is frequently observed, see, e.g., references [30-39]. Van Berkum et al [30] explained this departure by a model of infinitely narrow strain fields assuming that the whole irradiated volume of a (polycrystalline) specimen can be considered as a single coherently diffracting unit. In [31] and [35-39], it was shown that this phenomenon is related to the partial coherence of neighbouring nanocrystallites and that it is observed frequently in samples with a strong (local) texture. The typical effects of the partial coherence of nanocrystallites are a gradual increase of the XRD line broadening with increasing size of the diffraction vector at the smallest diffraction vectors, an abrupt increase of the XRD line in broadening at medium diffraction vectors and a constant line broadening at largest diffraction vectors. It will be explained below, how the limits of the above regions depend on the crystallite size, on the disorientation of neighbouring crystallites and on the kind and ordering of microstructure defects. The experimental examples given here illustrate the effect of the partial coherence of adjacent nanocrystallites on the XRD line broadening in single-phase nanocrystalline (Cr, Al) N coatings and in multi-phase BN nanocomposites. Results of XRD are verified and complemented by transmission electron microscopy (TEM) and transmission electron microscopy with high resolution (HRTEM).

2. Mathematical description of the partial coherence of nanocrystallites

The intensity of X-rays scattered on a material with the electron density $\rho(\mathbf{r})$ is given by

$$I(\mathbf{q}) \propto E^* E = \int_V \rho(\mathbf{r}) e^{-i\mathbf{q}\cdot\mathbf{r}} d\mathbf{r}^3 \cdot \int_V \rho(\mathbf{r}) e^{i\mathbf{q}\cdot\mathbf{r}} d\mathbf{r}^3 , \tag{1}$$

where $\mathbf{q}$ is the diffraction vector and $\mathbf{r}$ the position vector. The integrals in equation (1) that have the meaning of the scattered amplitudes are calculated over the irradiated volume of the sample. If the sample consists of individual crystallites, which can be represented by defect-poor domains in nanocrystalline materials, the intensity can be expressed in terms of the modulus of amplitudes that is averaged over all crystallites within the irradiated volume of the sample:

$$I(\mathbf{q}) \propto \left\langle \sum_{\text{Crystallite}} f_n e^{-i\mathbf{q}\cdot\mathbf{r}_n} \cdot \sum_{\text{Crystallite}} f_n e^{i\mathbf{q}\cdot\mathbf{r}_n} \right\rangle = \left\langle \sum_{n=-\infty}^{\infty} \Omega(\mathbf{r}_n) f_n e^{-i\mathbf{q}\cdot\mathbf{r}_n} \cdot \sum_{n=-\infty}^{\infty} \Omega(\mathbf{r}_n) f_n e^{i\mathbf{q}\cdot\mathbf{r}_n} \right\rangle \tag{2}$$

In equation (2), f_n means the atomic scattering factor of the atom n; $\Omega(\mathbf{r}_n)$ is the Heaviside function that describes the crystallite shape. Within the crystallites, $\Omega(\mathbf{r}_n)$ is equal to unity, outside to zero. If the diffraction and position vectors are written as linear combinations of the reciprocal and direct lattice vectors, **a***, **b***, **c*** and **a**, **b**, **c**, respectively:

$$\mathbf{q} = 2\pi \left(u\mathbf{a}^* + v\mathbf{b}^* + w\mathbf{c}^* \right) \quad \text{and} \quad \mathbf{r}_n = x_n\mathbf{a} + y_n\mathbf{b} + z_n\mathbf{c} + n_1\mathbf{a} + n_2\mathbf{b} + n_3\mathbf{c} , \tag{3}$$

where u, v and w are real numbers defining the position of the diffraction vector in the reciprocal space, x_n, y_n and z_n the atomic fractional coordinates and n_1, n_2 and n_3 integers defining the positions of the elementary cell in the crystallite, the summation over all atoms within individual crystallites can be split into the summation over the 3D periodic elementary cells (n_1, n_2, n_3) and into the summation over the first elementary cell (j):

$$\sum_{n=-\infty}^{\infty}\Omega(\mathbf{r}_n)f_n e^{i\mathbf{q}\cdot\mathbf{r}_n} = \sum_{n_1=-\infty}^{\infty}\sum_{n_2=-\infty}^{\infty}\sum_{n_3=-\infty}^{\infty}\Omega(n_1,n_2,n_3)e^{2\pi i(un_1+vn_2+wn_3)}\cdot\sum_{j=1}^{N}f_j e^{2\pi i(ux_n+vy_n+wz_n)} \quad (4)$$

The first part of the right-hand side of equation (4) has the meaning of the Fourier transformation of the shape function Ω; the second part can be understood as a generalised structure factor. At the reciprocal lattice points, i.e. for $h = u$, $k = v$, $l = w$, the generalised structure factor is equal to the standard structure factor:

$$F(hk\ell) = \sum_{j=1}^{N} f_j e^{2\pi i(hx_n+ky_n+\ell z_n)} \quad (5)$$

Thus, the intensity diffracted can be calculated from:

$$I \propto \sum_{\text{Crystallites}}[FT(\Omega)\cdot F(hk\ell)]^{*}\cdot\sum_{\text{Crystallites}}FT(\Omega)\cdot F(hk\ell), \quad (6)$$

where the summation is performed over all crystallites within the irradiated area of the sample. The asterisk in equation (6) means the complex conjugate. Taking into account the neighbouring crystallites of the same order (with the same or similar distance), equation (6) can be rewritten into the following form, which is more convenient for explanation of the partial coherence of neighbouring crystallites:

$$I \propto \sum_{n=1}^{N}[F_{hk\ell}\cdot FT(\Omega)]_n^{*}\cdot[F_{hk\ell}\cdot FT(\Omega)]_n + 2\sum_{m=1}^{N-1}\sum_{n=1}^{N-m}[F_{hk\ell}\cdot FT(\Omega)]_n^{*}\cdot[F_{hk\ell}\cdot FT(\Omega)]_{n+m}, \quad (7)$$

where N is the number of crystallites in the irradiated volume of the sample. The first sum in equation (7) describes the scattering of X-rays on individual crystallites, the second sum the interference of X-rays scattered on adjacent (non-identical) crystallites. As the kinematical diffraction theory neglects the coherence of different crystallites, the second term is usually omitted. However, this is only true if the reciprocal lattice points do not overlap each other. For overlapping reciprocal lattice points, the second term in equation (7) that describes the interference phenomena between different crystallites is larger than zero. Thus, the crystallites with overlapping reciprocal lattice points can be mutually coherent.

The requirements of the partial coherence of crystallites are defined both in the reciprocal and direct space. As it follows from equation (7), the reciprocal lattice points from partially coherent crystallites must partially overlap each other. This requirement is fulfilled if the reciprocal lattice points are large and their centres close to each other. The presence of considerably broadened reciprocal lattice points is a typical feature in nanocrystalline materials, as the size of the reciprocal lattice points is inversely proportional to the crystallite size. Small distances between the centres of the reciprocal lattice points can be reached by a small disorientation of crystallites with the same crystal structure. For crystallites with different crystal structures, a strong orientation correlation and small differences in the related interplanar spacings of both crystal structures are required. In the direct space, the distances between crystallites must be smaller than the coherence length of the X-ray radiation in order that the X-rays scattered by different crystallites can interfere and thus the crystallites can be seen as partially coherent [40]. This requirement can be fulfilled either through a good spectral quality and a negligible divergence of the primary beam [41] or through small distances between possibly partially coherent crystallites.

The product of the structure factor of individual crystallites and the Fourier transformation of their shape function in equation (7) can formally be replaced by:

$$[F_{hk\ell} \cdot FT(\Omega)]_n = |F_{hk\ell,n} \cdot FT(\Omega_n)| \cdot e^{i\mathbf{q}\cdot\mathbf{R}_n}, \tag{8}$$

where $\mathbf{R}_n$ are the position vectors of the crystallites in the irradiated volume of the sample. Using equation (8) and assuming crystallites with the same crystal structure and with approximately the same size, the scattered intensity from equation (7) can be rewritten into the following form (see also [31]):

$$I \propto \left\langle |F_{hk\ell,n} \cdot FT(\Omega_n)|^2 \right\rangle \cdot \left[N + 2\sum_{m=1}^{N-1} w_m (N-m) \cdot \left\langle \mathrm{Re}\left(e^{i\mathbf{q}\cdot\Delta\mathbf{R}_m}\right)\right\rangle \Gamma_m \right] \tag{9}$$

In equation (9), the modulus of the product of the structure factor with the Fourier transformation of the shape factor is averaged over the whole ensemble of crystallites within the irradiated volume of the sample. The same averaging must also be applied to the phase shift of the scattered waves, exp ($i\mathbf{q} \cdot \Delta\mathbf{R}_m$), which depends of the diffraction vector and on the distance between neighbouring crystallites, $\Delta\mathbf{R}_m = \mathbf{R}_{n+m} - \mathbf{R}_n$. N is the number of diffracting crystallites. In the reciprocal space, the degree of the coherence of adjacent crystallites is described by the amount of the overlap of the reciprocal lattice points, w_m. In the direct space, the mutual coherence of two waves scattered by different crystallites is characterised by the coherence factor Γ_m [40]. For nanocrystalline materials, Γ_m can be regarded to be equal to unity for volumes containing hundreds of nanocrystallites, as the coherence length of X-rays (approximately 1 μm) is much larger than the size of crystallites and thus it is much larger than the distances between the nanocrystallites.

On the contrary, the amount of the overlap of the reciprocal lattice points, w_m, is a crucial factor for the phenomenon of the partial coherence of crystallites, as it influences strongly the shape and the width of XRD lines. For non-overlapping reciprocal lattice points, i.e. for non-coherent crystallites, $w_m = 0$ and equation (9) reduces to the well-known kinematical form, in which the diffraction line broadening is given by the Fourier transformation of the shape function of crystallites (if lattice defects are neglected). For fully coherent crystallites, $w_m = 1$, which means that the Fourier transformation of the shape function of crystallites is multiplied by an oscillating function that makes the XRD lines narrower. The frequency of the oscillations depends mainly on the projection of the position vectors ($\Delta\mathbf{R}_m$) of crystallites into the direction of the diffraction vector ($\mathbf{q}$) and on the size of the diffraction vector. For partially coherent crystallites, w_m is between zero and unity. The dependence of the factor w_m on the diffraction vector is illustrated in figure 1 for reciprocal lattices of two slightly disoriented face centred cubic (fcc) crystallites. Obviously, w_m depends on the size of the reciprocal lattice points, on the distance of the reciprocal lattice point from the origin of the reciprocal space, i.e. on the size of the diffraction vector, and on the mutual disorientation of neighbouring crystallites. A simplified form of w_m was given in [31]. Alternatively, w_m can be calculated from the overlap of the reciprocal lattice points divided by the volume of the reciprocal lattice points. In a two-dimensional case (see figure 2), w_m is given by:

$$w_m = \frac{A}{\pi R^2} = \frac{R^2(\alpha_m - \sin\alpha_m)}{\pi R^2} = \frac{\alpha_m - \sin\alpha_m}{\pi} \quad (10)$$

The angle α is related to the size of the reciprocal lattice points, R, and to the shift s of the reciprocal lattice points from each other:

$$\alpha_m = 2\arccos(s_m/2R) \quad (11)$$

The shift of the reciprocal lattice points increases with the size of the diffraction vector q and with the disorientation of crystallites ω_m:

$$s_m = 2q\tan\frac{\omega_m}{2} \quad (12)$$

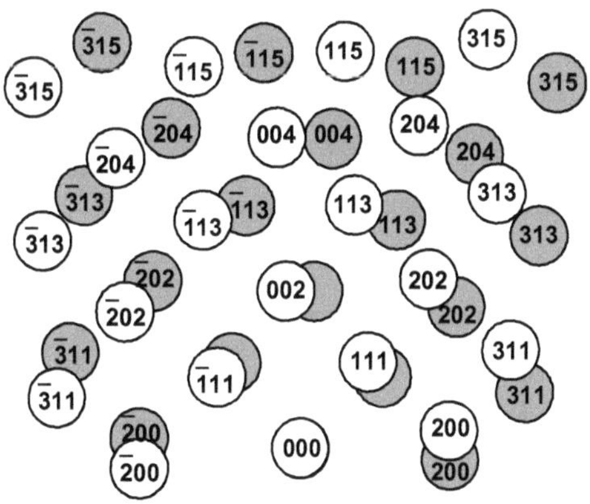

Figure 1. Projection of the reciprocal lattice points from two slightly disoriented fcc crystallites into the x-y plane. The reciprocal lattice points 111, 200, 220, 311 overlap partially in this particular case.

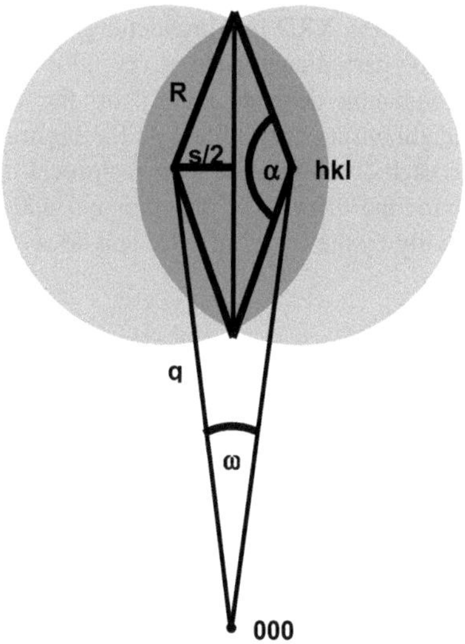

Figure 2. Partial overlap of the reciprocal lattice points shown for two crystallites with the disorientation ω. ***R*** *is the size of the reciprocal lattice points,* ***q*** *the diffraction vector.*

As indicated already by equation (9), w_m depends also on the order (or distance) of the next neighbours. Several models can be considered for description of the disorientation of the neighbouring crystallites: random disorientation, "zigzag" disorientation and cumulative disorientation. In compact materials, a small random disorientation of neighbouring crystallites is rather unlikely. The other two models can be taken into consideration, for instance as a consequence of a high concentration of lattice defects [42]. The consequence of the "cumulative" model is a steep decrease of w_m with increasing m (as compared at the same diffraction vector), which means that the more distant crystallites are rather non-coherent. An additional decrease of the degree of the partial coherence with increasing distance between crystallites is related to the coherence factor Γ_m in equation (9) that applies also for the "zigzag" model.

The dependence of the XRD line broadening on the sine of the diffraction angle is illustrated in figure 3 as calculated using equation (9). At the smallest diffraction angles, the neighbouring crystallites are partially coherent. As the degree of their partial coherence decreases with increasing size of the diffraction vector, the XRD line broadening increases with increasing diffraction vector. In the classical kinematical diffraction theory, the increase of the XRD line broadening with increasing diffraction vector can be related directly to the dislocation density. For partially coherent nanocrystallites, the increase of the XRD line broadening is related to the crystallites size, to the mean disorientation of neighbouring crystallites and to the disorientation model. Still, like in the classical kinematical diffraction theory, the XRD line broadening extrapolated to $q = 0$ can be assigned to the reciprocal size of clusters that are composed from slightly disoriented nanocrystallites and that therefore appears partially coherent to X-ray scattering at small diffraction angles. At a certain size of the diffraction vector, the XRD line broadening increases abruptly. This phenomenon is related to the loss of the partial coherence of crystallites. The position of the abrupt increase of the XRD line broadening depends mainly on the crystallite size and on the mutual disorientation of neighbouring crystallites. At the highest diffraction angles, the crystallites are non-coherent to each other, because their reciprocal lattice points do not overlap as it can be seen in figure 1 for the diffraction lines beyond 400. Thus, the XRD line broadening is given predominantly by the crystallite size and does change with increasing diffraction angle.

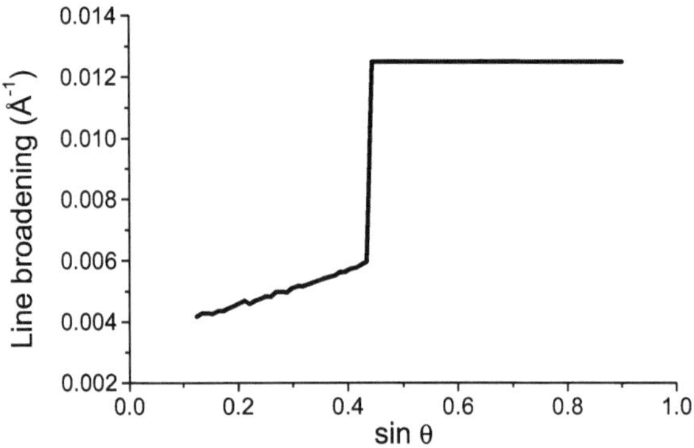

Figure 3. Effect of the partial coherence of neighbouring crystallites on the XRD line broadening.

3. Experimental details

The capability of the XRD approach introduced above is illustrated on examples of nanocrystalline (Cr, Al) N coatings and BN nanocomposites. The (Cr, Al) N coatings were deposited using the vacuum arc evaporation [35, 36, 38]. The BN nanocomposites were produced using the high-pressure and high-temperature (HP/HT) synthesis at the pressure 15 GPa and at the temperature 1660°C [39]. During the HP/HT process, *h*-BN crystallites (space group P 6_3/mmc) in the starting material transform through the metastable *w*-BN (space group P 6_3mc) into the stable *c*-BN (space group F -43m). The phase transformation is running concurrently from several seeds within individual *h*-BN crystallites, as it was confirmed through a smaller size of transformed crystallites and through the presence of partial coherence of *w*-BN and *c*-BN crystallites in the initial stages of the HP/HT transformation process. Due to the orientation relationships between *h*-BN, *w*-BN and *c*-BN, $\langle 11\text{-}20\rangle_h \parallel \langle 11\text{-}20\rangle_w \parallel \langle -110\rangle_c$ and $(0002)_h \parallel (0002)_w \parallel (111)_c$ (see e.g. [43, 44]), the mutual orientations of *h*-BN, *w*-BN and *c*-BN crystallites within the individual crystallites of the *h*-BN starting material are preserved during the phase transition.

X-ray diffraction experiments were performed in the glancing angle diffraction geometry (GAXRD) on a D8 diffractometer (Bruker AXS) that was equipped with a sealed tube with the copper anode, the parabolic Goebel mirror, the Soller collimator with the acceptance of 0.12°, the flat LiF analyser and a scintillation detector. The Goebel mirror was located in the primary beam, the Soller collimator and the LiF analyser in the diffracted beam. TEM and HRTEM were done on the high-resolution transmission electron microscope JEM 2010 FEF (JEOL) at the acceleration voltage of 200 kV. The high resolution of the microscope was reached by using an ultra-high-resolution objective lens (C_s = 0.5 mm) and an in-column energy filter. The latter was used to select only the elastic electrons for the image recording. The samples for TEM were prepared in the plane-view orientation.

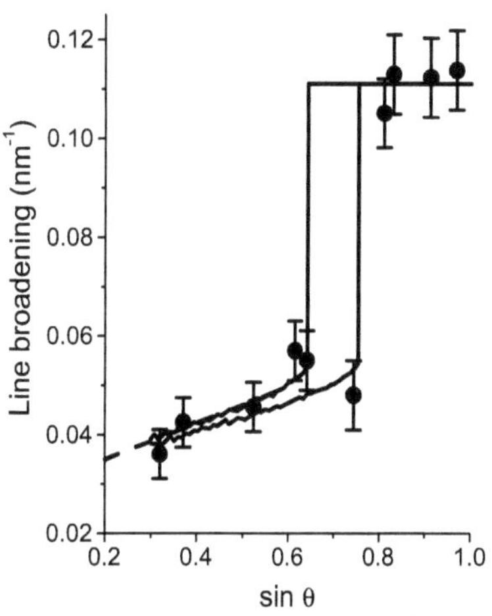

Figure 4. XRD line broadening (integral breadth) as measured in the single-phase sample with the chemical composition $Cr_{0.92}Al_{0.08}N$. Solid lines were calculated for the limiting disorientations of crystallites using equation (9).

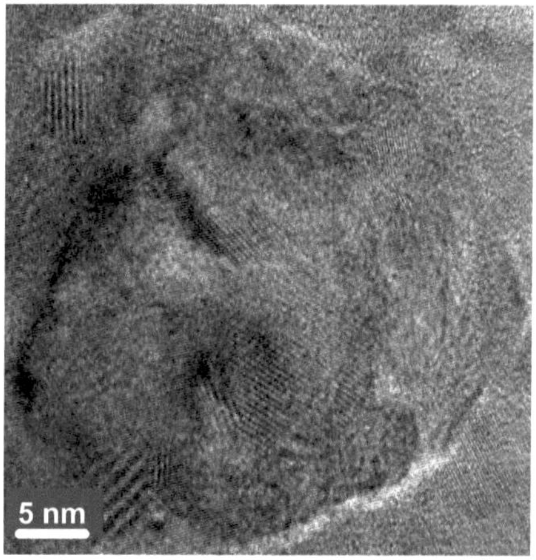

Figure 5. HRTEM micrograph of a nanostructured cluster in the sample from figure 4, i.e. $Cr_{0.92}Al_{0.08}N$ containing a single fcc phase.

4. Experimental results

The capability and reliability of the theory of partial coherence of crystallites was verified for the example of a single-phase (fcc) nanocrystalline coating with the chemical composition $Cr_{0.92}Al_{0.08}N$. The dependence of the XRD line broadening on the sine of the diffraction angle (figure 4) shows all features discussed above for partially coherent crystallites (see also figure 3): the increase of the XRD line broadening at small diffraction angles, a steep in-

visible for X-rays due to the very high defect density. Using our approach, the nature of the dislocation walls and/or other lattice defects can be concluded from the local disorientation of nanocrystallites.

5. Concluding remarks

It was shown that the analysis of the XRD line broadening, which takes into account the partial coherence of nanocrystallites, is a powerful experimental technique for the local microstructure analysis of nanocrystalline materials and nanocomposites. Primarily, this technique can reveal the size of nanocrystallites. If the disorientations of neighbouring nanocrystallites are small, the nanocrystallites are partially coherent, which implies occurrence of additional interference phenomena related to the X-rays scattering on partially coherent crystallites. From these interference effects, the disorientation of partially coherent crystallites and the size of clusters composed from partially coherent crystallites can be calculated. The maximum disorientation, at which the crystallites are still partially coherent, depends generally on the crystallite size, i.e. on the size of the reciprocal lattice points and on the crystal structure, i.e. on the distance of the reciprocal lattice points with the lowest low diffraction indices from the origin of the reciprocal space. Nevertheless, the maximum disorientation of partially coherent crystallites is approximately 1 – 2° for fcc nanocrystallites with the lattice parameter of approximately 4 Å and the size of about 10 nm.
Theoretical considerations and the comparison of results obtained from the XRD line profile analysis and TEM/HRTEM yielded useful information on the reliability of the microstructure parameters calculated from XRD data: the size and disorientation of nanocrystallites and the size of clusters of partially coherent crystallites. The most reliable microstructure parameter is the crystallite size, as it is calculated directly from the saturated XRD line broadening. A relatively reliable parameter is the cluster size. It is calculated from the extrapolation of the XRD line broadening in the partially coherent region to $q = 0$ like the crystallite size in the "classical" XRD line profile analysis. The main uncertainty of the cluster size originates from the general non-linearity of the XRD line broadening in the partially coherent region of the diffraction vectors, which moreover depends on the kind and ordering of the lattice defects. Nevertheless, numerous results have shown that the cluster size obtained from the XRD line broadening correlates well with the size of similar microstructural features observed by TEM/HRTEM if the number of diffraction lines affected by the phenomenon of the partial coherence is sufficiently large (five and more). On the other hand, relatively uncertain is the local disorientation of crystallites that strongly depends on the microstructure model used for simulation of the XRD line broadening. Furthermore, the local crystallite disorientation could be verified using indirect methods only at the moment.

References

1. Snyder, R.L., Fiala, J. & Bunge, H.J. (editors), 1999, *Defect and Microstructure Analysis by Diffraction* (Oxford University Press).
2. Mittemeijer, E.J. & Scardi, P. (editors), 2003, *Diffraction Analysis of the Microstructure of Materials* (Berlin: Springer).
3. Warren, B.E., 1990, *X-ray Diffraction* (New York: Dover).

4. Krivoglaz, M.A., 1996, *X-ray and Neutron Diffraction in Non-ideal Crystals* (Berlin: Springer).
5. Williamson, G.K. & Hall, A.H., 1953, *Acta Metall. Mater.*, **1**, 22.
6. Wilkens, M., 1970, *Phys. Status Solidi A*, **2**, 359.
7. Langford, J.I. & Wilson, A.J.C., 1978, *J. Appl. Cryst.*, **11**, 102.
8. Adler, T. & Houska, C.R., 1979, *J. Appl. Phys.*, **50**, 3282.
9. Houska, C.R. & Smith, T.M., 1981, *J. Appl. Phys.*, **52**, 748.
10. Langford, J.I. & Louër, D., 1982, *J. Appl. Cryst.*, **15**, 20.
11. Keijser, Th.H. de, Langford, J.I., Mittemeijer, E.J. & Vogels, A.B.P., 1982, *J. Appl. Cryst.*, **15**, 308.
12. Delhez, R., de Keijser, Th.H. & Mittemeijer, E.J., 1982, *Z. Anal. Chem.*, **312**, 1.
13. Vargas, R., Louër, D. & Langford, J.I., 1983, *J. Appl. Cryst.*, **16**, 512.
14. Groma, I., Ungár, T. & Wilkens, M., 1988, *J. Appl. Cryst.*, **21**, 47.
15. Klimanek, P. & Kužel, R., 1988, *J. Appl. Cryst.*, **21**, 59.
16. Kužel, R. & Klimanek, P., 1988, *J. Appl. Cryst.*, **21**, 363.
17. Kužel, R. & Klimanek, P., 1989, *J. Appl. Cryst.*, **22**, 299.
18. Balzar, D., Von Dreele, R.B., Bennett, K. & Ledbetter, H., 1998, *J. Appl. Phys.*, **84**, 4822.
19. Langford, J.I., Louër, D. & Scardi, P., 2000, *J. Appl. Cryst.*, **33**, 964.
20. Ungár, T., Gubicza, J., Ribárik, G. & Borbély, A., 2001, *J. Appl. Cryst.*, **34**, 298.
21. Scardi, P. & Leoni, M., 2001, *Acta Cryst. A*, **57**, 604.
22. Estevez-Rams, E., Martinez, J., Penton-Madrigal, A. & Lora-Serrano, R., 2001, *Phys. Rev. B*, **63**, 054109.
23. Popa, N. C. & Balzar, D., 2002, *J. Appl. Cryst.*, **35**, 338.
24. Estevez-Rams, E., Leoni, M., Scardi, P., Aragon-Fernandez, B. & Fuess, H., 2003, *Phil. Mag.*, **83**, 4045.
25. Leoni, M., Gualtieri, A.F. & Roveri, N., 2004, *J. Appl. Cryst.*, **37**, 166.
26. Ungár, T., Tichy, G., Gubicza, J. & Hellmig, R.J., 2005, *Powder Diffraction*, **20**, 366.
27. Balogh, L., Ribárik, G. & Ungár, T., 2006, *J. Appl. Phys.*, **100**, 023512.
28. Scardi, P. & Leoni, M., 2006, *J. Appl. Cryst.*, **39**, 24.
29. Ortiz, A.L., Cumbrera, F.L. & Sánchez-Bajo, F., 2006, *J. Appl. Cryst.*, **39**, 598.
30. Van Berkum, J.G.M., Delhez, R., de Keijser, Th.H. & Mittemeijer, E.J., 1996, *Acta Cryst. A*, **52**, 730.
31. Rafaja, D., Klemm, V., Schreiber, G., Knapp, M. & Kužel, R., 2004, *J. Appl. Cryst.*, **37**, 613.
32. Ribárik, G., Audebrand, N., Palancher, H., Ungár, T. & Louër, D., 2005, *J. Appl. Cryst.*, **38**, 912.

33. Comedi, D., Zalloum, O.H.Y., Irving, E.A., Wojcik, J., Roschuk, T., Flynn, M.J. & Mascher, P., 2006, *J. Appl. Phys.*, **99**, 023518.
34. Mandal, T.K., 2007, *Mater. Letters*, **61**, 850.
35. Rafaja, D., Poklad, A., Klemm, V., Schreiber, G., Heger, D., Šíma, M. & Dopita, M., 2006, *Thin Solid Films*, **514**, 240.
36. Rafaja, D., Dopita, M., Růžička, M., Klemm, V., Heger, D., Schreiber, G. & Šíma, M., 2006, *Surf. Coat. Technol.*, **201**, 2835.
37. Rafaja, D., Poklad, A., Klemm, V., Schreiber, G., Heger, D. & Šíma, M., 2007, *Mat. Sci. Eng. A*, **462**, 279.
38. Rafaja, D., Wüstefeld, Ch., Dopita, M., Růžička, M., Klemm, V., Schreiber, G., Heger, D. & Šíma, M., 2007, *Surf. Coat. Technol.*, **201**, 9476.
39. Rafaja, D., Klemm, V., Motylenko, M., Schwarz, M.R., Barsukova, T., Kroke, E., Frost, D., Dubrovinsky, L. & Dubrovinskaia, N., 2008, *J. Mater. Res.*, **23**, 981.
40. Born, M. & Wolf, E., 1999, *Principles of optics*, 7th edition (Cambridge: University Press).
41. Pietsch, U., Holý, V. & Baumbach, T., 2004, *High-Resolution X-ray Scattering*, 2nd edition (New York: Springer).
42. Mughrabi, H., 1983, *Acta Metall.*, **31**, 1367.
43. Kurdyumov, A.V., Britun, V.F. & Petrusha, I.A., 1996, *Diam. Relat. Mater.*, **5,** 1229.
44. Huang, J. & Zhu, Y.T., 2000, *Defect and Diffusion Forum*, **186-187**, 1.
45. Williams, D.B. & Carter, C.B., 1996, *Transmission Electron Microscopy III – Imaging* (New York: Plenum Press).
46. Randl, V., 1992, *Microtexture determination and its applications* (London: Inst. of Materials).
47. Klemm, V., Klimanek, P. & Seefeldt, M., 1999, *phys. stat. sol. a*, **175**, 569.
48. Frank, F.C., 1950, *Report on the Symposium on Plastic Deformation of Crystalline Solids, Carnegie Institute of Technology and Office of Naval Research* (Washington: U.S. Government Printing Office), page 150.
49. Wüstefeld, Ch., Rafaja, D., Dopita, M. & Klemm, V., *Z. Kristallographie*, this issue.
50. Motylenko, M., Klemm, V., Schreiber, G., Rafaja, D., Schwarz, M., Barsukova, T. & Kroke, E., *Z. Kristallographie*, this issue.

Acknowledgements. The authors acknowledge the financial support of this work through the German Research Council (DFG) in frame of the Priority Programme # 1181 and the projects # RA-1050/9 and KR 1739/13-1. JEM 2010 FEF was purchased from the funds of the Priority Programme # 1060. We thank also Dr. Růžička and Dr. Šíma (SHM Šumperk, Ltd.) for their assistance with the PVD deposition of nanocrystalline (Cr, Al) N coatings.

Optimisation of coherent X-ray diffraction imaging at ultrabright synchrotron sources

Ian K. Robinson[1,2,*]

[1] Diamond Light Source, Harwell Science and Innovation Campus, Didcot, OX11 0DE, United Kingdom
[2] London Centre for Nanotechnology, University College, Gower St, London, WC1E 6BT, United Kingdom
[*] i.robinson@ucl.ac.uk

Keywords: diffraction, coherence, imaging, nanocrystals, synchrotron radiation

Abstract. Coherent X-ray Diffraction Imaging (CXDI) is a powerful method of imaging single crystalline grains within a powder. Direct Fourier transformation of the oversampled diffraction pattern surrounding a Bragg peak is possible once the phases have been obtained using a 'support' constraint. The image is in general complex with the phase representing a projection of the atomic displacements allowing access to the internal strains inside the crystal. CXDI relies crucially on the production of a coherent beam of X-rays, which is one of the technical advances of the latest 3rd generation Synchroton Radiation (SR) sources. It is shown here how the use of a secondary source in the design of the X-ray beamline allows the coherence to be controlled continuously over a finite range without any loss of flux. The conclusions will have significant impact on instruments currently being designed at the Diamond Light Source and the new National Synchrotron Light Source (NSLS-2).

1. Imaging of crystals

Strain fields inside nanometre-size crystals can be imaged quantitatively using the method of Coherent X-ray Diffraction Imaging (CXDI) [1-5]. A coherent X-ray beam is focussed onto a nano-grained powder sample so that relatively few grains are illuminated. One of the diffraction spots of an individual grain can isolated on a Charge-Coupled Device (CCD) detector with sufficient resolution that it becomes oversampled relative to the spacing of the fringes that arise from the finite size of the grain. The angular spacing, $\Delta\theta$, of the fringes is given by the spacing formula, $\Delta\theta = \lambda / d$, where λ is the X-ray wavelength and d is the grain size. This simply requires sufficiently small pixels and sufficient distance between the sample and the detector. The diffraction pattern is recorded in three dimensions (3D) by rocking the sample through the Bragg peak and recording a sequence of frames [5].

The 3D diffraction pattern can be Fourier inverted to an image once it has been phased. The phasing method that is most widely used today is the Hybrid Input-Output (HIO) method of

Fienup [6], which repeatedly cycles back and forth between real and reciprocal spaces using discrete Fast Fourier Transforms (FFT), while applying constraints on each side. The reciprocal space constraint is the overwriting of the amplitude of the diffraction pattern, keeping the phases. The real space constraint is the cutting off of the image outside a 'support', a finite-sized region of space just big enough to contain the crystal. The method works because the diffraction pattern is oversampled relative to the spatial Nyquist frequency [7]. When there are more measurement points than image points within the support, a solution can be found. Mathematical analysis has shown that non-unique solutions in 3D are not possible in general [8], while the effects of experimental noise on the convergence are still being explored [9]. Except in the special case where the diffraction intensity function is locally symmetric about its centre (the exact Bragg point), the resulting image is a complex 3D function. It has been shown that the phase of this image at each point is equal to the scalar (dot) product of the momentum transfer vector with the local atomic displacement from the ideal crystal lattice [10,11]. This allows quantitative mapping of strain fields.

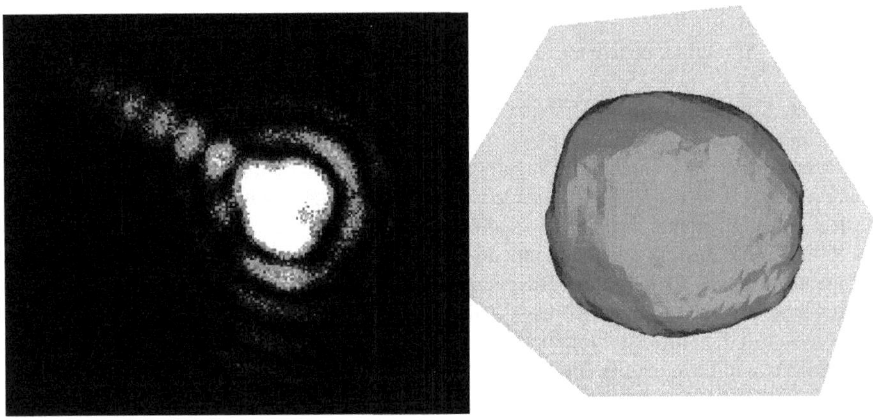

Figure 1. (a) One frame of the 3D diffraction pattern of gold crystal Au607-112, slightly off centre from the Bragg point. Circular fringes are attributed to the approximately spherical shape of the crystal, while the fringed flare on the top-left side is due to a prominent facet. (b) View of one isosurface of the inverted image of the crystal showing the expected features in the shape. A translucent image of the support used in the phasing calculation is overlaid.

An illustration of the current capabilities is shown in Figure 1. The sample is a 200nm sized crystal of gold prepared by dewetting an evaporated film of about 20nm thickness by heating in air at 1100C for about 10 hours. The substrate is the oxide layer on a silicon wafer. One issue addressed by the experiment is the extent to which the crystal becomes strained by the presence of the substrate. An analogous experiment on Pb showed a distinct pattern of strain inside the crystal that was attributed to its contact with the substrate [1]. One frame of the diffraction pattern, measured at sector 34-ID-C of the Advanced Photon Source (APS) is shown in Figure 1(a). One view of the inverted image is shown in Figure 1(b) as an isosurface (3D contour map) of the 3D density function. The phasing algorithm was a variant of HIO in which the phase outside the support was constrained to values in the range $\pm \pi/2$ [12].

2. Secondary slit for beamline optimisation

The brightness of a SR beamline arises directly from the strength of the quadrupole focussing magnets in the lattice of the storage ring and is limited by the horizontal betatron oscillations. In this paper we will ignore the vertical direction altogether, where the figures are two orders of magnitude better; at NSLS-2 the vertical direction is expected to be brightness limited in the hard X-ray part of the spectrum. The horizontal emittance, $\varepsilon = \sigma\sigma'$, is the quantity most directly relevant, where σ is the Gaussian half width of the electron source spatial distribution and σ' is the Gaussian half width of the electron source angular distribution. By Louiville's theorem applied to the electron orbit, ε is conserved around the ring, but the relative amount of the two components can vary. The ratio, $\beta = \sigma / \sigma'$ is the quantity that can be varied to match the needs of the experiments. At NSLS-2, the beta-function alternates between adjacent straight sections of the lattice. At Diamond, there are 8m 'long' and 5m 'short' straight sections around the ring with quite different beta functions. The latest-available relevant source parameters are tabulated in Table 1. There are smaller variations even within the straights, with beta rising from 10m at the end to 13m at the centre of Diamond's 'long' straight, but we will ignore these for simplicity here. Since a CXDI experiment could be implemented on any of these straights, it is necessary to choose the appropriate source parameters. The choice between these is important because it can have significant impact on the available flux and hence on the scientific outcome of the beamline.

Table 1. Design parameters for the two storage rings considered in this paper. Derived quantities refer to the symbols defined in the text.

	units	NSLS-2 Low-beta	NSLS-2 High-beta	Diamond Short	Dimond Long
σ	μm	28	99	122.9	178.4
σ'	μ-rad	19	5.5	24.2	16.5
σ_X'	μ-rad	25.30	17.58	29.40	23.48
ε	nm-rad	0.53	0.54	2.97	2.94
ε_X	nm-rad	0.71	1.74	3.61	4.19
β	m	1.47	18.00	5.08	10.81
D_1	m	150	150	250	250
D_2	m	25	25	25	25

Option I : direct use of source

s_1	µm	268	76	102	70
Coh frac		0.035	0.014	0.007	0.006

Option II: secondary slit at source

s_1 min	µm	268	76	102	70
s_1 max	µm	7589	5275	14701	11738

Option III: secondary slit at shield wall

C_{min}	µm	110	110	105	105
s_1 min	µm	-	83	107	72
s_1 max	µm	-	907	2105	3140

While the conserved emittance is defined for the electron beam, the generated X-rays no longer obey this simple conservation law. The reason is that there is a finite opening angle for the radiation emitted as it passes through an undulator, located in one of the straight sections of the ring. An electron passing through an undulator emits a radiation cone of half width [13], $1/\sqrt{N}\gamma$, where γ is the relativistic factor, 6000 for both 3GeV rings considered here, and N is the number of magnetic poles in the undulator, which we will take to be 100. The X-ray opening angle is then 16.7mrad. This opening angle has to be combined (quadratically) with the electron beam divergence to give the X-ray divergence σ_X'. As Table 1 shows, the product of X-ray source size and X-ray divergence, which we define as the "X-ray emittance" ε_X, now does vary around the ring. All numbers quoted in this paper are calculated for λ=0.1nm or 12 keV energy, which is typical for coherence-based diffraction experiments. The table shows there is a natural advantage of using the low-beta straights at NSLS-2 or the short straights at Diamond for their smaller X-ray emittance, which translates into a larger coherent flux, as seen below. The difference is more than a factor of two for NSLS-2.

Coherence-based experiments require that only part of the beam can be used. An aperture must be introduced into the beamline to limit the spatial extent of the beam to approximately one lateral coherence length. Exceptions can be made, for example in X-ray Photon Correlation Spectroscopy (XPCS), where the best signal-to-noise is found to occur for a slit size nearer to two coherence lengths [14]. The lateral correlation length of a wavefield at position P is defined from the van Cittert-Zernike theorem [15] as,

$$\xi = \lambda D / 2\sigma \tag{1}$$

where D is distance from the source to point P. Multiplicative factors close to unity that arise from the exact distribution of the source have been omitted here and elsewhere in this paper. Once a coherent part of the wavefield has been selected by the aperture, the beam can then be manipulated using optics to concentrate it onto a small sample [16]; perfect optics will not affect the degree of partial coherence, in principle. This paper will focus on the issues associated with the choice of aperture needed to define the coherence and not discuss the following optics further, simply assuming that one coherence length is required.

3. Design options

In both the Diamond and NSLS-2 projects, the option of 'long' beamlines for CXDI, extending outside the storage ring building, is being considered seriously. In neither case are the distances decided, but typical estimates of the position of the experiment and its coherence-defining entrance slit are given in Table 1, D_l=150m for NSLS-2 and 250m for Diamond. The conclusions of the paper are not very sensitive to the exact choice.

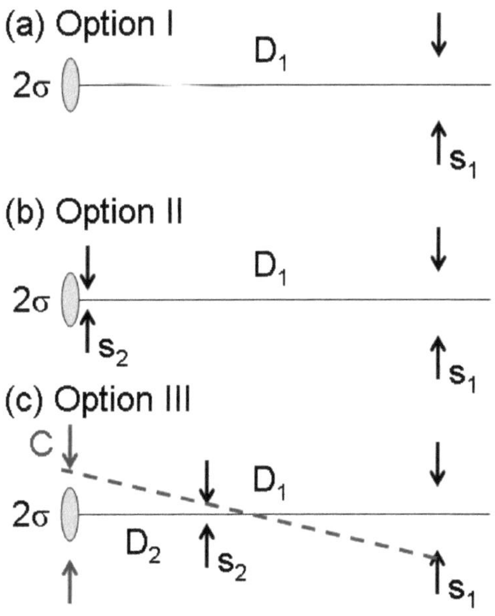

Figure 2. Schematic beamline layouts for consideration of secondary slits.
(a) Option I where the coherence at the location of the experiment entrance slit, s_1, is set by the synchrotron source itself. (b) Option II where a secondary slit is introduced at the source. (c) Option III where a secondary slit is introduced at the synchrotron shield wall. The dashed line is the ray-tracing discussed in the text to determine the projected size, C, of the slits onto the source.

The simplest option, denoted option I in Figure 2(a), is to place nothing in the beam, except perhaps windows and a monochromator that do not affect the light propagation in principle. The coherence-defining horizontal slit is then set to one coherence length, given by Equation 1 with $D=D_l$, the position of the experiment. The resulting slit/beam size, $s_l=\xi$, given in table 1, is in the range of 100 microns. This coherent beam can then illuminate the sample directly or be manipulated with optics to match the sample. In any case, the useful flux is given by the fraction of the X-ray opening angle subtended by the slit. Assuming this to be a small fraction of the divergence, we can write this as the coherent faction of the total flux,

$$f = \frac{s_1 / D_1}{2\sigma_X'} = \frac{\lambda}{4\sigma\sigma_X'} = \frac{\lambda}{4\varepsilon_X} \tag{2}$$

As can be seen, this depends only on the X-ray emittance. It is immediately obvious that this option favours the lowest possible beta for the storage ring because of the compounding effect of the X-ray opening angle on the horizontal divergence. The coherent fraction, calculated in Table 1 to be at most a few percent of the full beam, shows that the low-beta straight at NSLS-2 gives more than a factor of two more flux than the high-beta straight for this reason.

Option II in Figure 2(b) is a hypothetical variation showing the simplest kind of secondary source. Although obviously impractical, a secondary slit of width s_2 is placed directly on top of the electron source to limit its spatial extent. The value of s_1, again equal to one coherence length of the source, now defined by s_2, is given by $s_2 = \lambda D_1 / s_1$ and tabulated in table 1, is bigger than before and can indeed be made arbitrarily large. However the resulting coherent fraction,

$$f = \frac{s_1 / D_1}{2\sigma_X'} \times \frac{s_2}{2\sigma} = \frac{\lambda}{4\varepsilon_X} \tag{3}$$

is exactly the same as before. This shows the basic principle of the secondary source, the ability to manipulate the coherence beam size without cost in flux. The two limits on the range of values of s_1, listed in table 1, are given by the constraints that the secondary slit s_2 must be smaller than the source, $s_2<2\sigma$, and that the coherence defining slit s_1 must be filled by the X-ray divergence, giving $s_1<2D_1\sigma_X'$. This gives a wide range of adjustable coherence lengths that do not reduce the coherent flux.

Option III in Figure 2(c) is the practical version of the secondary source located at the closest accessible position in the beamline, the distance of the shield wall that isolates the storage ring from the experimental hall, D_2. A slit of width s_2 is placed at distance D_2 to determine the coherence length at the experiment. This acts as a virtual source in the sense that each point within the secondary slit is incoherently illuminated by multiple overlapping waves from the actual source. The coherence-defining slit at the experiment, s_1, selecting one coherence length, is again given by Equation 1, but now with $D=(D_1-D_2)$ and 2σ replaced by s_2., giving

$$s_1 = \lambda(D_1 - D_2) / s_2 \tag{4}$$

Like with option II, a range of coherence lengths can be achieved. Within that range, as discussed below, the coherent flux is unchanged from the expression in Equations (2) and (3). Deriving this is a little harder than before; it has been verified in the special symmetric case of $s_1/D_1=s_2/D_2$ but is expected to be generally true by the Louiville theorem.

Ray tracing is needed to find the working range of values of the slit sizes s_1 and s_2 that preserve the ideal coherent flux. For simplicity, we assume box-shaped (instead of Gaussian) functions for the spatial and angular distributions; only the extreme choices for s_1 and s_2 are likely to be much affected by this assumption. It turns out that the angular distribution is sufficiently wide that it is never a constraint. The critical ray that limits the range, shown in colour in figure 2(c), passes the extremes of the two slits and must stay within the source. The equation for this is,

$$C = s_1 \frac{D_2}{D_1 - D_2} + s_2 \frac{D_1}{D_1 - D_2} \leq 2\sigma \tag{5}$$

where the quantity C on the left-hand side of the inequality is the back-projected image of the slits and depends on both slit settings and their distances. C can be reduced to a function of a single variable by inserting the relation (4) above between s_1 and s_2 determined by the one-coherence length criterion. Expressed in terms of s_1,

$$C(s_1) = s_1 \frac{D_2}{D_1 - D_2} + \frac{\lambda D_1}{s_1} \tag{6}$$

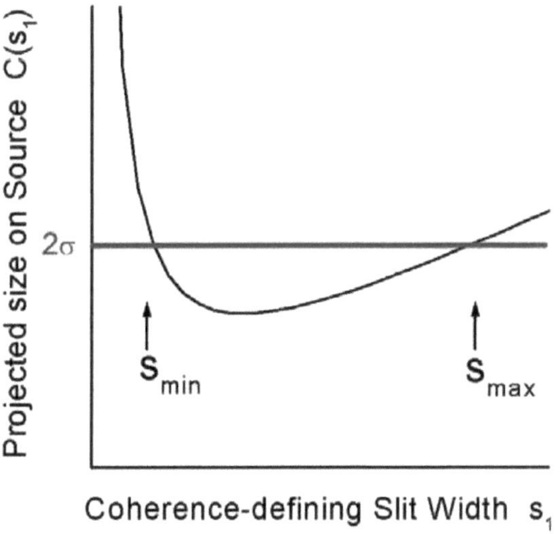

Figure 3. Variation of the projected size, C, of the slits onto the source (see Figure 2(c)) as a function of the final slit size s_1, subject to the constraint that the coherence size s_1 is set by the secondary slit s_2.

This expression is plotted in figure 3. The minimum value of $C(s_1)$, given by $C_{min} = 2\sqrt{\frac{\lambda D_1 D_2}{D_1 - D_2}}$, is tabulated in Table 1. When $C(s_1)$ falls below the source size, 2σ, the ray-tracing condition is satisfied and we have a viable configuration with a fully illuminated secondary source. But $C(s_1)$ rises on both sides giving a limited range of allowed slit settings. The values of s_1 bounding the allowed range are,

$$s_{max/min} = \frac{D_1 - D_2}{D_2} \left\{ \sigma \pm \sqrt{\sigma^2 - \frac{\lambda D_1 D_2}{D_1 - D_2}} \right\} \tag{7}$$

also tabulated for the various sources in Table 1. It is notable that there is no solution for the otherwise attractive low-beta option at NSLS-2: C_{min} is already greater than the source size. In order to obtain any solution, the secondary slit must be brought forward to D_2=7.45m. In the other three cases a useful range of values is apparent. The geometric mean of s_{max} and s_{min}, also listed, is the natural centre of the two ranges. This corresponds to the symmetric choice of $s_1/D_1=s_2/D_2$ mentioned above.

4. Conclusions

The technique of CXDI is illustrated as producing a 3D image of a 200nm Au nanocrystal. The phasing was achieved by a phase-constrained HIO algorithm using oversampled 3D diffraction data. The CXDI method depends on coherence of the X-ray beams used. It will be more powerful with the production of higher fluxes in the coherent beams at the new 3rd generation SR sources. The development of suitable X-ray optics to match the sample size to the coherent beam size, which is a subject outside the present discussion, will clearly profit from the ability to adjust the size of the incoming coherent beam. It is shown here that the introduction of a secondary slit at the position of the shield wall allows control of the coherent beam sizes without cost of flux. The available range of sizes depends on the source and the layout of the beamline, but excludes the low-beta option for NSLS-2.

The main advantage of using a secondary slit is the continuous control of coherent beam size, but there are other advantages also. Because the secondary source is at an accessible location and because only a small fraction of the source is subtended by the slit, the beam may be multiplexed into multiple branches operating in parallel. This can be achieved using mirrors as done presently at 34-ID and 8-ID of APS, or else using crystal optics. The multiplexing option would be possible to implement at Diamond or NSLS-2, but only on the high-beta straights of the latter. Another advantage is that it sets the coherence outside the storage ring without reliance on a stable source of the correct size. It therefore offers immunity to instabilities of the source which would lead to a drop in flux rather than loss of coherence.

References

1. Pfeifer, M.A., Williams, G.J., Vartanyants, I.A., Harder, R. & Robinson, I.K., 2006, *Nature*, **442**, 63.
2. Harder, R., Pfeifer, M.A., Williams, G.J., Vartaniants, I.A. & Robinson, I.K., 2007, *Phys. Rev. B*, **76**, 115425.
3. Williams, G.J., Pfeifer, M.A., Vartanyants, I.A. & Robinson, I.K., 2006, *Phys. Rev. B*, **73**, 094112.
4. Robinson, I. & Miao, J., 2004, *Mat. Sci. Bull.*, **29**, 177.
5. Williams, G.J., Pfeifer, M.A., Vartanyants, I.A. & Robinson, I.K., 2003, *Phys. Rev. Lett.*, **90**, 175501-1.
6. Fienup, J.R., 1987, *J. Opt. Soc., Am. A.*, **4**, 118.
7. Sayre, D., 1952, *Acta Cryst.*, **5**, 843.

8. Bates, R.H.T., 1982, *Optik*, **61**, 247.
9. Williams, G.J., Pfeifer, M.A., Vartanyants I.A. & Robinson, I.K., 2007, *Acta Cryst.*, A**63**, 36.
10. Robinson , I.K. & Vartanyants, I.A., 2001, *Appl. Surf. Sci.*, **182**, 186.
11. Vartanyants, I.A. & Robinson, I.K., 2001, *J. Phys. : Condens. Matter*, **13**, 10593-611.
12. Harder, R., to be published.
13. Als-Nielsen, J. & McMorrow, D., 2001, *Elements of Modern X-ray Physics* (Wiley).
14. Sandy, A., private communication.
15. Born, M. & Wolf, E., 1999, *Principles of Optics: Electromagnetic Theory of Propagation, Interference and Diffraction of Light* (Cambridge: University Press), 7th edition.
16. Robinson, I.K., Pfeiffer, F., Vartanyants, I.A., Sun Y. & Xia, Y., 2003, *Optics Express*, **11**, 2329.

Acknowledgements. The author wishes to thank Ulrich Wagner, Christoph Rau, Scott Coburn and Lonny Berman for valuable discussions about the latest developments at the two new SR sources. Ross Harder assisted with the measurements and his phase constrained HIO algorithm [12] was used for inversion of the data in Figure 1. The experimental work was carried out at the Advanced Photon Source (APS) beamline 34-ID-C, which was operated by the Materials Research Laboratory of the University of Illinois under DOE contract DEFG02-91ER45439. APS is operated by DOE contract No W 31 109 ENG 38. This research was supported by the EPSRC.

Capacitor discharge sintering of nanocrystalline copper

A. Fais[1,*], P. Scardi[2]

[1] Department of Materials Science and Chemical Engineering, Turin Polytechnic, 10129 Corso Duca degli Abruzzi 24, Turin, Italy
[2] Department of Materials Engineering and Industrial Technologies, University of Trento, 38100 via Mesiano, 77, Trento, Italy
[*] alessandro.fais@polito.it

Keywords: capacitor discharge sintering, electron back scattering diffraction, line profile analysis, whole powder pattern modelling, nanocrystalline materials

Abstract. High purity copper was ground by high energy milling and the resulting nanocrystalline power sintered by Capacitor Discharge Sintering to dense (93% of theoretical density) bulk components. The evolution of the nanostructure from the starting nanocrystalline powder to the sintered component was studied by Whole Powder Pattern Modelling, a state of the art X-ray Diffraction methods for Line Profile Analysis, and by Electron Back Scattering Diffraction. As it results from the present work, the Capacitor Discharge Sintering process is very effective in densifying nanocrystalline metal powders with limited effects on the domain size and defect density.

1. Introduction

Capacitor Discharge Sintering (CDS) [1] is a novel Field Assisted Sintering Technique (FAST), where pre-compacted metallic powders are rapidly densified by a single, high current pulse. As Resistive Sintering has been derived from Resistive Welding technology [2], CDS has been derived from Capacitor Discharge Welding, a relatively new process for automatic welding of homogeneous and heterogeneous metal and composite junctions. The main advantages with respect to other FASTs and more traditional sintering techniques are: very short discharge time (of the order of 10 ms), that potentially prevents (or at least limits) annealing and grain coarsening effects; large applied pressure levels (up to 1 GPa, considerably higher than in other FASTs); high repeatability of the current impulses; possibility of process automation, allowing for interesting industrial applications [3].
Besides showing the main features of the CDS, with the specific example of pure copper, the present work addresses a basic, fundamental problem in nanotechnology, which is the accurate determination of the crystalline domain size in materials at the nano scale. Both the ball milled Cu powder and the bulk component produced by CDS were analysed by two different, and somehow complementary techniques: X-ray Diffraction (XRD) Line Profile Analysis (LPA) and Electron Back Scattering Diffraction (EBSD).

The XRD analysis was carried out by state-of-the-art LPA, using a modern Whole Powder Pattern Modelling (WPPM) [4-6] for the simultaneous analysis of domain size distribution and lattice defects, which in the present case are dislocations and (deformation and growth) stacking faults in the fcc copper structure. The EBSD application follows a recent evolution of electron microscopy, where the introduction of high resolution field emission guns (FEG) for the electron beams has considerably improved the resolution of the technique down to a few nanometers. Nanocrystalline domains and strain field can directly be imaged providing additional information on grain texture.

2. Experimental

Copper powders with 99.999% purity (Sigma-Aldrich) were ground in a high energy planetary mill apparatus Retsch PM-100 with a 50 ml WC jar and WC balls. Milling was carried out with a ball-to-powder weight ratio of 10:1 and a 3 ml addition of ethanol as wetting agent at 300 rpm. The milling process consists of the following steps: (1) 5 minutes on at 300 rpm, (2) 1 minute off, (3) inversion and back to (1) for a total time of 48 hours.
After evaporation of the wetting agent in air, the powders were sintered in the CDS apparatus sketched in figure 1. Current and applied tension were externally monitored by a PC-controlled USB two-channel oscilloscope. Electrical power dissipated by Joule effect was obtained from the integration over time of the dissipated electrical power. The main process parameter, directly related to densification, in analogy to other techniques [7], is called Specific Energy Input (SEI) and is given in units of [J/g]. For the present work nominal pressure at the electrodes was 350 MPa while SEI was 3100 J/g.

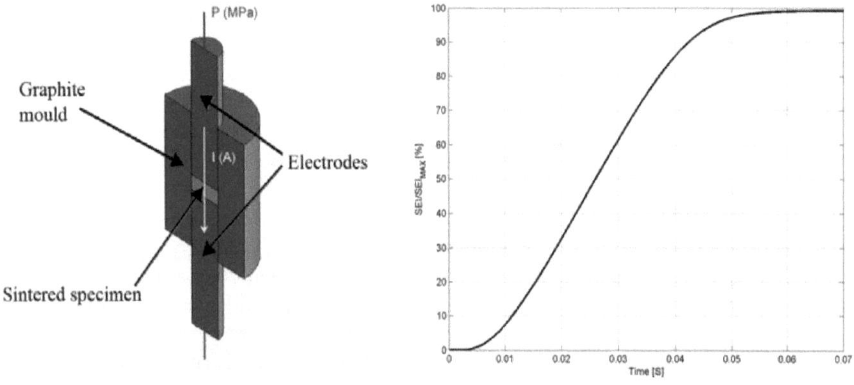

Figure 1. Schematics of the CDS mould [1,3] (left) and variation of normalized SEI (over maximum value) in time for the analysed sample (right).

In the CDS process the electrical energy varies sigmoidally in time. Maximum energy is usually reached at the end of the discharge (30-40 ms), although about 5 ms after the current starts flowing, 98% of the total energy has already been supplied to the powders (see figure 1). The sintered sample was removed from the mould, and its density measured with Archimedes' Principle. Vickers microhardness test was performed after proper polishing with

SiC papers and diamond pastes down to 1μm. For more accurate results, a final polishing step with colloidal silica was finally performed before EBSD.
EBSD patterns where measured on a Carl-Zeiss Supra 55 VP FEG SEM equipped with an Oxford Instruments – HKL Nordlys S type sensor and then analysed with Channel 5 software. The step size during the acquisition was set to 5 nm for high resolution and 100 nm for low resolution mapping.
XRD patterns on both powder and sintered component were collected on a Rigaku PMG-VH diffractometer, based on the conventional powder geometry, with narrow Soller slits (2°) and secondary graphite monochromator, especially optimized to produce highly symmetrical and narrow instrumental profiles, though at the expenses of relatively long counting time. CuKα radiation (produced at 40kV, 30mA) was used for both measurements. The Cu powder pattern was collected with 1° Divergence Slit (DS), 0.15 mm Receiving Slit (RS) and 105 s counting time per 0.05° step. The sintered component, being much smaller, required narrower slits (1/2° DS) to avoid sample illumination problems, and 90 s per 0.05° step.

3. Results and discussion

After ball milling SEM images of the ground powders evidence a plate like shape with a diameter of 30-40 μm and thickness of 5-15 μm, as shown in figure 2. After sintering a density of 8.265 g/cm^3, corresponding to 93% of theoretical density, was measured. Vickers microhardness testing over 19 different indentations revealed a normal distribution of values around 182 HV with a standard deviation of 12 HV. An estimate of the flow stress can then be obtained as $\sigma_y = HV/3 = 595\pm38$ MPa. This value, if used in the Hall-Petch relation reported in the literature [7], gives a mean grain size of 38±6 nm.

Figure 2. Morphology of ball milled Cu powder used in the CDS process.

The XRD patterns of ball milled Cu powder and CDS component were preliminarily studied by profile fitting (*MarqX* software, based on pseudo-Voigt peak shapes [8]), in order to separate overlapping peaks and background. Results were then analysed by the Williamson-Hall method [9,10], for a preliminary assessment of line profile effects. The result, shown in figure 3a for the powder, points out the important contribution to the diffraction line profile given by lattice defects, causing the steep slope in the plot.

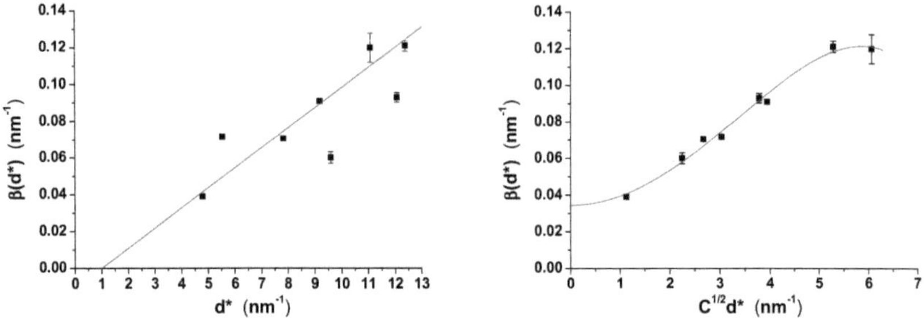

Figure 3. Williamson-Hall plot for the Cu powder: standard linear plot (a) and modified plot, accounting for dislocation anisotropy (b)

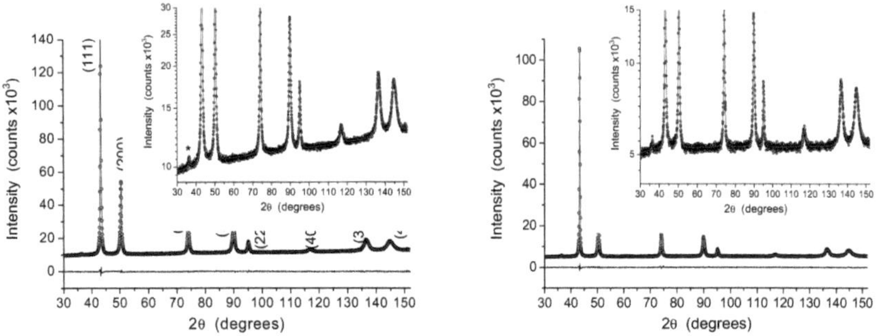

Figure 4. Results of WPPM for ball milled powder (a) and sintered component (b). The asterisk in the inset of (a) marks the most intense line of the Cu_2O phase.

The intercept, whose inverse should provide the volume-weighted mean column length $<L>_V$ [4,10], is negative because the line broadening anisotropy due to lattice defects was not considered. To account for dislocation anisotropy the data were reanalysed considering the Cu contrast factor ($\overline{C}_{hkl} = A + B \cdot H$, where $H = (h^2k^2 + k^2l^2 + l^2h^2)/(h^2 + k^2 + l^2)^2$, A_e=0.3046, B_e=0.5002 for edge and A_s=0.2983, B_e=0.7088 for screw dislocations respectively [11]). The result, shown in figure 3b, gives $<L>_V$=29(3) nm from the inverse of the intercept, whereas the refined B value is 0.735, slightly above the limit for screw dislocations. This suggests that faulting, not considered in the analysis of figure 3b, might have a role.
After the preliminary screening by the W-H plot, the patterns of powder and sintered components were both analysed by WPPM, following the approach described in the recent literature [4-6]. Since the main contribution is expected from lattice defects, crystalline domains can sufficiently well be accounted for by a lognormal distribution of spheres. Lattice defects considered for Cu were edge/screw dislocations and deformation/twin faulting. In fact, both defects are relatively low energy in copper and may be present simultaneously.
Figure 4a shows the modelling result for the nanocrystalline powder. As shown in the inset, in addition to the fcc Cu peaks, weak reflections of Cu_2O are also observed. As a consequence the WPPM analysis was made considering both phases.

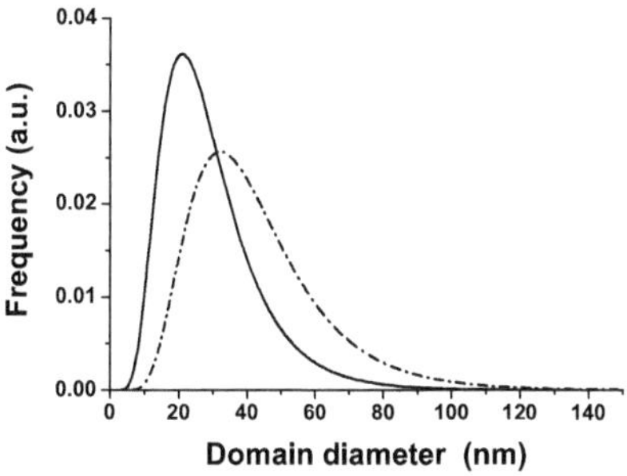

Figure 5. Domain size distribution for ball milled powder (line) and sintered component (dash).

A similar analysis was carried out for the CDS component, and results are shown in figure 4b. The two patterns are qualitatively similar, the Cu_2O small fraction being present in both. Figure 5 shows the size distributions obtained by WPPM, whereas additional results are reported in table 1. As expected, the high energy milling leads to very fine domains (D=29 nm) tending to grow during the CDS process (D=43 nm). Sintering broadens the size distribution, and the dislocation density, about 1.5 x10^{16} m^{-2} in the as milled powder, is nearly halved. Faulting is present in both samples, with a predominance of twins, and decreases slightly after the CDS. It is worth noting that, when faulting is considered, the refined contrast factor is well within the range for edge and screw dislocations: the effective fraction of edge dislocations (screw fraction $f_S=1-f_E$) is 0.46-0.47 for both powder and sintered component.
In both cases Re is relatively small, giving a Wilkens parameter ($R_e\rho^{1/2}$) below unity, formally outside the limits of validity of the theory [12]. This is quite a frequent observation in highly deformed metals, possibly pointing out a strong interaction between dislocations (e.g., see [13] and references therein). As is known, dislocations in nanocrystalline materials formed by severe plastic deformation tend to arrange in low-energy patterns (as excess dislocation boundaries), which are quite different from what observed for plastic deformation of materials in massive components and domain in the micrometer scale [14].
Beyond all those considerations, a basic evidence of this work is that the CDS process reached a high degree of compaction (density 8.3 g cm^{-3}) with a relatively limited loss of nanostructural properties: domain size is still well below 100 nm, and the dislocation density is just halved with respect to the starting powder, in any case close to 10^{16} m^{-2}.

Table 1. WPPM results: unit cell parameter (a_0), mean domain size (D) and standard deviation (s.d.) of the lognormal distribution; dislocation density (ρ) and effective outer cut-off radius (R_e); fraction of edge dislocations (f_E), deformation (α) and twin (β) fault probability. Estimated standard deviations are given in parenthesis and refer to the last significant digit.

	a_0 (nm)	D (nm)	s.d. (nm)	ρ (10^{16} m^{-2})	R_e (nm)	f_E	α (%)	β (%)
Ball milled powder	0.361600 (5)	29 (2)	15 (1)	1.5 (1)	2.9 (5)	0.46 (3)	0.41 (2)	0.60 (5)
CDS bulk component	0.361611 (4)	43 (4)	20 (3)	0.80 (5)	5.1 (4)	0.47 (4)	0.30 (2)	0.54 (5)

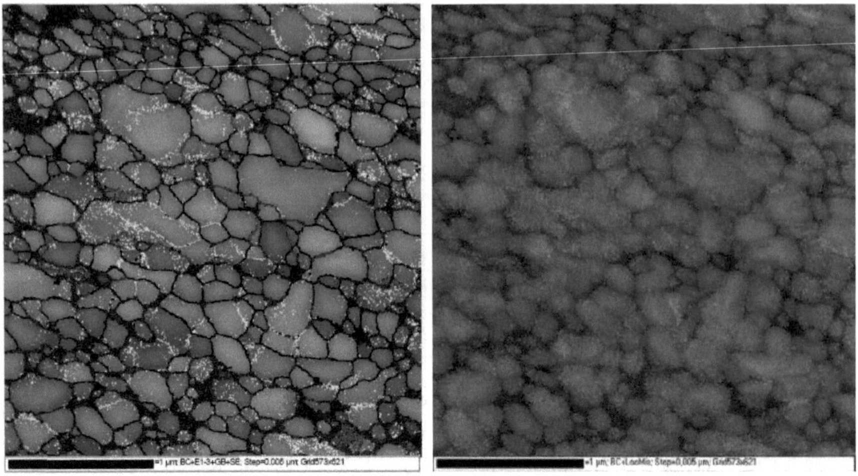

Figure 6. EBSD Map of CDS copper. On the left, grain boundaries are shown in black, twin grain boundaries in red, local misorientation due to dislocations in yellow. Local misorientation alone is shown on the right.

Indexing of the Kikuchi bands in the creation of EBSD maps was relatively successful and stayed above 60% of the whole map, with over 70% area of successful indexing (figure 6). The remaining 30% of non indexed area is attributable either to very small grains (<10-15 nm, two or three pixels in a 5 nm step size map) or to highly strained portions of material. Automatic mean grain size determination and plot of the grain size distribution has not given results in agreement with WPPM. This is probably due to the high percentage of single pixel or two-pixel "grains" detected by the software. Furthermore it is a well known issue that effective resolution in EBSD still needs some improvement to obtain reliable grain size data in nanocrystalline materials [15].

Local misorientation is also shown in figure 6 (right), relative to 3 by 3 pixels correlations. Even if most literature [16-17] correlates this parameter with plastic strain, no formalism has yet been developed to correlate these results to other microstructural, like, e.g., the dislocation density. Compared to EBSD maps of unstrained metals, figure 6 is a clear evidence, though on a semi-quantitative level, of a high dislocation density and residual strain in the sintered sample.

The following maps (figure 7) concern local texture, a feature hardly accessible by diffraction techniques with such a high level of detail and spatial resolution. Both pictures and orientation maps reveal a very fine grained and homogeneous microstructure and a local texturing due to the different orientation of the plate-like grains in the sintered compact. In fact, while the pole figures for the whole micrograph in figure 7 suggest a rather homogeneous microstructure, subsets taken in the two areas marked on the micrograph reveal a local texture. With respect to the cylindrical shape of the sintered component, subsets 2 and 3 refer to a plane, respectively parallel and perpendicular to the basis. All of this demonstrate that EBSD can complement the information from other microscopies and by XRD LPA to better understand the mechanisms of sintering of nanostructured powders.

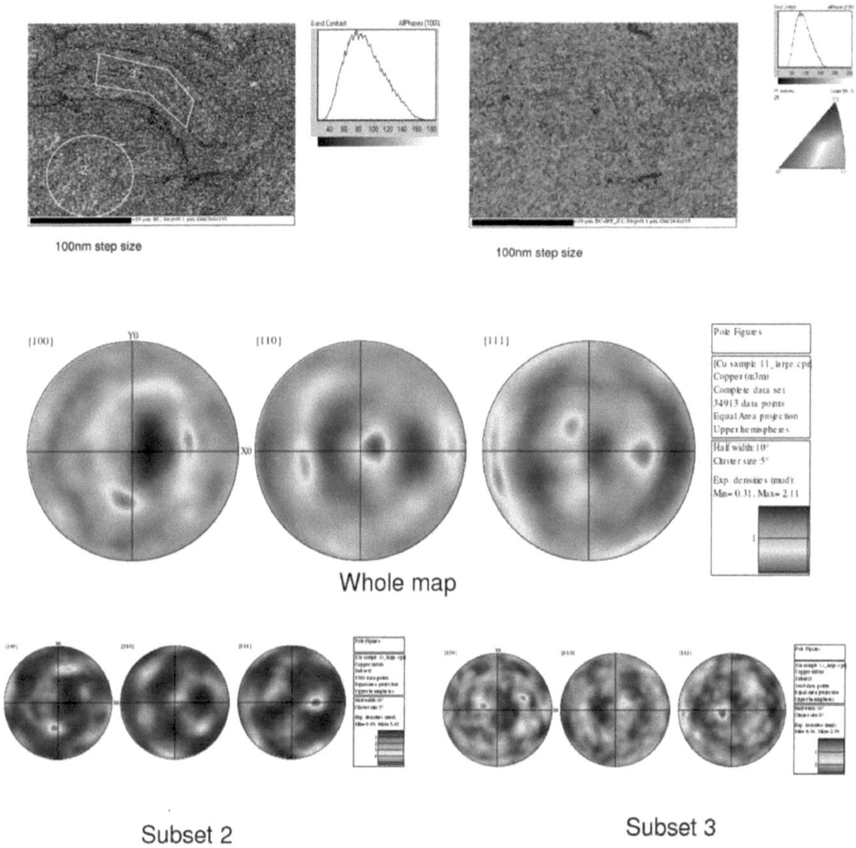

Figure 7. EBSD results of CDS copper: low resolution map and associated pole figures showing texture inside the particles after sintering. (100), (110) and (111) pole figures are shown for the whole area in the micrograph (upper corner, left), and as subsets 2 and 3, referred to the two marked areas, respectively.

4. Concluding remarks

The present work showed how a novel process, Capacitor Discharge Sintering, can be used to produce dense metallic components starting from highly defected, nanocrystalline metal powders. The main difference with respect to traditional sintering processes, is the possibility to preserve, at least partly, the nanostructure of the starting powders, a feature made possible by the very short length of the CDS process.

XRD Line Profile Analysis proved useful to follow the evolution of crystalline domain size and lattice defects from the ball milled starting powder to the CDS component. In particular, features of the domain size distribution and detailed information of lattice defect nature and amount could be obtained by Whole Powder Pattern Modelling, beyond the qualitative indications of integral breadth methods. EBSD showed several features on the distribution of strains and orientations in sintered component that could not be easily observed by diffraction or by other microscopy techniques. As such, XRD LPA and EBSD can be proposed as very practical and informative techniques to support sintering process modelling and optimization.

References

1. Fais, A., 2004, *La Sinterizzazione a Scarica Capacitiva – Il processo, i materiali e la loro caratterizzazione*, MoS thesis, Politecnico di Torino, Italy.
2. Lenel, F.V., 1955, *J.O.M. Transactions AIME*, **7**, 158.
3. Fais, A. & Maizza, G., 2007, *J. Mat.Proc.Tech.*, in press.
4. Scardi, P. & Leoni, M., 2001, *Acta Cryst. A*, **57**, 604.
5. Scardi, P. & Leoni, M., 2002, *Acta Cryst. A*, **58**, 190.
6. Leoni, M. & Scardi, P., 2004, *J. Appl. Cryst.*, **37**, 629.
7. Cheng, S., Ma, E., Wang, Y.M., Kecskes, L.J., Youssef, K.M., Koch, C.C., Trociewitz, U.P. & Han, K., 2005, *Acta Mater.* **53,** 1521.
8. Dong, Y.H. & Scardi, P., 2000, *J. Appl. Cryst.*, **33**, 184.
9. Williamson, G.K. & Hall, W.H., 1953, *Acta Metall.* **1**, 22.
10. Scardi, P., Leoni, M. & Delhez, R., 2004, *J. Appl. Cryst.*, **37**, 381.
11. Leoni, M. Martinez-Garcia, J. & Scardi, P., 2007, *J. Appl. Cryst.*, **40**, 719.
12. Wilkens, M., 1970, in *Fundamental Aspects of Dislocation Theory*, NBS Special Publication No. 317, Vol. II, edited by Simmons, J.A., de Wit, R. & Bullough, R. (Washington, DC: National Bureau of Standards), pp. 1195-1221.
13. Scardi, P., 2005, *Z. Metallkd*, **96**, 698.
14. Valiev, R.Z., Islamgaliev, R.K. & Alexandrov, I.V., 2000, *Progr. Mater. Sci.* **45**, 103.
15. Humphreys, F.J., 2004, *Scr. Mater.,* **51,** 771.
16. Maitland, T. & Sitzman, S., 2007, in *Scanning Microscopy for Nanotechnology Techniques and Applications*, edited by Zhou W. & Wang Z.L., (Berlin: Springer). pp. 41-75.
17. Sutliff, J.A., 1999, *Microsc. Microanal. (Suppl. 2)*, **5**, 236.

Acknowledgements. Authors thank Dr M. D'Incau for assistance in XRD data collection and Dr Ali Gholinia for EBSD data.

Microstructure study on BN nanocomposites using XRD and HRTEM

M. Motylenko[1,*], V. Klemm[1], G. Schreiber[1], D. Rafaja[1], M. Schwarz[2], T. Barsukova[2], E. Kroke[2]

[1] Institute of Materials Science, TU Bergakademie Freiberg, Germany
[2] Institute of Inorganic Chemistry, TU Bergakademie Freiberg, Germany
[*] motylenk@ww.tu-freiberg.de

Keywords: boron nitride, nanocomposites, HP/HT synthesis, XRD, HRTEM

Abstract. X-ray diffraction and high-resolution transmission electron microscopy were used for microstructure analysis of the bulk boron nitride (BN) nanocomposites containing the graphitic, wurtzitic and sphaleritic modifications of BN that were produced via high-pressure and high-temperature synthesis. The main tasks of the microstructure analysis were to quantify the phase composition of the nanocomposites, to determine the crystallite size and to visualise the atomic ordering at the interfaces between individual nanocrystallites. A particular task of the transmission electron microscopy was to look for possible orientation relationships between the nanocrystallites with different crystal structures in order to be able to explain the origin and the role of the high density of microstructure defects in the BN nanocomposites.

1. Introduction

Boron nitride (BN) nanocomposites are considered as alternative ultra-hard materials for production of cutting tools, particularly in the applications, in which diamond tools would react chemically with the work pieces. An important property of such materials is their high hardness, which increases with decreasing crystallite size according to the well-known Hall-Petch relationship [1, 2] and with increasing density of the internal interfaces that hinder the movement of microstructure defects. In [3], it was shown that the nano-structuring of the BN nanocomposites can be done via phase transformation, which is running during high-pressure and high-temperature (HP/HT) synthesis. During the HP/HT synthesis, graphitic BN (*h*-BN, space group $P6_3/mmc$) usually transforms via the meta-stable wurtzitic BN (*w*-BN, $P6_3mc$) in the sphaleritic BN (*c*-BN, F-43m) by maintaining the following orientation relationships between the individual phases [4, 5]: $(0002)_h \parallel (0002)_w \parallel (111)_c$ and $\langle 11\bar{2}0\rangle_h \parallel \langle 11\bar{2}0\rangle_w \parallel \langle \bar{1}10\rangle_c$. Another transformation path is a direct *h*-BN to *c*-BN conversion [6]. In the samples, which were synthesised from hot-pressed *h*-BN with the crystallite size of approximately 100 nm, the presence of the above orientation relationship between *h*-BN, *w*-BN and

c-BN led to a very strong local preferred orientation of transformed nanocrystallites of the same phase [7]. The transfer of the preferred orientation between adjacent nanocrystallites caused their partial coherence, which was visible via the interference of X-rays scattered on different nanocrystallites [8]. The mutual disorientation of neighbouring nanocrystallites as calculated from their partial coherence for X-ray scattering was approximately 1.5° [9].
However, the partial coherence of nanocrystallites for X-ray scattering is observed in a very narrow range of their disorientations that depends on the crystallite size and on the minimum size of the reciprocal lattice vector. For large (> 15 nm) or mutually disoriented (> 2.5°) crystallites, no partial coherence effects can be observed [9]. The aim of this work was to show possible orientation relationships between different phases in the BN system, the effect of these orientation relationships on the disorientation of neighbouring nanocrystallites and the effect of these crystallites disorientation on the X-ray diffraction (XRD) line broadening.

2. Experimental details

The BN nanocomposites under study were synthesised from the pyrolytic hexagonal boron nitride (PBN) that contained less than 0.8 wt. % of oxygen as determined by the electron probe microanalysis with wavelength-dispersive spectroscopy. XRD measurements on the starting material confirmed the expected strong preferred orientation of crystallites. For the synthesis, cylindrical pieces of PBN with the diameter of 2.7 mm and the height of 4.4 mm were cut out, so that the cylinder axis was parallel to the {001} texture. The HP/HT synthesis was performed in a hydraulically driven 6-8 type multi-anvil press [3]. The synthesis parameters that were employed in individual experiments are summarized in Table 1.

Table 1. Parameters of the HP/HT synthesis

Sample	Pressure	Temperature	Dwell time
PBN-1	15 GPa	1650°C	74 min
PBN-2	15 GPa	1550°C	0.5 min
PBN-3	9 GPa	1560°C	1 min

For microstructure investigations on the starting material and on the products of the HP/HT synthesis, the combination of XRD and high-resolution transmission electron microscopy (HRTEM) was used. XRD was employed mainly for the quantification of the phase composition and for the determination of the mean crystallite size. The XRD measurements were performed on a Bruker D8 diffractometer equipped by a sealed tube with copper anode and by two Goebel mirrors. The first one was located in the primary beam, the second in the front of a scintillation detector. The diffraction patterns were taken in the symmetrical diffraction geometry. The advantage of the full parallel-beam optics against the Bragg-Brentano geometry was its non-sensitivity to the surface roughness of the specimens. The phase analysis was done using the Fullprof suite [10], which is one of the implementations of the Rietveld method [11]. The analysis of the crystallite size was performed using the Williamson-Hall analysis [12] after correcting the measured XRD line broadening for instrumental effects. The instrumental line broadening was measured with the LaB_6 standard from NIST. The crystallite size obtained from the XRD line profile analysis was verified by HRTEM. HRTEM was done on a 200kV analytical high-resolution TEM JEM 2010 FEF from Jeol equipped with ultra-high-resolution objective lens (C_s = 0.5 mm) and an in-column energy filter, which selected only the elastic electrons for the image formation. The most important task for HRTEM was the determination of the orientation relationships between neighbouring crystallites with the aid of the Fast Fourier Transform (FFT) of selected regions in the HRTEM micrographs. This method is known to have a rather low accuracy, but the more accurate method that is using the Kikuchi patterns for the disorientation measurements [13] was not applicable because of too low diffraction contrast of small crystallites in the BN nanocomposites.

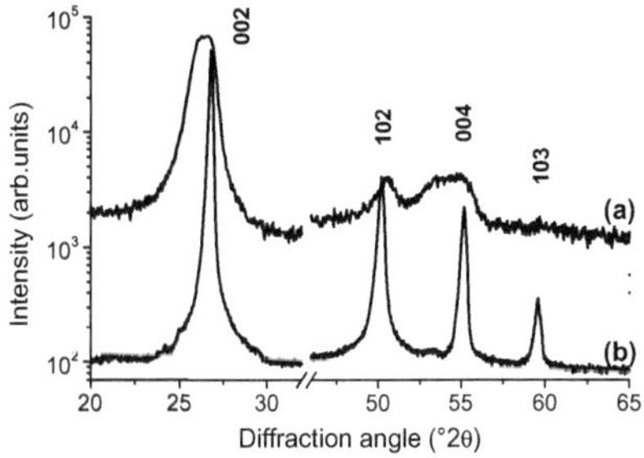

Figure 1. A part of the diffraction pattern of the PBN starting material (a) as compared with the diffraction pattern of a hot-pressed graphitic BN (b).

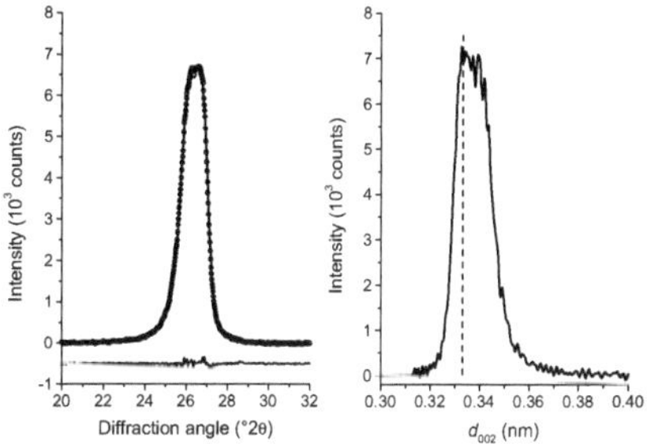

Figure 2. Left: measured (circles), re-convoluted (upper line) intensity of the diffraction line (002). Bottom line shows the difference between them. Right: Distribution of the interplanar spacing as calculated from the deconvoluted profile. Dashed line indicates the intrinsic interplanar spacing of h-BN taken from [14].

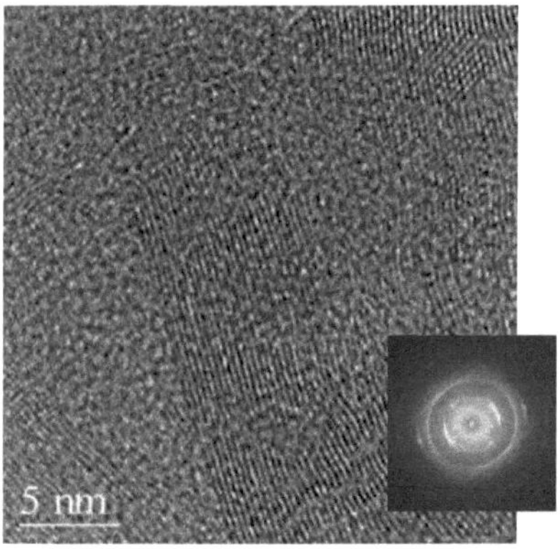

Figure 3. HRTEM micrographs of the h-BN starting material and its FFT (inset) showing the strong {001} preferred orientation of crystallites and a large disorientation of crystallites in the {hk0} directions.

2.1 Microstructure analysis of the starting material

XRD measurements done on the PBN starting material have shown both the {001} texture and a large expected variation of the interplanar spacings $d_{00\ell}$ due to the puckering effects [4]. The large variation of the interplanar spacings $d_{00\ell}$ was indicated by the huge broadening of the diffraction lines (002) and (004) in figure 1 and confirmed by the shape of the distribution of the interplanar spacing d_{002}, which is shown in figure 2. The distribution of the interplanar spacing d_{002} was calculated from the deconvolution of the diffraction line (002). In order to visualise the differences between different starting materials (pyrolytic and hot-pressed *h*-BN), the hot-pressed *h*-BN (figure 1b) was used as a "standard" for the deconvolution. HRTEM confirmed the {001} texture in the axis of the cylindrical sample and illustrated high disorientation of crystallites in the {*hk*0} directions (figure 3). Because of a large spread of the interplanar spacings $d_{00\ell}$ due to the puckering, the size of crystallites in the pyrolytic *h*-BN could only be calculated from the broadening of the diffraction lines (*hk*0) to be approximately 7 nm. This crystallite size was confirmed by HRTEM (figure 3).

2.2 XRD analysis of the BN nanocomposites

According to XRD, *c*-BN was the dominant phase in all three samples under study, i.e. in PBN-1, PBN-2 and PBN-3. Besides, XRD identified the metastable *w*-BN and the remnants of *h*-BN in all samples. The phase compositions of individual samples as obtained from the Rietveld analysis [11] are summarised in Table 2 together with the crystallite size in *c*-BN and *w*-BN, which was determined from the XRD line broadening. Figure 4 shows the Williamson-Hall plots for *c*-BN in all samples under study. It follows from the comparison of the phase compositions and the crystallite sizes given in Table 2 with the parameters of the HP/HT synthesis shown in Table 1 that the increase of the pressure at the same temperature (samples PBN-2 and PBN-3) during the HP/HT synthesis accelerates the conversion of starting *h*-BN to *c*-BN. Also the increase of the temperature and the dwell time at the constant pressure of 15 GPa (compare samples PBN-1 and PBN-2) forces the phase transformation from *h*-BN to *c*-BN. However, the increase of the *c*-BN amount in the sample PBN-1 as compared with the amount of *c*-BN in the sample PBN-2 is much lower than the increase of the amount of *c*-BN between the samples PBN-3 and PBN-2. This effect looks like a saturation of the phase transformation rate for *c*-BN amounts approaching 100 %. As a possible reason for the retardation of the phase transformation rate, the microstructure defects in *w*-BN and in the bits of *h*-BN that both hinder the phase transformation were discussed in [7].

Table 2. Phase composition and crystallite size as calculated from the XRD data. The crystallite size in h-BN could not be determined from the XRD line broadening because of the small amount of this phase in the samples.

Sample	Phase composition, wt%			Crystallite size, nm		
	PBN-1	PBN-2	PBN-3	PBN-1	PBN-2	PBN-3
c-BN	96 ± 4	92 ± 4	75 ± 4	45 ± 7	34 ± 2	14 ± 3
w-BN	3 ± 1	6 ± 2	15 ± 3	7 ± 2	19 ± 3	10 ± 4
h-BN	~ 1	~ 2	10 ± 3			

Other microstructure defects that can obstruct both the *h*-BN (→ *w*-BN) → *c*-BN phase transformation and the growth of the *c*-BN crystallites can be the presence of the original grain boundaries in the starting *h*-BN. In the BN nanocomposites that were synthesised from the hot-pressed *h*-BN [7], the maximum size of the *c*-BN crystallites did not exceed the size of the crystallites in the starting material, which was approximately 100 nm. Thus, the crystallites boundaries in the starting hot-pressed *h*-BN seemed to act as powerful barriers for the growth of the *c*-BN crystallites [7].

The size of the *h*-BN crystallites in PBN was much smaller than in the hot-pressed *h*-BN – approximately 7 nm as determined using XRD and confirmed using HRTEM. In PBN, HRTEM revealed additionally amorphous boron nitride (*a*-BN) between *h*-BN crystallites (figure 3). Therefore, the pyrolytic *h*-BN is an appropriate starting material for investigation of the growth of the *c*-BN crystallites during the HP/HT synthesis, especially if the growth of *c*-BN crystallites over the original crystallites boundaries in the starting material is the subject of the study. The crystallites in the starting material are sufficiently small, thus the barriers for the growth are very near to each other. The mutual disorientation of crystallites in PBN is advantageous for investigation of the possible orientation relationships between *h*-BN, *w*-BN and *c*-BN.

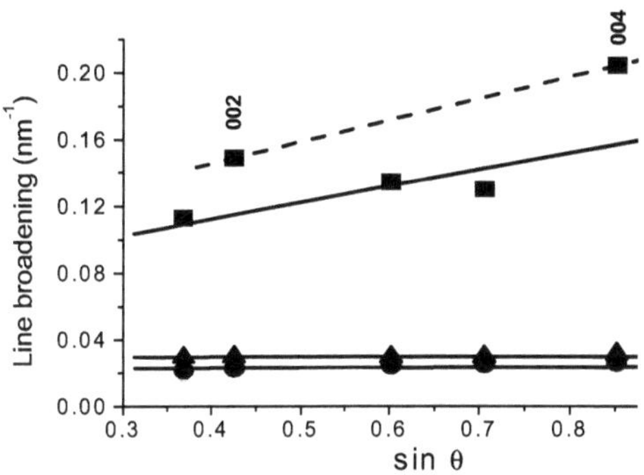

Figure 4. Williamson-Hall plots of the samples PBN-1 (circles), PBN-2 (triangles) and PBN-3 (boxes). The solid lines show the respective linear fit of the physical line broadening. For the sample PBN-3, the broadening of the diffraction lines (002) and (004) was excluded from the fit as these data are apparently influenced by the residual puckering of the basal planes.

The first insight into the microstructure of the *c*-BN crystallites in the samples synthesised from PBN was obtained from the dependence of the XRD line broadening on the size of the diffraction vector (from the Williamson-Hall plot) shown in figure 4. The relatively small and almost constant XRD line broadening in the samples PBN-1 and PBN-2 indicated the presence of defect-free *c*-BN crystallites having the size of 47 and 34 nm, respectively. This crystallite sizes exceed the crystallite size in the starting material several times, thus the crystallites boundaries in the starting PBN could not hinder the phase transformation. However, the phase transformation and the growth of the *c*-BN crystallites across the original crystallites boundaries must be connected either with the re-ordering of the atoms like in the recrystallisation process or with the transfer of the crystallites orientation through alternative crystallographic directions in the neighbouring crystallites. In the sample PBN-3, an increase of the XRD line broadening with increasing size of the diffraction vector was observed that is characteristic for the presence of microstructure defects within crystallites, although the *c*-BN crystallites in this sample were much smaller than the *c*-BN crystallites in the samples PBN-1 and PBN-2. Moreover, the XRD line broadening observed on the *c*-BN nanocrystallites from the sample PBN-3 did not show the steep increase, which is typical for the partial coherence of nanocrystallites and which was observed for this crystallite size in the samples produced from the hot-pressed *h*-BN [9]. A possible explanation of this phenomenon would be a very small disorientation of neighbouring nanocrystallites that does not destroy their partial coherence within the accessible range of the diffraction vector. This means that the steep increase of the XRD line broadening would first be observed at the diffraction vectors, which are not accessible within the reachable 2θ range of the diffractometer (at the wavelength of the radiation used). Therefore, only the size of domains consisting of partially coherent crystallites can be obtained from XRD [15], not the size of individual crystallites. These domains had the size of approximately 14 nm as it can be seen from Table 2. However, the results of XRD must be verified by electron microscopy, particularly in this disputable case.

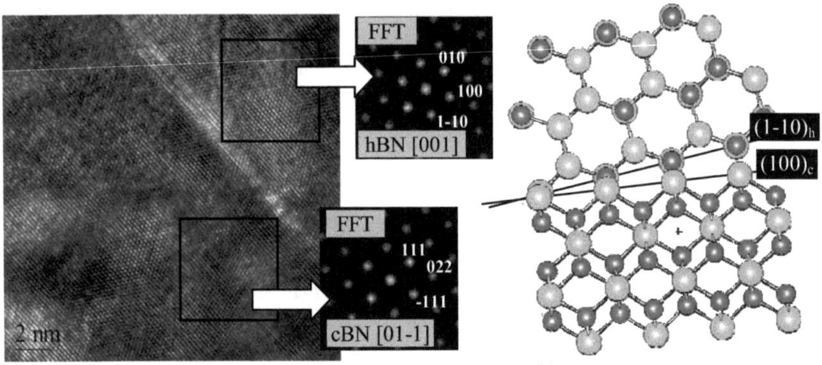

a) b)

Figure 5. An interface between h-BN/w-BN [001] and c-BN [01-1] in PBN-3: HRTEM and FFTs of selected regions (a); suggested atomic positions in both phases (b).

2.3 HRTEM analysis of the BN nanocomposites

The HRTEM micrograph of the sample PBN-3 (figure 5a) revealed a negligible disorientation of nanocrystallites in *c*-BN even at the boundary to *h*-BN, which is expected to accommodate the maximum density of microstructure defects. The presence of the nanocrystallites can only be guessed from the fluctuations of the contrast in the HRTEM micrograph. Their size is approximately 5 nm, which is less than the size of partially coherent domains (~ 14 nm) that was obtained from XRD. The reverse calculation of the XRD line broadening from the size of nanocrystallites estimated using HRTEM yields that the maximum XRD line broadening would be 0.2 nm^{-1}.This is not in contradiction to the results shown in figure 4, where the experimental XRD line broadening did not exceeded this maximum. The Fast Fourier Transform of the HRTEM micrograph from figure 5a was used for determination of the mutual orientation of *c*-BN and *h*-BN at the phase boundary. The indexation of the FFT patterns (the centre of figure 5) was done by means of the measurement of the interplanar distances and the angles between the lattice planes. A possible atomic arrangement at the *c*-BN/*h*-BN interface is proposed in figure 5b. The corresponding orientation relationship between *h*-BN and *c*-BN is $(01\bar{1}0)_h \parallel (111)_c$ and $\langle 0001 \rangle_h \parallel \langle \bar{1}10 \rangle_c$.

One example for another orientation relationship between *h*-BN and *c*-BN, $(01\bar{1}1)_h \parallel (111)_c$ and $\langle 21\bar{3}\bar{1} \rangle_h \parallel \langle \bar{1}10 \rangle_c$, is given in figure 6. The angle between the lattice planes $(01\bar{1}0)_h$ and $(01\bar{1}1)_h$ is ~18° as calculated for the lattice parameters $a = 0.250$ nm and $c = 0.666$ nm. This angle is probably compensated through the orientation relationship between *h*-BN crystallites, $(10\bar{1}1)_h \parallel (01\bar{1}\bar{1})_h$ and $\langle \bar{1}011 \rangle_h \parallel \langle \bar{2}\bar{1}3\bar{1} \rangle_h$, that is shown in figure 7. The angle between the lattice planes $(10\bar{1}1)_h$ and $(01\bar{1}\bar{1})_h$ in the elementary cell of *h*-BN with the above lattice parameters is ~69°, which matches well with the complement of the angle between $(01\bar{1}0)_h$ and $(01\bar{1}1)_h$ to 90°. On the other hand, the angle of ~69° agrees well, for instance, with the angle between the lattice planes $(111)_c$ and $(11\bar{1})_c$ in *c*-BN. Still, the lattice mismatch, which is a consequence of the slight disagreement of the angles and atomic positions in figures 5 – 7, leads to the generation of microstructure defects at the phase or grain boundaries. The energy accumulated in such microstructure defects prevents probably the phase transformation in its late stage and contributes substantially to the preservation of *h*-BN in samples with high *c*-BN contents.

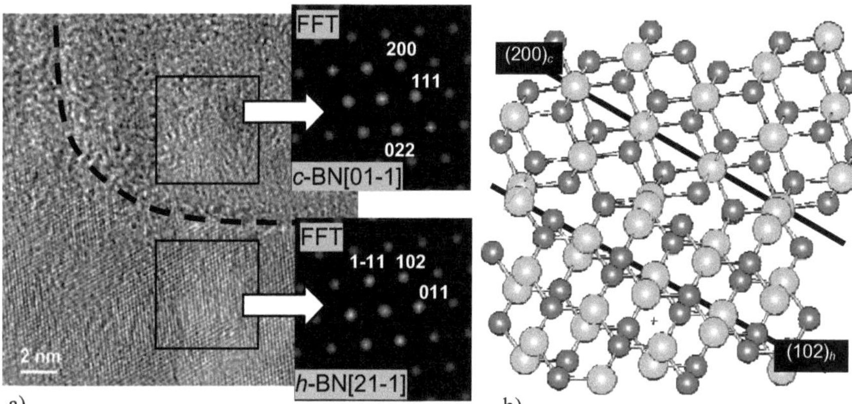

a) b)

Figure 6. Partial coherence boundary (dashed line) between c-BN and h-BN phases in PBN-1: HRTEM micrograph with FFTs of selected regions (a); possible atomic positions in both phases.

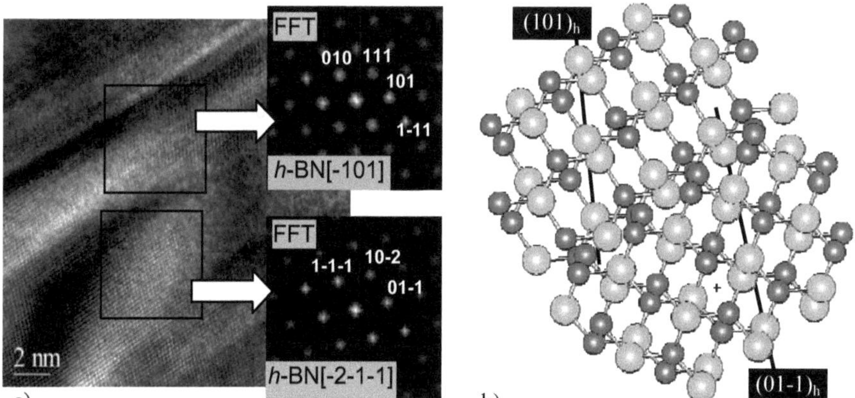

a) b)

Figure 7. Interfaces between h-BN nanocrystallites in PBN-1.

3. Conclusions

Bulk BN nanocomposites synthesised from the pyrolytic boron nitride using HT/HP conversion process were investigated using XRD and HRTEM. XRD revealed the crystallite size and the distribution of the interplanar spacing d_{002} in the starting material. Furthermore, XRD was applied to get the phase composition and the crystallite size in the nanocomposites. It was found that the transformation process is retarded in nanocomposites containing over 90 % of *c*-BN, which was explained by the presence of microstructure defects mainly in *h*-BN, which hinder the phase transformation, like stacking faults or twins. The presence of these defects was correlated to the orientation relationships found using HRTEM and FFT of the HRTEM micrographs.

References

1. Hall, E.O., 1951, *Proc. Phys. Soc. London Sect. B*, **64**, 747.
2. Petch, N.J., 1953, *J. Iron Steel Inst.*, **174**, 25.
3. Rubie, D.C., 1999, *Phase Transitions*, **68**, 431.
4. Kurdyumov, A.V., Britun, V.F. & Petrusha, I.A. 1996, *Diamond Relat. Mater.*, **5**, 1229.
5. Huang, J. & Zhu, Y.T., 2000, *Defect and Diffusion Forum*, **186-187**, 1.
6. Horiuchi, S., He, L.-L., Onoda, M. & Akaishi, M., 1996, *Appl. Phys. Lett.*, **68**, 182.
7. Rafaja, D., Klemm, V., Motylenko, M., Schwarz, M. R., Barsukova, T., Kroke, E., Frost, D., Dubrovinsky, L. & Dubrovinskaia, N., 2008, *J. Mater. Res.*, 23, 981.
8. Rafaja, D., Klemm, V., Schreiber, G., Knapp, M. & Kužel, R., 2004, *J. Appl. Cryst.*, **37**, 613.
9. Rafaja, D., Klemm, V., Wüstefeld, C., Motylenko, M., Dopita, M., Schwarz, M., Barsukova, T. & Kroke, E., *this issue*.
10. Roisnel, T. & Rodriguez-Carvajal, J., 2001, *Mater. Sci. Forum*, **378-3**, 118.
11. Rietveld, H.M., 1967, *Acta Cryst.*, **22**, 151; Rietveld, H.M., 1969, *J. Appl. Cryst.*, **2**, 65.
12. Williamson, G.K. & Hall, A.H., 1953, *Acta Metall.*, **2**, 22.
13. Straube, H., Klemm, V., Klimanek, P. & Motylenko, M., 2003, *in:* Problems of materials science, Eds: Rybin V. & Teodosiu C., N1 (**33**), 124.
14. Inorganic Crystal Structure Database, 2007, FIZ Karlsruhe.
15. Rafaja, D., Wüstefeld, C., Dopita, M., Růžička, M., Klemm, V., Schreiber, G., Heger, D. & Šíma, M., 2007, *Surf. Coat. Technol.*, **201**, 9476.

Acknowledgements. The authors acknowledge gratefully the financial support of the present work through the German Research Council (DFG), project # RA 1050/9 and KR 1739/13, within the Priority Programme Nr. 1181 „Nanoskalige anorganische Materialien durch molekulares Design".

PLASTIC DEFORMATION

Impact of dislocation cell elastic strain variations on line profiles from deformed copper

L. E. Levine[1,*], B. C. Larson[2], J. Z. Tischler[2], P. Geantil[3], M. E. Kassner[3], W. Liu[4], M. R. Stoudt[1]

[1] Materials Science & Engineering Laboratory, National Institute of Standards and Technology, Gaithersburg, MD 20899-8553, USA
[2] Materials Science & Technology Division, Oak Ridge National Laboratory, Oak Ridge, TN 37831, USA
[3] Department of Aerospace and Mechanical Engineering, University of Southern California, Los Angeles, CA 90089, USA
[4] Advanced Photon Source, Argonne National Laboratory, Argonne, IL 60439-4800, USA
[*] Lyle.Levine@nist.gov

Keywords: micro-diffraction, dislocations, residual stress, plastic deformation

Abstract. Energy scanned, sub-micrometer X-ray beams were used to obtain diffraction line profiles from individual dislocation cells in copper single crystals deformed in compression. Sub-micrometer depth resolution was provided by translating a wire through the diffracted beams and using triangulation to determine the depths of the diffracting volumes. Connection to classic volume-averaged results was made by adding the line profiles from 52 spatially resolved dislocation cell measurements. The resulting sub profile is smooth and symmetric, in agreement with early assumptions; the mean strain and full width half maximum are consistent with the average of the parameters extracted from the more exact individual dislocation cell measurements.

1. Introduction

When metals are plastically deformed, their mechanical properties change dramatically in a process called work hardening. The physical process that underlies this behaviour is the formation and propagation of dislocations through these metals. As deformation proceeds, these dislocations strongly interact through long range ($1/r$) stress fields and self-organize into three-dimensional cellular structures that impede the motion of mobile dislocations. These cellular arrangements are composed of cell walls with high dislocation density surrounding cell interiors with much lower dislocation density.

Since dislocations interact primarily through their stress fields, the mesoscopic distribution of stresses within a cellular dislocation structure plays a critical role in the work hardening process, and this topic has received considerable attention in the literature over the past few decades [1-9]. On the theoretical side, the composite model originally proposed by Mughrabi is the most widely accepted [1]. Here, the dislocation cell walls and cell interiors are considered as strong and weak components, respectively, that deform compatibly under an externally applied load. Thus, under an applied compressive load, the axial compressive stress within the cell walls would be larger than that within the cell interiors. After elastic unloading to zero applied stress, the cell walls would therefore still be in compression and the cell interiors would be in tension.

Until recently, the strongest experimental evidence for the composite model came from X-ray diffraction line profiles from plastically deformed metal single crystals (primarily copper) [1-6]. As a crystal deforms, the line profiles broaden considerably and exhibit a pronounced peak shape asymmetry, which is interpreted as the superposition of two approximately symmetric line profiles corresponding to diffraction from the cell walls and cell interiors. The separation between these sub-profiles in reciprocal space corresponds to the difference in elastic strain (and thus stress).

Although this explanation is generally accepted, definitive and quantitative measurements were not possible before the recent development of intense sub-micrometer X-ray beams at third-generation synchrotrons. In a recent study, Levine *et al.* [9] used energy scanned differential aperture X-ray microscopy (DAXM) with 0.5 μm X-ray beams [10,11] to measure the elastic strains within individual, spatially resolved dislocation cells in single crystal Cu samples deformed in uniaxial tension and compression along a <001> axis. The measurements by Levine *et al.* had a spatial resolution of 0.5 μm in all three dimensions. These spatially resolved measurements found large elastic strains (and thus stresses) that are compressive in the unloaded tension samples and tensile in the unloaded compression samples. These results are qualitatively consistent with Mughrabi's composite model. However, the measured strains also exhibited large cell-to-cell variations corresponding to axial stresses ranging from zero to 50 % of the flow stress.

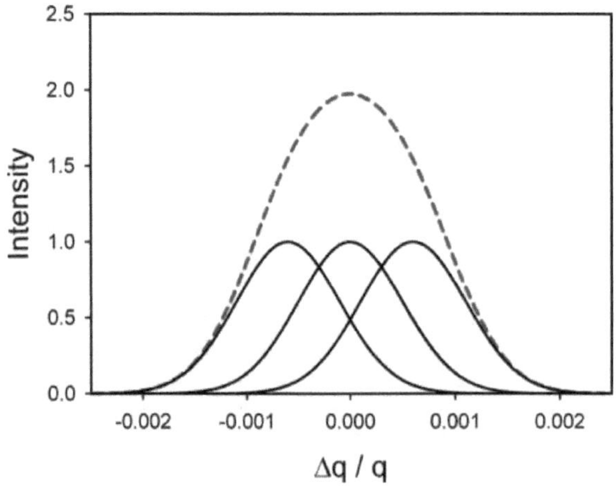

Figure 1. The solid curves are Gaussian functions representing idealized diffraction line profiles with different mean elastic strains. The dotted curve is the sum of the three Gaussians and is analogous to a classical cell interior sub peak.

The broad distribution of elastic strains found in the previous DAXM paper raised an important question about the validity of using a classical sub-profile analysis to draw conclusions about the distribution of dislocations in plastically deformed metals [9]. In such studies, dislocation cell interior and cell wall sub peaks are extracted using semi-empirical techniques, and the resulting profiles are then analyzed using a theoretical description of diffraction from dislocation structures [2,3,6,12-18]. However, the dislocation diffraction theories used in these analyses generally assume that all of the observed broadening originates directly from the dislocations and not from the recently observed, large cell-to-cell variations in the elastic strain. Such strains are sometimes referred to as elastic strains of the 2nd kind [19, 20, 21] and are often neglected in the analysis of dislocation line broadening. The question then becomes: does the shape of such a sub peak reflect the dislocation broadening behaviour in some average sense? It is not immediately clear that this is true. For example, the cell interior sub peak corresponds to a direct sum of the line profiles from all of the dislocation cell interiors. A simple geometrical construction in figure 1 illustrates the basic problem. Here, the three smaller peaks are identical Gaussian functions that are offset along the X axis. The peaks can be thought of as idealized line profiles from individual dislocation cell interiors, and the offsets therefore correspond to three different values of the elastic strain. The larger peak is the direct sum of the three smaller peaks and is therefore analogous to a classical dislocation cell interior sub peak. The full width half maximum (FWHM) of the composite peak is 1.9×10^{-3} which is very different from the average FWHM of the individual peaks which is 1.2×10^{-3}. In this case, the width and shape of the composite peak does not correspond to an average behaviour of the individual cell interior peaks, and a conventional dislocation diffraction analysis would be invalid.
To investigate this problem for real dislocation cell interior sub profiles, we directly construct such a sub profile by summing 52 diffraction line profiles obtained from individual, spatially resolved dislocation cells. We then compare the shape of the constructed profile to those measured from individual dislocation cells.

2. Sample and X-ray methods

The 99.999 + % pure single crystal Cu sample was cylindrical with nominal dimensions of 10 mm diameter by 20 mm long. The cylinder axis had a <001> crystallographic orientation. The sample was deformed in uniaxial compression at an approximate strain rate of 4×10^{-4} s^{-1} to a true strain of 0.28 at a true stress of 210 MPa. After deformation, the sample was cut in half perpendicular to the strain axis using wire electrical discharge machining (EDM). To remove the EDM damage layer, the top 120 μm of the sample was removed by electropolishing in a solution of ortho-phosphoric acid with a nominal applied voltage of 1.7 V. All X-ray measurements were conducted on the polished face from a region close to the centre of the original gauge section.
Spatially resolved axial elastic strains were measured using scanning monochromatic DAXM as described previously [9,22,23]. Briefly, a monochromatic X-ray beam from an undulator on XOR/UNI beam line 34-ID at the Advanced Photon Source is focused using crossed Kirkpatrick-Baez focusing mirrors with focal lengths of 130 mm and 60 mm. Since the dislocation microstructure contains numerous dislocation cells with different crystallographic orientations and elastic strains, only part of the sample illuminated by the monochromatic X-ray beam will diffract. The photon energy is therefore scanned over a large enough energy range to extract diffraction data from most of the probed sample volume.

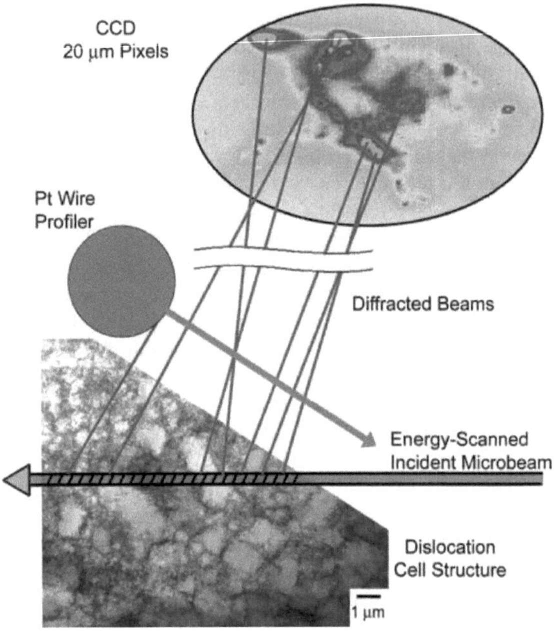

Figure 2. An X-ray microbeam is incident upon a plastically deformed Cu single-crystal and diffracted beams intersect a CCD area detector. Incrementally translating a Pt wire profiler parallel to the sample surface breaks diffracted beams allowing the depth of the originating sample volume to be extracted by triangulation.

The X-ray CCD area detector shown in Fig. 2 displays a false colour map of the incident diffracted X-ray intensity summed over all energies for a single sample position. Depth resolution is provided by incrementally translating a Pt wire parallel to the sample surface (for each energy) and using triangulation to determine the sample depth from which the observed diffracted X-ray intensity originates (see figure 2). Since the beam and detector pixel positions are known from the microscope calibration procedure, the elastic strains from individual dislocation cells can be directly extracted using Bragg's law with a strain resolution (one standard deviation) of about 10^{-4}. The sampled volume, as determined by the beam size and the depth profiling wire step size, must be small enough so that diffraction from individual dislocation cells can be unambiguously identified. Adjacent sample volumes (perpendicular to the incident X-ray beam) can be probed by translating the sample perpendicular to the beam and repeating the scans of energy and wire position.

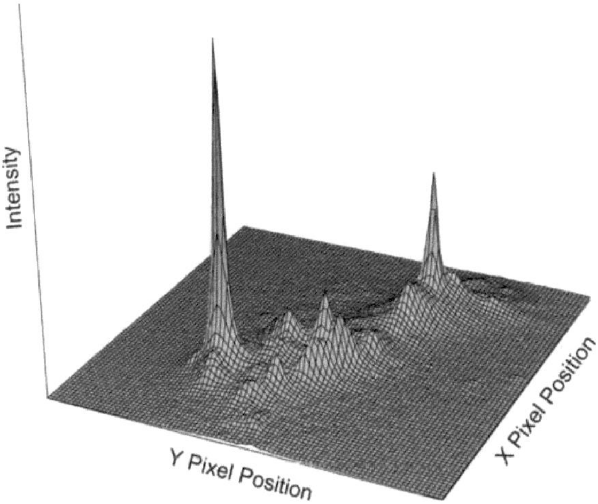

Figure 3. Distribution of diffracted X-ray intensity incident upon the detector from 0.5 μm³ of material buried several micrometers below the sample surface. The peaks originate from individual dislocation cells within this sample volume.

3. Results and conclusions

In these experiments, the axial 006 reflection was used to probe the local elastic strains (and thus stresses) within individual dislocation cells. To acquire diffraction data from a wide range of dislocation cell crystallographic orientations, the photon energy was scanned from 14.085 keV to 14.535 keV in 3 eV steps. The beam size for these measurements was approximately 1.0 μm × 0.7 μm; the wire step size was 1.0 μm giving a projected depth resolution of 0.7 μm and a depth resolved sample volume of about 0.5 μm^3. The transmission electron microscope image in figure 2 shows the dislocation cell structure of a separate Cu compression sample deformed along a <001> axis to a true strain of 0.25. Although a typical resolved sample volume of 0.5 μm^3 would contain more than one dislocation cell interior, the relative rotations between these dislocation cells produce well separated peaks on the detector, allowing independent diffraction line profiles to be extracted from each of these dislocation cell interiors. Figure 3 shows a typical pattern of X-ray intensity incident upon the detector from 0.5 μm^3 within the sample and integrated over all energies. The two large peaks originate from the interiors of two individual dislocation cells; the peak positions on the detector primarily reflect the relative crystallographic orientations of the cells. The diffuse scattering comes from a combination of much smaller coherently scattering volumes and dislocation cell walls.

A region of the sample with dimensions 5.6 μm × 1 μm × 7 μm (80 separate ≈ 0.5 μm^3 sample volumes) was examined, and a total of 52 clear, isolated peaks (from different dislocation cell interiors) were selected for detailed study. Diffraction line profiles were measured from all 52 cell interiors. The mean elastic strains from all of the cells were tensile, in agreement with the Mughrabi composite model and with our earlier DAXM measurements. The elastic strains ranged from $(0.3 \pm 1.0) \times 10^{-4}$ to $(16.0 \pm 1.0) \times 10^{-4}$. Converting these axial elastic strains to axial stresses using the elastic modulus in the <001> direction, $E_{001} = 66.6 \pm 0.5$ GPa [24], the stresses range from zero (within our measurement uncertainty) up to 106 MPa ± 7 MPa, or approximately 50 % of the macroscopic flow stress, again in agreement with our earlier work.

The instrument resolution function (due mainly to the pixel spacing, the sample-to-detector distance, and the convergence of the focused X-rays) was determined by measuring the 006 diffraction line profile from an unstrained, nearly dislocation free, <001> Cu single crystal under the same experimental conditions. The instrument function is almost perfectly Lorentzian with a FWHM of (0.065 ± 0.002) nm^{-1}. Scaling this by the magnitude of the 006 scattering vector, q (where $q = |\boldsymbol{q}| = (4\pi/\lambda)\sin(\theta)$, λ is the X-ray wavelength, and θ is the Bragg angle) gives a FWHM of 6×10^{-4} in $\Delta q/q$. Since the FWHM of the instrument profile is not negligible when compared with the FWHMs of the measured line profiles, intrinsic peak widths must be determined using a deconvolution algorithm. The measured profiles are intermediate between Gaussian and Lorentzian line shapes and may therefore be approximated as Voigt functions. Thus, the FWHMs of the approximately Gaussian intrinsic line shapes were determined using the formula [25]

$$F_G = \sqrt{(F_V - 0.5346F_L)^2 - 0.2166F_L^2}\,, \tag{1}$$

where F_G, F_V and F_L are the FWHMs of the Gaussian (intrinsic line profiles), Voigt (measured line profiles) and Lorentzian (measured instrument function) line shapes, respectively.

The FWHMs of the 52 $\Delta q/q$ line profiles range from 7.2×10^{-4} to 2.5×10^{-3}. The intensities of the peaks vary by a factor of 17, but this intensity only reflects the part of the dislocation cell volume that resides within a given 0.5 μm^3 sample volume and therefore is only statistically connected with the dislocation cell size. Figure 4a shows three intrinsic line profiles (Gaussian) selected to demonstrate the broad range of behaviors found in different dislocation cells.

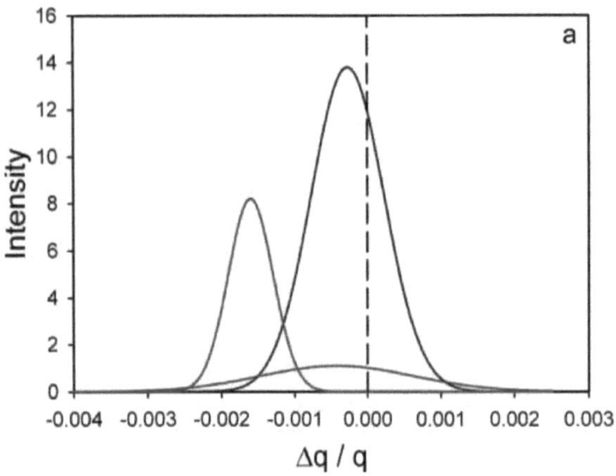

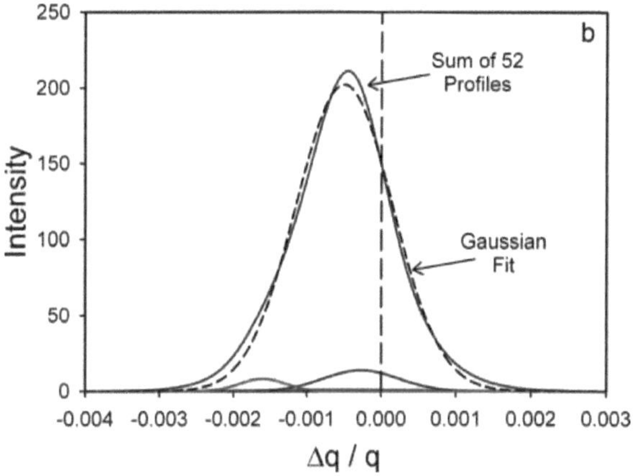

Figure 4. a) Range of (Gaussian) intrinsic line profiles obtained from three individual dislocation cell interiors. b) The same 3 line profiles plotted with the sum of 52 such profiles. The Gaussian fit to the composite peak demonstrates that this peak is nearly symmetric.

It is clear that a classic cell interior sub profile derived from X-ray measurements that average over many dislocation cells cannot provide information on the wide range of conditions present in the ensemble of individual dislocation cells. At best, it may provide information on the dislocation cell interior stresses and broadening in an average sense. To explore this possibility, we constructed a dislocation cell interior sub profile by adding the 52 measured line profiles together. We emphasize that, unlike sub profiles found in the literature, this sub profile was not empirically determined but, instead, directly reflects the summed diffracted intensity from a collection of real dislocation cell interiors. Figure 4b shows the constructed sub profile plotted with the same individual line profiles shown in figure 4a. The peak is very smooth and exhibits no pronounced irregularities, thus demonstrating that our sample size of 52 line profiles is statistically adequate for capturing the general behavior. In agreement with the early assumptions made by Mughrabi, this constructed sub profile is remarkably symmetric. The FWHM is 1.5×10^{-3} and the mean elastic strain is 5.0×10^{-4}. These values compare well with those obtained by directly averaging the FWHMs and elastic strains of the individual line profiles. This average gives a FWHM of $(1.58 \pm 0.06) \times 10^{-3}$ and a mean elastic strain of $(5.3 \pm 0.3) \times 10^{-4}$. The reason for this good agreement is that the three peaks shown in figure 4a represent extreme conditions in the sample. The majority of cell interiors fall between these extremes, and the average is dominated by the dislocation broadening rather than the distribution of elastic strains. Of course, the detailed shape of the sub profile depends upon both factors, and their relative importance likely depends upon both the material and the deformation conditions. Additional measurements on other systems are required to explore how general these results are. For the case explored in this study, we have confirmed that, if a dislocation cell interior sub profile is accurately extracted from volume averaged X-ray data, it can provide a good approximation for the mean stress condition and broadening behavior of an average dislocation cell.

An important issue that is not addressed in this paper is an evaluation of the various procedures that have been proposed for extracting sub profiles from volume-averaged line profile data [2,3,6,12-18]. This could be accomplished by measuring a volume-averaged line profile from the same sample used in this microbeam study and extracting sub profiles using published procedures for comparison with our composite sub profile.

Finally, the measurements and analyses conducted here only address the sub profile associated with dislocation cell interiors. Although it is likely that a similar study of the cell wall sub profile would show analogous results, this has not yet been confirmed. The diffuse nature of diffraction from cell walls would make such a study considerably more difficult than the present work.

In summary, we have used scanning monochromatic DAXM with sub micrometer sized X-ray beams to measure diffraction line profiles from 52 individual dislocation cells within a <001> Cu single crystal deformed in compression to 0.28 true strain. The distribution of elastic strains is consistent with previously published results. The intrinsic (deconvolved) widths of the line profiles were extracted from the data, and the resulting Gaussian profiles were added to produce a composite profile analogous to a cell interior sub profile extracted from volume averaged X-ray data. The sub profile is both smooth and nearly symmetric, confirming Mughrabi's early assumption about this sub profile shape. The mean elastic strain and FWHM of the constructed sub profile accurately reflect the average of the parameters extracted from the more exact individual dislocation cell measurements.

References

1. Mughrabi, H., 1983, *Acta. Metall.,* **31**, 1367.
2. Ungár, T., Mughrabi, H., Roennpagel, D. & Wilkens, M, 1984, *Acta Metall.,* **32**, 333.
3. Mughrabi, H., Ungár, T., Kienle, W. & Wilkens, M., 1986, *Phil. Mag. A*, **53**, 793.
4. Ungár, T., Mughrabi, H., Wilkens, M. & Hilscher, A., 1991, *Phil. Mag. A*, **64**, 495.
5. Ungár, T., 1994, *Mater. Sci. Forum*, **166-169**, 23.
6. Straub, S., Blum, W., Maier, H.J., Ungár, T., Borbély, A. & Renner, H., 1996, *Acta Metall.*, **31**, 4337.
7. Kassner, M.E., Perez-Prado, M.-T., Vecchio, K.S. & Wall, M.A., 2000, *Acta Mater.*, **48**, 4247.
8. Mughrabi, H, 2006, *Acta Mater.*, **54**, 3417.
9. Levine, L.E., Larson, B.C., Yang, W., Kassner, M.E., Tischler, J.Z., Delos-Reyes, M.A., Fields, R.J. & Liu, W., 2006, *Nat. Mater.*, **5**, 619.
10. Yang, W., Larson, B.C., Tischler, J.Z., Ice, G.E., Budai, J.D. & Liu, W., 2004, *Micron*, **35**, 431.
11. Ice, G.E., Larson, B.C., Yang, W., Budai, J.D., Tischler, J.Z., Pang, J.W.L., Barabash, R.I. & Liu, W., 2005, *J. Synchrotron Radiat.*, **12**, 155.
12. Krivoglaz, M.A. & Ryaboshapka, K.P., 1963, *Fiz. Metal. Metalloved,*. **15**, 18. Engl. transl: 1963, *Phys. Metals Metallogr.*, **15**, 14.
13. Wilkens, M., 1970, *Phys. Status Solidi A*, **2**, 359.
14. Wilkens, M. Herz, K. & Mughrabi, H., 1980, *Z. Metallkd.*, **71**, 376.
15. Ungár, T, Groma, I. & Wilkens, M., 1989, *J. Appl. Crystallogr.*, **22**, 26.

16. Ungár, T. & Borbély, A., 1996, *Appl. Phys. Lett.*, **69**, 3173.
17. Levine, L.E. & Thomson, R., 1997, *Acta Crystallogr., A*, **53**, 590.
18. Groma, I., 1998, *Phys. Rev. B*, **57**, 7535.
19. Krivoglaz, M.A., 1961, *Fiz. Metal. Metalloved,* **12**, 465.
20. Kuzel, R. 2007, *Z. Kristallorgr.*, **222**, 136.
21. Breuer, D., Klimanek, P. & Pantleon, W., *J. Appl. Crystallogr.*, **33**, 1284.
22. Larson, B.C., Yang, W., Ice, G.E., Budai, J.D. & Tischler, J.Z., 2002, *Nature*, **415**, 887.
23. Yang, W., Larson, B.C., Ice, G.E., Tischler, J.Z., Budai, J.D., Chung, K.-S. & Lowe, W.P., 2003, *Appl. Phys. Lett.*, **82**, 3856.
24. Ledbetter, H.M. & Naimon, E.R., 1974, *J. Phys. Chem. Ref. Data*, **3**, 897.
25. Olivero, J.J. & Longbothum, R.L., 1977, *J. Quant. Spectrosc. Ra.*, **17**, 233.

Acknowledgements. M. E. K. and P. G. acknowledge support from the Office of Basic Energy Sciences (BES), DOE, and thank the National Center for Electron Microscopy at LBNL for access to the TEM. Research at ORNL is supported by DOE BES Division of Materials Sciences and Engineering. The XOR/UNI facilities on sector 34 at the APS are supported by the US DOE, Office of Science.

Z. Kristallogr. Suppl. **27** (2008) 65-72 / **DOI** 10.1524/zksu.2008.0009

Determination of stored elastic energy in plastically deformed copper

A. Baczmański[1,*], N. Hfaiedh[2], M. François[2], K. Saanouni[2], K. Wierzbanowski[1]

[1] Faculty of Physics and Applied Computer Science, AGH- University of Science and Technology, al. Mickiewicza 30, 30-059 Kraków, Poland
[2] LASMIS, Universite de Technologie de Troyes, 11 rue Marie Curie,. B.P. 2060, 10010 Troyes, France
[*] baczman@ftj.agh.edu.pl

Keywords: diffraction, self-consistent model, residual stresses, crystallographic texture

Abstract. The quantitative analysis of stresses in plastically deformed copper samples was performed using the self-consistent model and X-ray diffraction measurements of lattice strains. The {220} strain pole figures were determined for the samples subjected to different magnitude of tensile deformation.
Using diffraction data and self-consistent model, the tensor of plastic incompatibility stress, was found for each orientation of a polycrystalline grain. The distribution of elastic stored energy was presented in the Euler space and correlated with preferred orientations of grains. Moreover, using the model prediction, variation of the critical resolved shear stress with grain orientation was shown.

1. Introduction

The internal stresses and strains in polycrystalline aggregate depend on the history of the material, and the final state of the stress field can be described on different scales. On the macroscopic scale the macrostress (first order stress) is defined as the average stress over a sample volume containing a large number of polycrystalline grains, such as the gauge volume in a diffraction experiment. The second level of the study concerns the grain-size scale. The incompatibility stress (second order) characterises the deviation of the stress for a particular grain from the macrostress (first order) due to the difference in the behaviour of grains with varying lattice orientations. Finally, third order stresses characterise stress heterogeneity around dislocations or other defects of the crystal lattice within a single grain.
The $sin^2\psi$ X-ray diffraction method [1] of stress analysis is based on the measurement of peak positions for a given *hkl* reflection and for various directions of the scattering vector (see figure 1a). The interplanar spacings are calculated using Bragg's law, and the macrostresses (first order stresses) are determined by applying a simple linear or elliptical regression procedure [1]. The effect of the intergranular second order stresses has been observed as nonlinearities on the $sin^2\psi$ plots measured for polycrystalline samples subjected to different

elastoplastic deformations [2,3]. The influence of sample anisotropy (crystallographic texture) on the measured interplanar spacings was studied and the calculations based on the crystallographic texture and single crystal elastic constants facilitated a prediction of the nature of the $sin^2\psi$ plots for the sample under applied (or residual) macrostresses [4,5]. However, the elastic anisotropy was insufficient to interpret nonlinearities of the $sin^2\psi$ plots. A modified $sin^2\psi$ method for quantitative evaluation of the stresses generated due to plastic incompatibilities (second order stresses) has been proposed by Baczmanski *et al.* [6,7,8]. The non-standard interpretation of experimental results allowed the separation of the macrostresses and microstresses created due to the anisotropy of plastic deformation. The information about tensor of plastic incompatibility stresses, which cannot be obtained by other methods of stress analysis, is useful in evaluating such physical properties of the material as: value of critical stress for slip, surface of plasticity and hysteresis of mechanical response during cycling loading of the sample. Also, the distribution of the stored energy is important in the study of texture transformation during the recrystallisation process [9]. Hence, two types of stored energy created during elastoplastic deformation are considered in this work, i.e., stored elastic energy caused by plastic incompatibility stresses (determined using modified $sin^2\psi$ method) and the stored energy, connected with the dislocation which can be estimated using the self-consistent model.

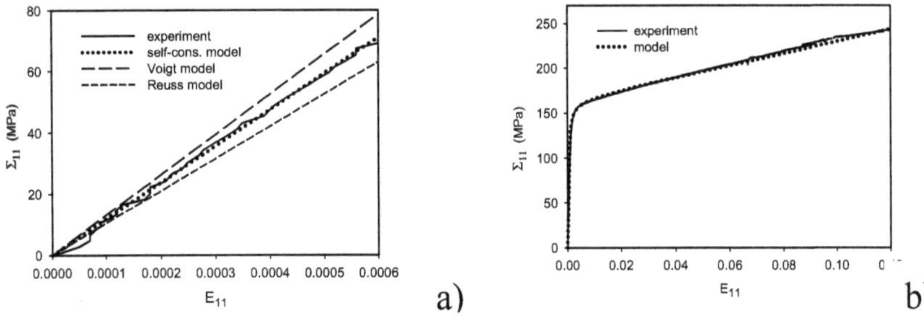

Figure 1. Results of mechanical tensile test (doted line) compared with the self-consistent prediction (solid line) for elastic (a) and elastoplastic (b) deformation. The macroscopic sample strain (E_{11}) vs. macrostress (Σ_{11}) applied to the sample is shown. In the case of elastic deformation the results of Reuss and Voigt models are also presented.

2. Model of elastoplastic deformation

The material chosen for the present study was a high-purity copper (99%). The self consistent model used in this work has been proposed by Berveiller and Zaoui [10] and Lipinski and Berveiller [11,12]. The tensile test was performed in transverse direction (TD) for the sample prepared from a copper sheet, obtained by hot rolling followed by cold rolling processes. The model calculations were carried out for 10^5 spherical inclusions (grains) embedded in a homogeneous matrix. The single crystal elastic constants of the grains are: C_{11} =170 GPa, C_{12} =123 GPa and C_{22} = 64.5 GPa. The crystal lattice orientations were assigned according to the initial orientation distribution functions measured by X-ray diffraction (see figure 3). It was assumed that the <110>{111} slip systems were active during plastic deformation of copper crystals (f.c.c. structure).

The initial critical resolved shear stress τ_c^{init} (MPa) = 50 MPa and the work hardening parameter H = 80 MPa were determined comparing model and experimental stress-strain curves obtained for a tensile test (see figure 1). As shown in figure 1 b, the linear hardening can be assumed until about 12% of sample deformation and in this range the model calculations perfectly agrees with experiment.

3. Determination of residual stresses by diffraction

As it was shown in many papers [6,7,8] only the orientation evolution of theoretical $\overline{\sigma_{ij}^{g(ic)}(\mathbf{g})}$ stresses (calculated by self consistent model), but not their amplitude, agrees with the experimental results. As proposed in [6], to relate the magnitude of theoretical stresses to the actual ones, an unknown scaling factor q can be introduced. This factor does not depend on the grain orientation $\mathbf{g}$ and it rescales the amplitude of the stress tensor, i.e. the second order incompatibility stress $\sigma_{ij}^{g(ic)}(\mathbf{g})$ in the real sample is equal to:

$$\sigma_{ij}^{g(ic)}(\mathbf{g}) = q\ \overline{\sigma_{ij}^{g(ic)}(\mathbf{g})} \tag{1}$$

where q is the scaling parameter and $\overline{\sigma_{ij}^{g(ic)}(\mathbf{g})}$ is the model predicted second order stress for a grain with $\mathbf{g}$ orientation.

The standard methods of stress determination are based on the measurement of interplanar spacing for various directions of the scattering vector, defined by the ϕ and ψ angles [1]. Using diffraction, the mean interplanar spacing $<d(\phi,\psi)>_{\{hkl\}}$, averaged only over reflecting crystallites, is measured. The measured interplanar spacings $<d(\phi,\psi)>_{\{hkl\}}$ can be related with the macrostresses σ_{ij}^{M} and theoretical incompatibility stresses $\overline{\sigma_{ij}^{g(ic)}(\mathbf{g})}$ [6,7,8]:

$$< d(\phi,\psi) >_{\{hkl\}} = \left\{F_{ij}\,(hkl,\phi,\psi)\,\sigma_{ij}^{M} + < \gamma_{3m}\gamma_{3n}s_{mnij}\,\overline{\sigma_{ij}^{g(ic)}(\mathbf{g})} >_{\{hkl\}}\right\} d_{\{hkl\}}^{0} + d_{\{hkl\}}^{0} \tag{2}$$

where s_{mnij} are the single crystal elastic constants, γ_{ij} are the directional cosines transforming strains from sample system to the system defined by scattering vector [1,6,7,8] system, $F_{ij}\,(hkl,\phi,\psi)$ are the diffraction elastic constants [4,5].

The solution of equation 2 can be found using the least square method in which the theoretical interplanar spacings $< d(\phi,\psi) >_{\{hkl\}}$ are fitted to the experimental ones. Measuring the $< d(\phi,\psi) >_{\{hkl\}}$ spacing for a sufficient number of scattering vector orientations the unknown macrostresses σ_{ij}^{M} and the q factor can be found. When the values of q parameter are determined by a fitting procedure, the real values of the plastic incompatibility stresses $\sigma_{ij}^{g(ic)}(\mathbf{g})$ can be calculated for all grain orientations $\mathbf{g}$ using equation 1. [1,6,7,8].

4. Texture and residual stresses in copper sample

Due to plastic deformation, the studied material (copper) was significantly textured. The diffraction measurements were performed for the samples subjected to different magnitude of tensile deformation along TD (transverse direction).

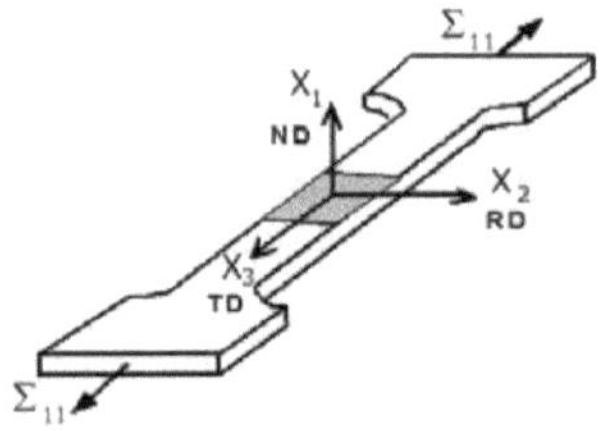

Figure 2. Orientation of the directions of rolling with respect to the X coordinates system used for definition of Euler angles and Euler space.

The X-ray patterns were recorded using a Seifert diffractometer, equipped with a chromium anode ($\lambda = 2.28970$). The 2θ angular range, between 30° and 140°, enabled to measure diffraction peaks corresponding to the reflections 111, 200 and 220. The angular step of the diffractograms was 0.05°, with a counting time of 63 s. The diffracted rays were recorded using a position sensitive proportional counter. The orientation distribution functions (ODF), characterising crystallographic textures, were calculated from measured incomplete pole figures ({111}, {200} and {220}) by using the harmonic method. Because of easier sample symmetry analysis, the ODF-s were presented in the Euler space defined for sample system with x_3 axis parallel to the tensile load (this sample system is shown figure 2). The measured and model ODF-s for different degrees of tensile deformations are shown in figure 3. Model functions were calculated by the self-consistent model starting from initial experimental texture or from random grains orientations. In the case of initial random orientations the ODFs show axial symmetry about the axis of the applied load (direction of Σ_{11}, i.e., x_3 in figure 2). The main texture maximum occurs for all grain orientations for which the applied load is perpendicular to {111} crystallographic planes. The ODF-s predicted from initial experimental texture do not exhibit axial symmetry and two separated maxima are created.
As it is shown in figure 3 also in this case both maxima belong to the group of orientations for which the applied load is normal to the {111} planes (these orientations are stable during tensile test). It can be concluded that the rotations of the grains are defined mostly by the mode of plastic deformation process and on the type of active slip systems. However, the final ODF strongly depends on the initial distribution of grains orientations. Finally, it should be underlined that the texture evolution during tensile test was perfectly predicted when the self-consistent calculations were performed starting from initial experimental texture (see figure 3).

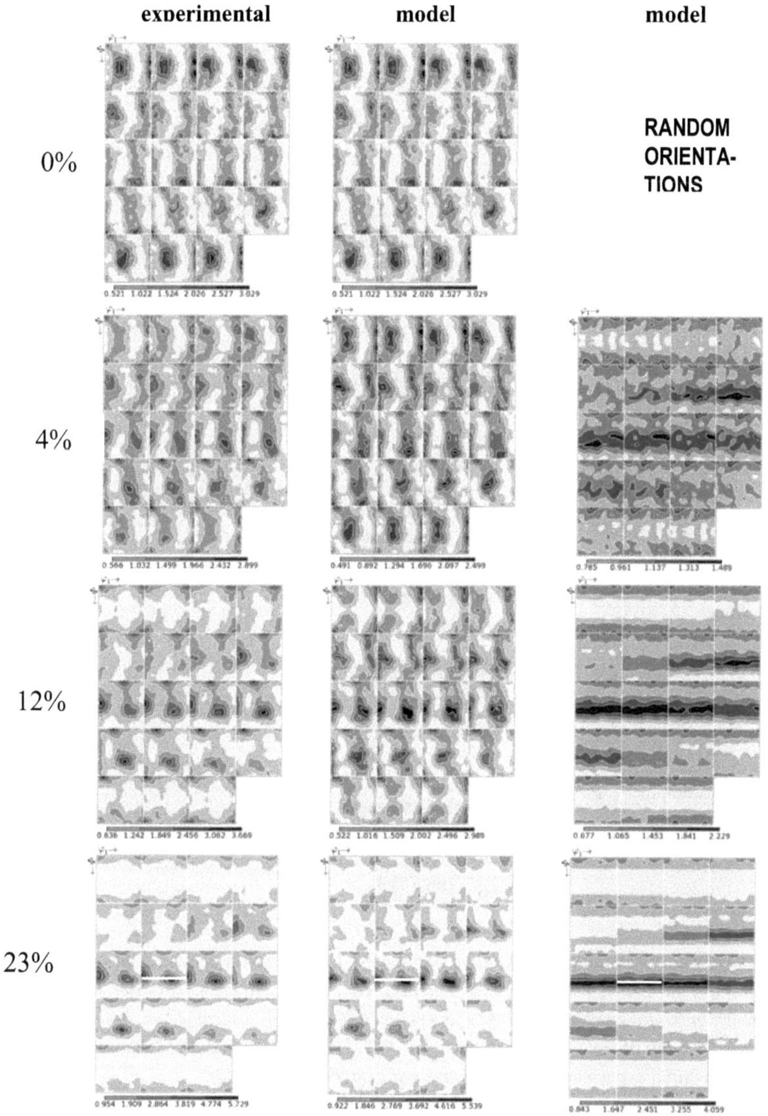

Figure 3. Orientation distribution function of the copper sample: experimental and model results. Model calculations performed from initial experimental texture (model 1) and from random grains orientations (model 2). Euler space is defined with respect to the sample system shown in figure 2 and the grains orientations having planes {111}$\perp$ Σ_{11} are indicated by solid lines.

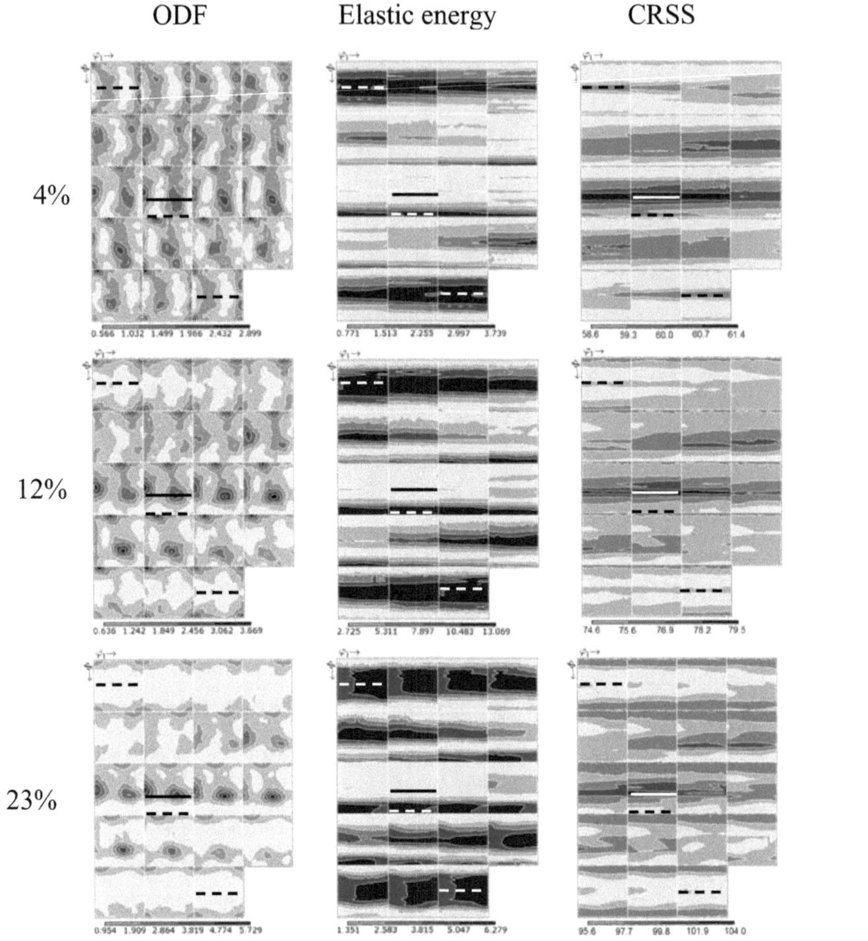

Figure 4. Experimental orientation distribution functions (ODF), stored energy caused by grains incompatibilities (specific elastic energy in kJ/m^3) and critical resolved stresses (CRSS in MPa) during tensile test shown in the Euler space (defined with respect to the sample system shown in figure 2). The grain orientations having planes {111}$\perp \Sigma_{11}$ and planes {110}$\perp \Sigma_{11}$ are indicated by solid and dashed lines, respectively. Initial experimental texture was used in model calculations.

5. Evolution of stored energy during elastoplastic deformation

In the case of 220 reflection, the positions of the diffraction peaks were determined and corresponding interplanar spacings were found for all orientations in the measured incomplete pole figure {220}. For the first time the quantitative analysis of stresses [6,7,8] was performed using fitting procedure based on equation 1 simultaneously for all interplanar spacings in the incomplete pole figure. The theoretical incompatibility stresses $\overline{\sigma_{ij}^{g(ic)}(\boldsymbol{g})}$ were calculated using self-consistent model for parameters established in section 2. As a result of the analysis, plastic incompatibility stresses $\sigma_{ij}^{g(ic)}$ as well as the macrostresses σ_{ij}^{M} were determined quantitatively.

The evolution of stored elastic energy corresponding to incompatibility stresses $\sigma_{ij}^{g(ic)}$ and to the dislocations during elastoplastic tensile test is considered. In figure 4 the experimental ODFs are set together with maps showing the variation of mean stored elastic energy caused by the $\sigma_{ij}^{g(ic)}$ residual stresses in Euler space. That energy is defined for each grain orientation $\mathbf{g}$, i.e.:

$$W^{g(ic)}(\mathbf{g}) = \frac{1}{2}\sigma_{ij}^{g(ic)}(\mathbf{g})\, s_{ijkl}^{g}(\mathbf{g})\, \sigma_{kl}^{g(ic)}(\mathbf{g}). \tag{3}$$

Also in figure 4, the variation of average critical resolved shear stress (CRSS denoted by τ_c) is presented. The later quantity is calculated as the average value over all slip systems for a given grain having orientation $\mathbf{g}$ and it is strictly connected with the stored energy caused by dislocations [13], i.e.:

$$W^{g(dis)}(\mathbf{g}) \sim \tau_c^2(\mathbf{g}) \tag{4}$$

As shown in figure 4, the axial symmetry around applied load was found and the minima correspond to the orientations for which the crystallographic planes {111} fulfil condition $\{111\}\perp\Sigma_{11}$, while maxima are close to the orientations for which $\{110\}\perp\Sigma_{11}$. Thus the grains having preferred texture orientations (maxima of ODFs) possess minimum elastic energy caused by grains incompatibilities ($W^{g(ic)}(\mathbf{g})$).

Analysing distributions of CRSS in Euler space, it was found that maximum energy caused by dislocations is stored in the grains having crystallographic planes $\{111\}\perp\Sigma_{11}$. In this case the variation of $\tau_c(\mathbf{g})$ with grain orientation is very small (about 3% of average value) and in the real deformation process it would strongly depend on the CRSS in the initial sample. However, the variation of CRSS in Euler space is unknown and can not be experimentally determined for that sample. Thus, in beginning of calculations the same value of $\tau_c^{init}(\mathbf{g})$ =50 MPa was attributed to all grain orientations. Consequently, the results of model calculations (see figure 4) show only the tendency of CRSS evolution during tensile test neglecting important influence of initial $\tau_c^{init}(\mathbf{g})$.

6. General conclusions

In the present work the ODFs and residual incompatibility stresses created during tensile test in polycrystalline copper were determined using the self-consistent model and X-ray diffraction measurements. It was shown that evolution of the measured texture agree very well with the model results when the experimental initial texture is used in model calculations. During elastoplastic deformation texture slowly changes and the final ODF depends strongly on the initial distribution of orientations. In contrast the distribution stored elastic energy caused by grain incompatibilities in Euler space rapidly changed when different type of deformation was applied (rolling was followed by tensile test). The distribution of stored elastic energy is almost independent on the initial texture and it shows symmetries depending mostly on the type of deformation. It was found that the grains having preferred orientations created during tensile process possess minimum of stored elastic energy caused by grain incompatibilities.

The model distribution of critical resolved shear stresses (related to stored energy caused by dislocations) was plotted in Euler space showing very small variation of CRSS with orientation after 23% of tensile deformation. The tendency of CRSS evolution was shown, however, the final distribution of CRSS in the real sample was not determined because it could be strongly affected by unknown initial state of the sample.

References

1. Noyan, I.C. & Cohen, J.B., 1987, *Residual Stress* (New York: Springer Verlag).
2. Hauk, V.M., 1986, *Adv. X-ray Anal.* **29**, 1.
3. Van Acker, K, Van Houtte, P. & Aernoudt, E., 1994, *Proc. of Int. Conf. on Residual Stresses*, ICRS4, Baltimore, Meryland, pp. 402.
4. Dolle, H., 1979, *J. Appl. Cryst.*, , **12**, 489.
5. Brakman, C.M., 1987, *Phil. Mag. A* , **55**, 39.
6. Baczmański, A., Wierzbanowski, K., Lipiński, P., Helmholdt, R.D., Ekambaranathan, G. & Pathiraj, B., 1994, *Phil. Mag. A*, **69**, 437.
7. Baczmański, A., Braham, C. & Seiler, W., 2003, *Phil. Mag.*, **83**, 3225..
8. Wroński, S., Baczmański, A. Dakhlaoui, R., Braham, C., Wierzbanowski, K. & Oliver, E.C., 2007, *Acta Mater.* , Acta Mater., **55**, 6219.
9. Baczmanski A., Wierzbanowski, K., Benmarouane, A., Lodini, A., Lipinski, P. & Bacroix, B., 2007, *Mat. Sci. Forum*, **539-543**, 3335.
10. Berveiller, M. & Zaoui, A.., 1979, *J. Mech. Phys. Solids*, **26**, 325.
11. Lipinski, P. & Berveiller, P., 1989, *Int. J. of Plasticity*, **5**, 149.
12. Lipinski, P., Berveiller, M., Reubrez, E. & Morreale, J., 1995, *Arch. Appl. Mech.*, **65**, 291.
13. Baczmanski, A., Wierzbanowski, K., Benmarouane, A., Lodini, A., Lipinski P. & Bacroix, B., 2007, *Mater. Sci. Forum,* **539-543,** 3335.

Acknowledgements. This research project has been partially supported by the Polish Ministry of Science and Higher Education (MNiSW) and by the European Network of Excellence: Complex Metallic Alloys (EU 6-th Frame Program). M.F. and N.H would like to thank Région Champagne Ardenne and the French government for supporting part of this research through Pôle Mécanique et Matériaux de Champagne Ardenne. N.H. would like to thank Région Champagne Ardenne and Fond Social Européen for her PhD allowance.

Structural studies of submicrocrystalline copper and copper composites by different methods

R. Kužel[1,*], V. Cherkaska[1], Z. Matěj[1], M. Janeček[2], J. Čížek[3], M. Dopita[4]

[1] Department of Condensed Matter Physics, Faculty of Mathematics and Physics, Charles University, Ke Karlovu 5, 121 16 Praha 2, Czech Republic
[2] Department of Physics of Materials, Faculty of Mathematics and Physics, Charles University, Ke Karlovu 5, 121 16 Praha 2, Czech Republic
[3] Department of Low Temperature Physics, Faculty of Mathematics and Physics, Charles University, Ke Karlovu 5, 121 16 Praha 2, Czech Republic
[4] Technical University Freiberg, Gustav Zeuner Str., Freiberg, Germany
[*] kuzel@karlov.mff.cuni.cz

Keywords: line profile analysis, transmission electron microscopy, electron backscatter diffraction, positron annihilation spectroscopy

Abstract. Microstructure of Cu, Cu + Al_2O_3, Cu + Zr samples deformed by high pressure torsion and equal channel angular pressing was studied by several methods: X-ray diffraction (XRD), transmission electron microscopy (TEM), positron annihilation spectroscopy (PAS) and electron back-scatter diffraction (EBSD). Line broadening analysis showed that the HPT samples have smaller crystallite size compared to the ECAP samples while the dislocation densities were similar – of the order of $1\times10^{15}m^{-2}$. Only small changes in the mean dislocation density with increasing number of ECAP passes were found by XRD. On the other hand, significant microstructural changes and grain fragmentation were revealed by TEM and EBSD. With increasing strain due to ECAP elongated dislocation cells with low-angle boundaries developed into more equiaxed subgrains and grains separated by high-angle grain boundaries.

1. Introduction

Single- or multiphase polycrystals with the mean grain size of about 100 nm are often called ultra fine-grained (UFG) or submicro-grained (SMG) materials.

The specific structure of these materials leads to a number of unusual physical and mechanical properties. In particular, the SMG materials exhibit anomalous high diffusion activity, unusual changes in Curie temperature, saturation magnetization and elastic properties. Moreover, the SMG materials are characterized by combination of the high strength and the ductility [1, 2].

It is well known that grain refinement and high density of dislocations enhance the mechanical strength of metal materials but also decrease ductility. Preparation of materials having exceptionally high strength and ductility requires new approaches. One of the examples reported in [3] is a nanostructured copper with a ''bimodal'' structure of micrometer-sized grains embedded in a matrix of nanocrystalline grains. The nanocrystalline grains provide high strength while the embedded larger grains stabilize the tensile deformation of the material. Another approach suggested [e.g. 4] is based on the *formation of second-phase particles* in the nanostructured metallic matrix, which modify shear-band propagation during straining, thereby increase the ductility. The third approach is based on the formation of ultrafine-grained structures with high-angle and non-equilibrium grain boundaries capable of grain boundary sliding [5]. It is well known that the low-angle GB networks usually suppress the ductility by glide. *The importance of high-angle grain boundaries* was reported in [6]. The authors compare the mechanical behaviour of metals subjected to different degrees of SPD which result in the formation of various types of grain boundaries (GBs).

SPD allows producing of bulk SMG materials with no porosity and impurities. At present, the majority of experiments are connected with application of two SPD methods [e.g. 1, 6]: high-pressure torsion (HPT) and equal channel angular pressing (ECAP). During ECAP pressing the rectangular specimen is repeatedly pressed through a special die having two mutually intercepting channels in an angle which usually equals to 90°. The number of passes and the way of specimen rotation between individual passes (so called route) are very important for the final microstructure refinement.

In this paper, microstructural investigations of Cu, Cu + Al_2O_3, Cu + Zr are presented. Different complementary techniques were applied: X-ray diffraction life-time positron spectroscopy transmission electron microscopy, electron backscatter diffraction.

2. Experiment

Both HPT and ECAP samples were studied. The HPT samples were obtained from Ufa Aviation Technical University, Russia. Torsion (5 rotations) of the initial coarse-grained samples was performed up to the true logarithmic strain of $e = 7$ under the pressure of 6 GPa. All the HPT specimens were disk-shaped with a diameter of 10÷12 mm and thickness of 0.2÷0.5 mm. A set of the SMG Cu and Cu+Zr samples were prepared by the ECAP technique (route Bc; the pressed sample is rotated clockwise by 90° after each pass). ECAP specimens were cut from the billet in the transversal direction (plane X). In order to remove the possible surface stress the as-obtained SMG samples where etched. Some measurements were also done in longitudinal direction of the deformation (plane Y).

XRD studies were performed with the aid of XRD7 (FPM-Seifert) and with Panalytical X'Pert Pro powder diffractometers using Cu_{K_α}-radiation filtered with a Ni foil and Soller slits placed in the diffracted beam. Measurements were conducted with the time of 20-200 s/step depending on the sample and step size 0.02-0.05°. On Panalytical diffractometer, variable divergence and variable anti scatter slits were used in order to keep the analyzed specimen area constant in the whole 2θ range. XRD profiles were fitted with the Pearson VII or pseudo-Voigt function. The correction for instrumental broadening was performed with the aid of a NIST LaB_6 standard and using the Voigt function. Rough estimation of the *preferred grain orientation* was done from the θ-2θ scans in terms of the Harris texture index [7]. *Line*

profile analysis was done by simplified method and total pattern fitting. The former method - the modified Williamson-Hall plots (integral breadth β (in 1/d) vs. sin θ) can be appropriately used for the visualization of basic features of line broadening – (the main reason of broadening, broadening anisotropy, basic trends in series of samples). For most of the as-deformed HPT and ECAP samples the strain broadening was dominant. In this case the dislocation density can be estimated from the line width [see 8, 9 for more].
In order to avoid rough and often inappropriate simplification of the above method the total pattern fitting is preferred nowadays [e.g. 10-12]. Essentially 4-5 free parameters are used – the dislocation density, the dislocation correlation factor, the mean crystallite size and the variance of the size distribution. Dislocation types can be estimated in some cases as well. Despite the description is more physically realistic than the simplified approximate method, the fitting must be carefully performed. In particular, it cannot be done in a fully automatic mode because of a high correlation between some parameters. Our own extension of FOX program based on the algorithms used in [10-11] was used for the fitting [12].
Pole figures of selected lattice planes (111), (200), (220) were measured with Seifert X-ray diffractometer XRD 3000 PTS in (Technical University Freiberg) and with the Philips X'Pert MRD Pro system equipped with the Eulerian cradle and polycapillary in the primary beam.
TEM observations were performed for selected samples using a Jeol 2000FX analytical microscope operating at 200 kV with EDX system LINK AN 10000. Small disks (of 3 mm diameter) were cut by spark erosion and mechanically polished down to a thickness of 0.1 mm and then thinned using a standard electrolytic double jet technique (Struers Tenupol).
EBSD measurements were performed on the high-resolution scanning electron microscope LEO-1530 equipped with field emission cathode with EDX and EBSD system and the software package Channel 5 (HILL Technology). The scan direction was chosen in the direction transversal to the ECAP pass. In positron annihilation experiments the BaF_2 spectrometer [13] was used with a time resolution of 150÷170 ps (FWHM) for ^{22}Na and a coincidence count rate of ~ 100 events per second. The diameter of radioactive spot was ≈ 3 mm.

3. Results

XRD

Lattice parameters of all specimens investigated (0.36149 – 0.361580 nm) agreed quite well with the theoretical values (a_{Cu}^{PDF2} = 0.3615 nm, ICDD PDF4 database). This indicates the absence of high stresses. Crystallite sizes, dislocation densities and texture indices of individual specimens are summarized in table 1.
All the analyzed HPT Cu samples show typical line broadening anisotropy $\beta_{h00} >> \beta_{hhh}$ (fig. 1 a, b). This can be explained by the orientation factors calculated assuming only dislocations with the Burgers vector $b \parallel <110>$ that are typical for the f.c.c. structures. The anisotropy is closely related to the elastic anisotropy of copper. The correspondence between calculated and experimental values is good in most cases. In the first approximation, the fractions of edge and screw dislocations can be taken as equal. However, in some cases, better fits can be obtained for uneven dislocation fractions. The differences are not significantly high but systematic trends can be seen (e.g. for samples with addition of Al_2O_3 or Zr) and related to the variation of line broadening anisotropy.

In most cases, the strain broadening was dominant and the crystallite size significantly exceeded 100 nm. Since for conventional powder diffraction this value is close to the limit of sensitivity and therefore only rough estimation of the mean crystallite size could be done. In any case, samples with different content of Al_2O_3 show smaller crystallites below 100 nm. No significant differences in microstructural parameters in samples differing in the amount of Al_2O_3 were found. Sample prepared using smaller pressure (3 GPa) has also lower dislocation density. All samples prepared with 6 GPa pressure exhibited preferred orientation of the type (111) with the texture index of 2-2.5. From measurements of pole figures it follows that textures are usually not too strong and of fibre type. In some samples, mixed – two component texture (111) – (100) was be found.

Table 1. Structure parameters for HPT Cu samples prepared at two different pressures, different amount of Al_2O_3, for ECAP Cu samples at longitudinal (plane Y) and transversal (plane X) directions and different number of passes (1, 2, 4, 8) and for Cu + Zr samples, D... crystallite size, ρ...dislocation density, T_{hkl}... texture indices, w... fractions of edge dislocations, 1 – w ... fractions of screw dislocations.

	Sample	D, nm	ρ, 10^{15} m^{-2}	w	T_{hhh}	T_{h00}	T_{hh0}	T_{311}
HPT samples	HPT Cu 1 – 6 GPa	200	1.2	0.5	2.1	1.1	0.66	1.1
	HPT Cu 2 – 3 GPa	190	0.82	0.8	0.78	1.4	0.85	1.0
	Cu + 0.3 wt.%Al_2O_3	92	1.1	1	2.5	0.49	0.40	1.0
	Cu + 0.5 wt.%Al_2O_3	55	1	0.5	2.4	0.95	0.37	0.9
	Cu + 1.1 wt.%Al_2O_3	60	1.9	0.2	2.3	0.78	0.54	1.2
ECAP samples	Cu-1-Y	190	1.1	0.5	2.4	1.3	0	0.7
	Cu-2-Y	110	1.6	0.5	1.7	1.9	0.1	0.7
	Cu-4-Y	210	0.93	0.5	1.5	1	0.8	0.7
	Cu-8-Y	160	0.66	0.5	2.1	1	0.3	0.7
	Cu-1-X	170	0.80	0.5	0.1	0.1	1.6	0.5
	Cu-2-X	> 500	1.2	0.5	0.3	0.5	1	0.8
	Cu + Zr 0.18-1-X	> 500	0.96	1	0.2	0.8	0.9	0.9
	Cu + Zr 0.18-2-X	> 500	1.6	1	0.6	0.3	0.9	0.8
	Cu + Zr 0.18-4-X	> 500	1.9	0.4	1.2	0.6	0.8	0.9
	Cu + Zr 0.18-8-X	> 500	2.1	0.2	1.0	0.8	0.8	0.8

In ECAP specimens XRD measurements (fig. 2) were performed in two different planes, parallel and perpendicular to the axis of deformation (plane Y) and (plane X), respectively. In general, in ECAP specimens crystallite size was found larger than in HPT specimens, especially in the direction parallel to the deformation axis. The mean dislocation density for pure Cu initially increases with number of ECAP passes, reaches its maximum in the specimen after 2 passes and declines with further straining. However, it must be noted that the dislocation correlation parameter was fixed at a single value for all specimens in these evaluations. More careful studies of profile shapes which are related to this factor and require high-quality data are under way. Mean dislocation densities in Cu 0.18 % Zr alloy are of the same order as those in Cu and Cu+Al_2O_3 samples. However, the density increases monotonously with increasing strain up to 8 passes. According to TEM observations (see below),

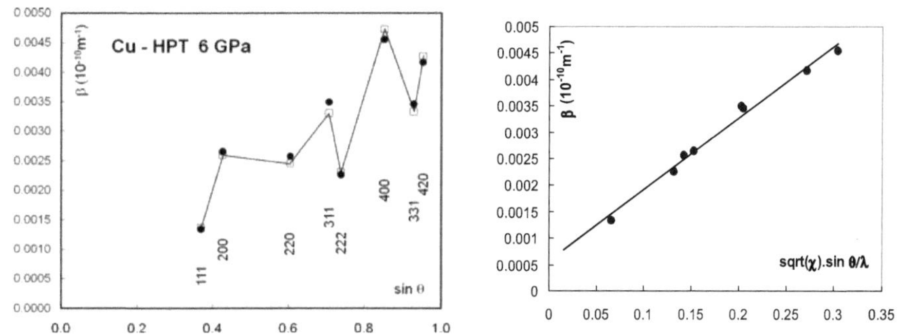

a) b)

Figure 1. WH plot with indicated hkl indices for sample HPT6 (a), modified WH plot of β vs. $\sqrt{\chi}$ sin θ/λ which should be linear, if correct orientation factors were used (b), for a) ● – experimental data; □ and line calculated values, in b) experimental values only.

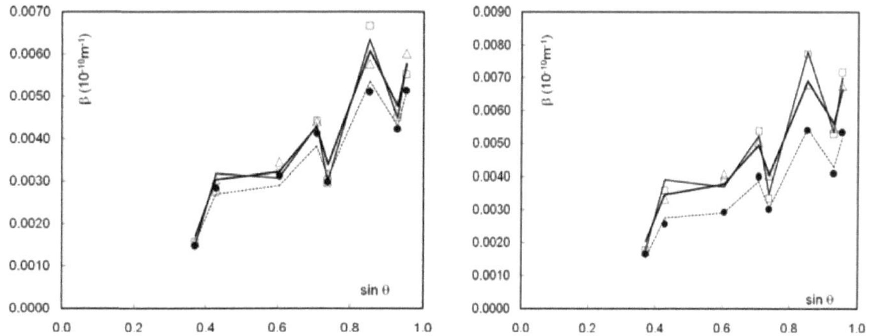

a) b)

Figure 2. WH plots for the measured Cu (a) and Cu 0.18 % Zr (b)samples in transversal direction after different number of ECAP passes: ●, dashed line – 1 pass; □, thicker line– 2 passes; △, line – 8 passes showing good agreement of experimental and calculated data (lines).

one may assume grain elongation (differences between Y and X, Cu-2) and partial recrystallization after four or more passes to occur in Cu specimens.

Harris texture indices relating to Y- and X- surfaces of ECAP samples are clearly different. However, the index is not a good parameter for texture characterization in this case. Pole figures (figure 3) indicate that the texture is not of simple fibre type and show its evolution with the number of passes. After one pass, the grains with the planes (111) slightly inclined to the sample surface prevail, a component close to the texture (110) with sufficient intensity is also present. After the second pass, the second texture component appears (two groups (111) with different inclinations). After the fourth pass the texture is smeared out and the texture after 8 ECAP passes is characterized by (110) component only. The pole figures for

Cu-1-X sample are in good agreement with theoretical simulations of [14]. Pole figures of ECAP Cu + Zr samples have shown sharper texture than pure Cu after one pass o and partially also after two ECAP passes. On the other hand, after 4 and 8 passes the pole figures of Cu and CuZr look nearly identical.

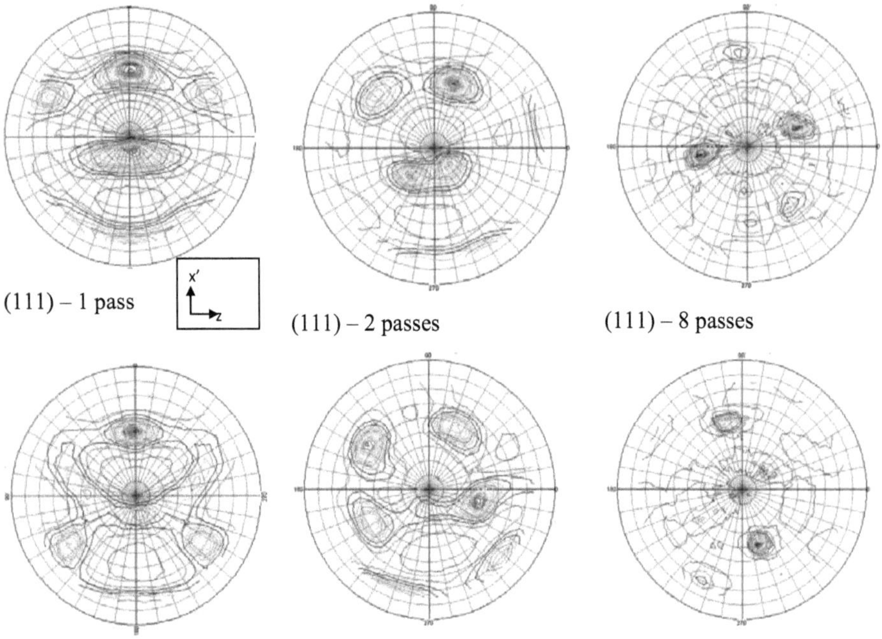

(111) – 1 pass (111) – 2 passes (111) – 8 passes

(100) – 1 pass (100) – 2 passes (100) – 8 passes

Figure 3. Pole figures for Cu samples after ECAP processed by different number of passes.

TEM, EBSD, PAS

TEM pictures of HPT samples show highly defected structure with inhomogeneous distribution of defects (regions with high and low dislocation density). Dislocations are situated in distorted regions along GBs, while grain interiors are almost free of dislocations.

More detailed studies were performed on the ECAP samples. Bright field TEM micrographs showing the evolution of the microstructure are presented in figure 4. After 1 pass (a) the microstructure consists mainly of strongly elongated dislocation cells and/or subgrains of average cross-sectional size of 300–400 nm. Two types of boundaries can be recognized in this micrograph. The thick dark lines can be identified as lamellae boundaries or dense dislocation walls (DDW) while the lighter and wider ones are individual dislocation cell boundaries. The DDWs usually enclose several cells and are classified as geometrically necessary boundaries as they are necessary to accommodate lattice rotations in adjoining volume. Selected area diffraction observations showed that these boundaries are mainly *low angle* ones and are oriented along the trace of a {111} plane, which indicates that they are parallel to a

<110> shear direction. They are in a nonequilibrium state as they contain a high density of extrinsic dislocations. After the second pass bands of elongated subgrains were found in all areas of the specimen. The average size of individual boundaries slightly decreased to 200–300 μm and more equiaxed subgrains with higher misorientation indicate the activity of additional slip systems during the second pass. After four passes (figure 4 b), the fraction of equiaxed subgrains increased and the larger proportion of high angle grain boundaries was observed in the structure as an indication that many new slip systems which are not parallel to the original slip system became active between the second and fourth pass of pressing reducing the length of the original subgrain boundaries. After 8 passes an almost homogeneous microstructure with equiaxed subgrains separated mostly by *high angle* grain boundaries was observed. The individual boundaries are straight with sharp contrast and very few dislocations in the grain interior [15].

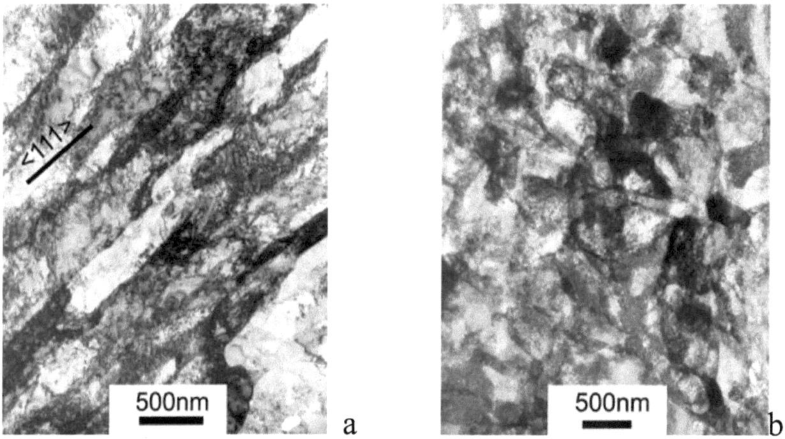

Figure 4. Typical bright field TEM pictures of ECAP Cu sample after 1 pass (a) and 4 passes (b).

As-prepared HPT Cu+Al_2O_3 samples may be characterized by a diffuse microstructure, with hardly distinguishable grains. Two kinds of Al_2O_3 particles were observed in the sample: isolated coarse particles with a size of approximately 100 nm, i.e. comparable with the grain size, and clusters of very fine Al_2O_3 nanoparticles with diameter less than 10 nm. TEM observations showed that grain size distribution in Cu+0.5wt.%Al_2O_3 sample is approximately lognormal. Parameters of the grain size distribution were L_{mean} = 140 nm, L_{median} = 112 nm, L_{td} = 95 nm. These values are in good agreement with XRD measurements.

EBSD analysis, in principle, confirmed the XRD (texture) and TEM (grain size, evolution of microstructure) studies. The grain size was found to decrease with increasing number of ECAP passes.. After 1 pass, almost the whole volume is deformed, while after 2 passes the fraction of recrystallised area increases and deformed part decreases. 4 passes resulted in about 76% of recrystallised volume and 21% of deformed one and after 8 passes approximately 20 % of substructured grains awere observed. With higher number of passes the deformation energy is accumulated and enhances the recrystallization. A successive evolution of grain boundary character from predominantly low-angle boundaries to high-angle boundaries with increasing numbers of passes was observed both in Cu and CuZr specimens *Positron annihilation spectroscopy* is very sensitive to small-sized open-volume defects as microvoids and dislocations. The majority of *positrons* are trapped at dislocations inside the distorted layers along GBs and contribute to the component with lifetime $\tau_2 \approx 164$ ps. A longer component with lifetime τ_3 comes from positrons trapped at small vacancy clusters, the so called microvoids situated inside grains. From the lifetime τ_2, one can deduce that size of microvoids, corresponds to approximately 4-5 vacancies. The size of the microvoids is the same for pure Cu and composites but their concentration is higher for the latter. These defects have not been observed by other technique.

4. Concluding remarks

For characterization of microstructure evolution of materials after severe plastic deformation the application of different complementary methods is inevitable, both because of their sensitivity to different microstructural features and the verification of the results obtained. Significant changes of microstructure of ECAP samples with number of passes from strongly deformed grains with low-angle boundaries to partially recrystallized grains with high-angle boundaries observed by TEM and EBSD have only small effect on XRD line broadening and the dislocation density evolution obtained from these measurements. Careful studies of line profile shapes for these samples are under way.

References

1. Valiev, R.Z., Alexandrov, I.V. & Islamgaliev, R.K., 2000, *Prog. Mat. Sci.*, **45**, 103.
2. Valiev, R.Z., 1997, *Mater. Sci. Eng. A.*, **234–236,** 59.
3. Wang, Y.M. & Ma, E., 2004, *Acta Mater.,* **52**, 1699.
4. Koch, C.C., 2003, *Scripta Mater.*, **49,** 657.
5. Wang, Y., Chen, M., Zhou, F. & Ma, E., 2002, *Nature*, **419,** 912.
6. Valiev, R.Z., Krasilnikov, N.A. & Tsenev, N.K. 1991, *Mater. Sci. Eng. A.*, **137,** 35.
7. Harris, G.B., 1952, *Phil. Mag.,* **43,** 113.
8. Krivoglaz, M.A., Martynenko, O.V. & Ryaboshapka, K.P., 1983, *Fiz. Metal. Metalloved.*, **55**, 5.
9. Kužel, R., 2007, *Z. Kristallogr.*, **222**, 136.
10. Scardi, P., Leoni, M. & Dong, Y.H., 2000, *Eur. Phys. J.,* **B18,** 23.
11. Ribárik, G., Ungár, T. & Gubicza, J., 2001, *J. Appl. Cryst.,* **34,** 669.
12. Matěj, Z., Nichtová, L. & Kužel, R., 2008, *Mater. Struct.,* **15,** 46.
13. Bečvář, F., Čížek, J. & Lešták, L., 2000, *Nucl. Instr. Meth. A.,* **443,** 557.
14. Li Saiyi, V., Beyerlein I.J. & Bourke M., 2005, *Mater. Sci. Eng. A.*, **394**, 66.
15. Janeček, M., Hadzima, B., Hellmig, R.J. & Estrin, Y., 2005, *Metallic Mater.*,**43**, 258.

Acknowledgements. The work is a part of the research plan MSM 0021620834 financed by the Ministry of Education of the Czech Republic and supported partially also by the Academy of Sciences of the Czech Republic under the number KAN400720701 and IAA101120803

Grain stresses and elastic energy in ferritic steel under uniaxial load

A. Baczmański[1,*], A. Tidu[2], P. Lipinski[3], K. Wierzbanowski[1]

[1] Faculty of Physics and Applied Computer Science, AGH University of Science and Technology, al. Mickiewicza 30, 30-059 Kraków, Poland
[2] LETAM, UMR CNRS 7078, ISGMP, Ile du Saulcy, F57045 METZ Cedex 1, France
[3] LPMM, UMR CNRS 7554, ENIM, Ile du Saulcy, F57045 METZ Cedex 1, France
[*] baczman@ftj.agh.edu.pl

Keywords: diffraction, self-consistent model, residual stresses, crystallographic texture

Abstract. The main purpose of this work is the study of grain stresses in plastically deformed textured samples under uniaxial loads where superposition of residual incompatibility stresses and imposed stresses occurs. The methodology for interpretation of diffraction data enabling determination of the grain stresses and elastic energy in polycrystalline materials was developed. Using X-ray diffraction, the lattice strains were measured in as received and "in situ" loaded samples produced from a cold rolled ferritic steel.
The self-consistent model of elastoplastic deformation was used for analysis of experimental data and the stress tensor was found for each orientation of polycrystalline grains. The stresses and elastic energy in the initial state were successfully determined and presented in Euler space.

1. Introduction

In a polycrystalline material, the incompatibility residual stresses arise during plastic deformation due to the anisotropy of plastic flow of different grains. The incompatibility stresses can be theoretically modeled by the self-consistent method [1,2] and determined from diffraction methods.
The methodology of stress determining based on the analysis of physical behaviour of crystallites in a polycrystalline material was proposed and developed by Baczmański et al. [3,4,5]. This method allows to evaluate quantitatively the macrostresses (the first order stresses) and the plastic incompatibility stresses (the second order stresses) using diffraction data and the self-consistent model. Analysis of the non-linear $<d(\phi,\psi)>_{\{hkl\}}$ vs. $sin^2\psi$ curves (measured using one reflection *hkl*) allowed to estimate the second order stresses in cold rolled ferritic steel [3,4] and in duplex steel [4,5]. The analysis of experimental data is based on the self-consistent calculations in which the interaction of ellipsoidal inclusion with a homogeneous matrix is considered. In this work, the method for determining the first and the

second order stresses is applied to the plastically deformed sample under "in situ" applied load.

2. Self-consistent model

The self-consistent model used in this work was proposed by Lipinski and Berveiller [1] and developed by Zattarin et al. [2]. The model calculations are performed on two different scales: the macro-scale, where the average quantities (Σ_{ij}, E_{ij} - applied to the sample) are defined, and the grain-scale, on which the behaviour of each crystallite under local stress σ_{ij}^{g} is analysed. On the grain-scale, plastic deformation occurs due to slip on the crystallographic planes. According to Schmid's law, slip can be activated only on a slip system *[uvw] (hkl)* (the slip plane and slip direction are specified) for which the resolved shear stress reaches a critical value τ_c, i.e.:

$$\sigma_{[uvw](hkl)} = \tau_c \tag{1}$$

where τ_c is the critical shear stress for slip, $\sigma_{[uvw](hkl)} = \frac{1}{2}(m_i n_j + n_i m_j)\sigma_{ij}^{g}$ is the resolved shear stress, $\boldsymbol{n} = [n_1, n_2, n_3]$ is the unit vector normal to the (*hkl*) plane and $\boldsymbol{m} = [m_1, m_2, m_3]$ is the unit vector parallel to the slip *[uvw]* direction.

During plastic deformation, the multiplication of dislocations and evolution of their spatial distribution inside grains leads to the hardening of slip systems (τ_c^s increases with deformation). In general, the dislocation interactions should be considered at the microscopic scale, i.e., inside the grain volume. However, in the kinetic description of slip systems, their latent hardening can be approximately described by the work hardening matrix H^{st} (i.e., the matrix reflecting the interaction between slip systems). Consequently, the rate of the critical shear stress on the *s*-th system $\dot{\tau}_c^s$ is equal to:

$$\dot{\tau}_c^s = \sum_t H^{st} \dot{\gamma}^t \tag{2}$$

where $\dot{\gamma}^t$ is the rate of the plastic glide on the *t*-th active system and dot means time derivative.

In the simplest approach, used in this work, a constant value of the *H* parameter assumed for all components of the hardening matrix (i.e., $H^{st} = H$), leads to the isotropic linear hardening law [1,2].

To predict elastoplastic deformation, these processes should be considered on the scale of a grain size. Moreover, the influence of boundary conditions and global variables on the behaviour of grains must be established. To this aim the idea of scale transition theory can be used for a polycrystalline material. The theory is based on the hypothesis of a concentration tensor $\boldsymbol{A}$ (or $\boldsymbol{B}$) which relates the macrostrain rate $\dot{E}_{kl}$ (or macrostress rate $\dot{\Sigma}_{kl}$) to the local strain rate $\dot{\varepsilon}_{ij}$ (or stress rate $\dot{\sigma}_{ij}$) defined for a polycrystalline grain, i.e.:

$$\dot{\sigma}_{ij}^{g} = B_{ijkl}^{g} \dot{\Sigma}_{kl} \quad \text{or} \quad \dot{\varepsilon}_{ij}^{g} = A_{ijkl}^{g} \dot{E}_{kl} \tag{3}$$

According to the self-consistent approach [1,2], the polycrystalline material is approximated by a macro-homogeneous medium at the macroscopic scale and a micro-heterogeneous medium at the microscopic scale. The relation between the stress rate and the strain rate can be written for both scales as follows:

$$\dot{\sigma}_{ij} = l_{ijkl}\dot{\varepsilon}_{kl} \qquad \text{and} \qquad \dot{\Sigma}_{ij} = L_{ijkl}\dot{E}_{kl} \tag{4}$$

where: l_{ijkl} is the local elastoplastic tangent modulus for the grain and L_{ijkl} is the macroscopic tangent modulus for a fictional average homogeneous medium.

The $\boldsymbol{L}$ tensor can be calculated only if the $\boldsymbol{A}^g$ tensor is known, and *vice versa*. To find both tensors the self-consistent procedure must be used [1,2].

The plastic incompatibility stresses can be theoretically modelled by the self-consistent method. In the calculations, the sample is represented by a number of grains, having a distribution of orientations reproducing the initial experimental textures. The model sample is subjected to elastoplastic deformation and next the external stresses are unloaded. Finally, the theoretical values of the second order plastic incompatibility stresses $\overline{\sigma_{ij}^{g(ic)}}$ (where the bar denotes the model-predicted quantities), defined as the stresses remaining after unloading of the macrostresses, can be calculated for polycrystalline grains.

3. Method of determining plastic incompatibility stresses

Let us suppose, that the anisotropy of the incompatibility stresses is well predicted by the model if the experimental texture is used as the input data. However, the absolute values of the stresses depend on the hardening process occurring during plastic deformation. To relate the magnitude of theoretical stresses to the actual one, an unknown scaling factor q is introduced. This factor does not depend on the grain orientation $\boldsymbol{g}$ and it rescales the amplitude of the stress tensor, i.e. the second order incompatibility stress $\sigma_{ij}^{g(ic)}(\boldsymbol{g})$ in the real sample is equal to:

$$\sigma_{ij}^{g(ic)}(\boldsymbol{g}) = q\ \overline{\sigma_{ij}^{g(ic)}(\boldsymbol{g})} \tag{5}$$

where q is the scaling parameter, and $\overline{\sigma_{ij}^{g(ic)}(\boldsymbol{g})}$ is the model predicted second order stress for a grain with $\boldsymbol{g}$ orientation.

The standard $sin^2\psi$ method of stress determination is based on the measurement of interplanar spacing for various directions of the scattering vector, defined by the ϕ and ψ angles [6-9]. It was shown by Baczmański [3,4,5] that the measured interplanar spacing can be expressed through:

$$< d(\phi,\psi) >_{\{hkl\}} = [F_{ij}(hkl,\phi,\psi)\sigma_{ij}^{M} + q\ \overline{<\gamma_{3m}\gamma_{3n}s_{mnij}\ \sigma_{ij}^{g(ic)}(\boldsymbol{g})>_{\{hkl\}}}\,]\, d^{o}_{\{hkl\}} + d^{o}_{\{hkl\}} \tag{6}$$

where $< d(\phi,\psi) >_{\{hkl\}}$ is the mean interplanar spacing for *{hkl}* planes in the direction of scattering vector, $d^{o}_{\{hkl\}}$ is the interplanar spacing for strain free material, $F_{ij}(hkl,\phi,\psi)$ are the diffraction elastic constants [7,8,9], $\overline{<\gamma_{3m}\gamma_{3n}s_{mnij}\ \sigma_{ij}^{g(ic)}(\boldsymbol{g})>_{\{hkl\}}}$ are the model predicted strains caused by the plastic incompatibility second order stresses, s_{mnij} are the single crystal elastic constants, $\boldsymbol{g}$ denotes grain orientation and γ_{ij} are the directional cosines transforming strains from the sample system to the system defined by scattering vector.

Using the least square method for equation 6, the fitting parameters (i.e., σ_{ij}^{M}, q and $d_{\{hkl\}}^{o}$) can be determined. The additional term $\overline{<\gamma_{3m}\gamma_{3n}s_{mnij}\ \sigma_{ij}^{g(ic)}(\boldsymbol{g})>_{\{hkl\}}}$ characterising the nonlinearities of the $sin^2\psi$ plot is adjusted to the experimental data. When the value of q parameter is determined by a fitting procedure, the real values of the plastic incompatibility second order stresses $\sigma_{ij}^{g(ic)}(\boldsymbol{g})$ can be calculated for all grain orientations $\boldsymbol{g}$ using equation 5. Thus the first order macrostresses σ_{ij}^{M} as well as the second order incompatibility stresses ($\sigma_{ij}^{g(ic)}(\boldsymbol{g})$) can be estimated by applying our method for the standard diffraction experiment.

4. Analysis of experimental data using the self-consistent model

Two samples were prepared from low carbon steel (ULC), cold rolled to the reduction of 96%. For the same steel, the diffraction elastic constants were determined using the self-consistent surface-free method [9]. The layers of *200 μm* and *100 μm* were removed by electropolishing in ULC96_1 and ULC96_2 samples, respectively. It was checked that the orientation distribution function was practically the same for both samples. The $<d(\phi,\psi)>_{\{211\}}$ vs. $sin^2\psi$ curves were measured using an improved Rigaku X-ray Stress Analyser (MSF-2M) equipped with a parallel beam optics and filtered $Cr_{K\alpha}$ X-ray radiation ($\lambda_{K\alpha}$ = 0.2291 nm). The measurements were carried out using the side-inclination method (ψ-goniometer geometry). To perform the measurements on the sample under a load, a tensile device was used. The samples were instrumented with strain gauges on both sides and were calibrated in an INSTRON universal test system. In addition, the tensile device had its own load cell to measure the applied forces.

To analyse the experimental results, at first, the cold rolling process was simulated using the self-consistent model. The polycrystalline sample was represented by *10000* grains having (initially) zero values of the stress components and the distribution of lattice orientations ($\boldsymbol{g}$) corresponding to the experimentally determined texture [3,10]. The same initial critical resolved stresses were attributed to each glide from the potentially active slips <111>{211} and <111>{110}. The cold rolling process was modelled using parameters given in table 1, up to sample strain E_{11} = 7.5% [10]. Next the applied macrostress was unloaded ($\Sigma_{kl} \rightarrow 0$) and the theoretical residual incompatibility stresses $\overline{\sigma_{ij}^{g(ic)}(\boldsymbol{g})}$ and lattice strains $\overline{<\gamma_{3m}\gamma_{3n}s_{mnij}\ \sigma_{ij}^{g(ic)}(\boldsymbol{g})>_{\{hkl\}}}$ were determined. The fitting procedure based on equation 6 was applied to adjust the theoretical $<d(\phi,\psi)>_{\{211\}}$ vs. $sin^2\psi$ functions to the experimental results for both samples (as received and loaded during tensile test) and the unknown adjusting parameters (i.e., σ_{ij}^{M}, q and d^0) were determined. The experimental orientation distribution function was used to calculate diffraction elastic constants [9] from single crystal elastic constants given in table 1.

Table 1. Parameters used for model prediction [10]

Assumed slip systems:	Single crystal elastic constants (GPa)			Initial critical resolved shear stress - τ_c (MPa)	Hardening parameter - H (MPa)
	c_{1111}	c_{1122}	c_{1212}		
<111>{211} and <111>{110}	231	134	116	250	80

In figure 1 the measured and theoretical (fitted) curves for as received and for loaded samples are shown. The theoretical plots are presented for two assumptions, i.e., the plastic incompatibility stresses are not present (dashed lines for *q=0*), and when the influence of $\overline{\sigma_{ij}^{g(ic)}(\boldsymbol{g})}$ stresses is taken into account and the *q* parameter is determined from equation 6 (continuous lines). As seen in figure 1, the calculated lines fit definitely better to experimental points when non-zero plastic incompatibility stresses are taken into account in the data analysis.

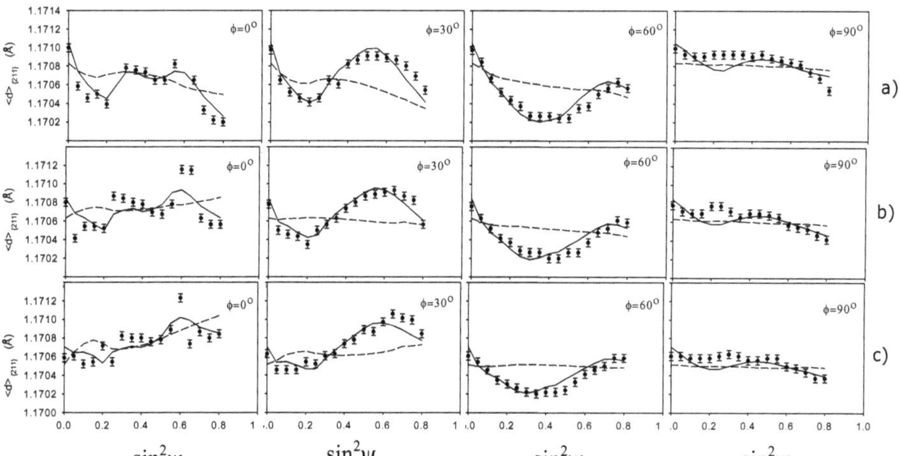

Figure 1. Measured interplanar spacings (points) and theoretical results of fitting (continuous lines for $q \neq 0$ and dashed lines for $q = 0$) for ULC96_1 sample. The results are presented for: (a) as received sample, superimposed tensile stress (b) Σ_{11} = 90 MPa and (c) Σ_{11} =120 MPa. The self-consistent surface free method was used for calculation of the diffraction elastic constants [9].

For a known *q* factor, the $\sigma_{ij}^{g(ic)}(\boldsymbol{g})$ incompatibility stresses were estimated using equation 5. It was found that the $\sigma_{ij}^{g(ic)}(\boldsymbol{g})$ stress are almost independent on the applied load (the equivalent von Mises stresses calculated from $\sigma_{ij}^{g(ic)}(\boldsymbol{g})$ is between 90 MPa and 125 MPa for all applied stresses).

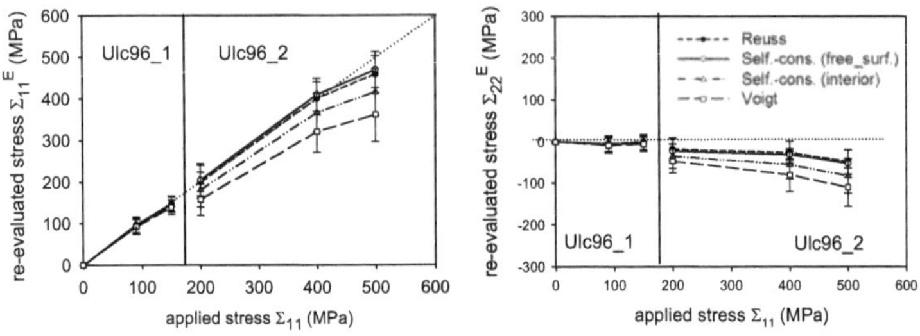

Figure 2. Values of re-evaluated stresses Σ_{ij}^E determined using different types of the diffraction elastic constants [9].

Finally, the possibility of determining the macrostress tensor was tested using the results obtained for different applied loads. In fact, it is impossible to verify the absolute values of the residual macrostresses for the as received samples, however, the differences between the determined macrostresses can be compared with the corresponding applied stresses. In figure 2 the re-evaluated stresses Σ_{ij}^E as a function of the applied stress Σ_{11} is presented. The re-evaluated stress is defined as:

$$\Sigma_{ij}^E = \sigma_{ij}^M(\Sigma_{11}) - \sigma_{ij}^M(0) \tag{7}$$

where $\sigma_{ij}^M(0)$ and $\sigma_{ij}^M(\Sigma_{11})$ are the macrostresses determined for as received sample and when the Σ_{11} stress is applied, respectively. It has been verified that the values of the applied stresses are precisely re-evaluated when the Reuss [8,9] or the self-consistent method for free surface [9] were used to calculate the diffraction elastic constants, i.e., the values of the applied stresses were correctly re-evaluated (below $\Sigma_{11} = 400$ *MPa*). Only for $\Sigma_{11} = 500$ *MPa* the re-evaluated stress is smaller than the applied one, probably due to local plasticity occurring when the superposed stresses exceed the elastic limit. Consequently, it can be concluded that the present method can be used to determine macrostresses in anisotropic samples (textured) in the presence of plastic incompatibility stresses.

5. Stored energy and stress orientation distribution functions

As mentioned, knowing the q factor from fitting procedure and applying equation 2 the values of the tensor of plastic incompatibility stresses $\sigma_{ij}^{g(ic)}(\boldsymbol{g})$ and next corresponding equivalent von Mises stress $\sigma_{eq}^{g(ic)}(\boldsymbol{g})$ for each orientation ($\boldsymbol{g}$) of a polycrystalline grain were determined. Moreover, the stored elastic energy $W^{g(ic)}(\boldsymbol{g})$ was calculated from the formula:

$$W^{g(ic)}(\boldsymbol{g}) = \frac{1}{2}\sigma_{ij}^{g(ic)}(\boldsymbol{g})\, s_{ijkl}^{g}(\boldsymbol{g})\, \sigma_{kl}^{g(ic)}(\boldsymbol{g}) \tag{8}$$

$s_{ijkl}^{g}(\boldsymbol{g})$ is the compliance tensor of a grain having the lattice orientation $\boldsymbol{g}$

In figure 3 the equivalent $\sigma_{eq}^{g(ic)}(\mathbf{g})$ stress, the stored elastic energy $W^{g(ic)}(\mathbf{g})$ together with the orientation distribution function f(**g**) are shown.

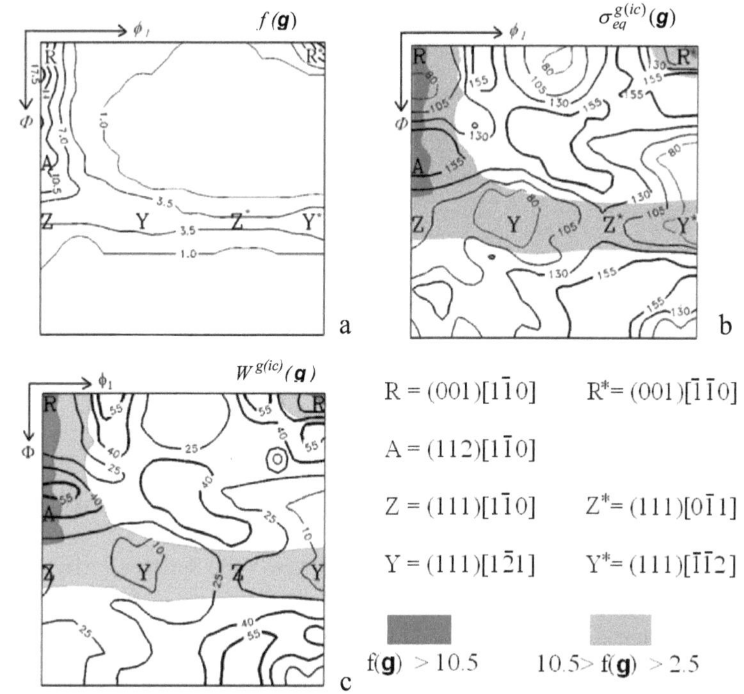

*Figure 3. Maps of orientation distribution function f(**g**) - (a), equivalent von Mises stress $\sigma_{eq}^{g(ic)}(\mathbf{g})$ [MPa]- (b) and stored elastic energy $W^{g(ic)}(\mathbf{g})$ [kJ/m^3] - (c) presented in the cross sections of the Euler space for $\phi_2=45^O$ ($0^O \leq \phi_1 \leq 90^O$ and $0^O \leq \phi_1 \leq 90^O$) [11]. The following notation is used for the main texture components: (hkl) is the crystallographic plane parallel to the surface of the sheet and [uvw] is the crystallographic direction parallel to the rolling direction (RD). The shades of grey (sections b and c) indicate area of high density of orientations.*

6. General conclusions

In the present study, the quantitative stress analysis based on the self-consistent model was performed using the X-ray diffraction data and modified $sin^2\psi$ method. As a result of the analysis the plastic incompatibility second order stresses as well as the macrostresses were determined quantitatively. The method was tested on samples for which known external loads were applied. It was found that for strongly textured samples the applied macrostresses can be accurately re-evaluated if the elastic anisotropy is properly taken into account. Moreover, the values of determined incompatibility stresses do not depend on the applied external load. Thus, it has been proved that the nonlinearities of the $<d(\phi,\psi)>_{\{hkl\}}$ vs. $sin^2\psi$ curves

cannot be explained only by the anisotropy of the diffraction elastic constants.
The presented analysis of the diffraction data is a quantitative verification of the self-consistent model. It has been shown that the deformation of different grains can be quantitatively predicted assuming the mechanism of plastic deformation based on slip systems and elastoplastic interaction between grains. The plastic incompatibility stress $\sigma_{ij}^{g(ic)}(\boldsymbol{g})$ was found for each orientation of a polycrystalline grain and the $\sigma_{eq}^{g(ic)}(\boldsymbol{g})$ equivalent von Mises stress was presented in the Euler space. The distribution of elastic energy $W^{g(ic)}(\boldsymbol{g})$ (caused by plastic incompatibilities between grains having various lattice orientations) was also shown.

References

1. Lipinski, P. & Berveiller, P., 1989, *Int. J. Plast.*, **5**, 149.
2. Zattarin, P., Baczmański, A., Lipiński P. & Wierzbanowski, K., 2000, *Arch. Metall.* **45**, 163.
3. Baczmański, A., Wierzbanowski, K., Lipiński, P., Helmholdt, R.D., Ekambaranathan, G. & Pathiraj, B, 1994, *Phil. Mag. A*, **69**, 437.
4. Baczmański, A., Braham, C. & Seiler, W., 2003, *Phil. Mag.*, **83**, 3225.
5. Wroński, S., Baczmański, A. Dakhlaoui, R., Braham, C., Wierzbanowski, K. & Oliver, E.C., 2007, *Acta Mat.,* **55**, 6219.
6. Noyan, I.C. & Cohen, J.B., 1987, *Residual Stress* (New York: Springer Verlag).
7. Brakman, C.M., 1987, *Phil. Mag. A* , **55**, 39.
8. Barral, M., Lebrun, J.L., Sprauel, J.M. & Maeder, G., 1987, *Met. Trans.*, **18A**, 1229.
9. Baczmanski, A., Tidu, A., Lipinski, P., Humbert, M. & Wierzbanowski, K., 2006, *Mat. Sci. Forum*, **524-525**, 235.
10. Baczmański, A., Lipinski, P., Tidu, A., Wierzbanowski, K. & Pathiraj, B., Quantitative Estimation of Incompatibility Stresses and Elastic Energy Stored in Ferritic Steel, submitted to *J. Appl. Cryst.*
11. Bunge, H.J., 1982, *Texture Analysis in Material Science: Mathematical Methods*, (London: Butterworths).

Acknowledgements. This research project has been supported by the Polish Ministry of Science and Higher Education (MNiSW) and by the European Network of Excellence: Complex Metallic Alloys (EU 6-th Framework Program).

Stress and hardness in surface layers of shot-peened steels

J. Drahokoupil[1,2,*], N. Ganev[2], M. Čerňanský[1], P. Boháč[1], R. Čtvrtlík[1,3], M. Stranyánek[1,3]

[1] Institute of Physics of the ASCR, v. v. i., Na Slovance 2, 182 21 Prague 8, Czech Republic
[2] Faculty of Nuclear Sciences and Physical Engineering, Czech Technical University, Trojanova 13, 120 00, Prague 2, Czech Republic
[3] Joint Laboratory of Optics of Palacky University and Institute of Physics of the ASCR, v.v.i., 17. listopadu 50, 772 07 Olomouc, Czech Republic
[*] jandrahokoupil@seznam.cz

Keywords: microstrain, hardness, macroscopic stress, shot peening, steels

Abstract. The main goal of the research is to study gradients of stresses and hardness in surface layer after shot peening and to look for correlations between these quantities, and this way to contribute to understanding of the onset of residual stress formation in engineering materials. X-ray diffraction was applied as a main technique for microscopic and macroscopic residual stress determination. Combination of X-ray diffraction with electrolytic polishing enables to study the depth profile of aforesaid quantities. Indentation on polished cross sections was used to study hardness depth profiles. It could be state that the shot peening produce the compressive macrostresses and the depth of the compressive layer is primarily influenced by intensity of blasting.

1. Introduction

Shot peening process consists of the controlled bombardment of the metal surface by spherical shot including steel shot, steel and stainless steel pieces of wire, ceramic or glass beads [1]. The shots may be driven by a high velocity stream of air or liquid or by mechanical device in which the shots are fed into a rotating wheel and thrown at the desired velocity. The treatment causes plastic flow of the surface layers, thereby inducing surface compressive stresses, change of microstructure and may cause phase transformation in the surface layers.
Although shot peening have been used widely in machined industry for a long time, it is still in the focus of interest of scientists and technologists [1]. The mechanism of the creation of compressive residual stress by shot peening has been recently clarified by Kobayashi et al. in [2] and a lot of papers dealing with the improvement of fatigue life of products caused by this treatment is still appearing [3 - 6]. Ultrasonic shot peening method is successfully used for generation of a nanostructured surface layer, i.e. surface nanocrystalisation of steels [7 - 8].

X-ray diffraction is widely used as the best developed experimental method of quantifying residual stresses. When the beneficial effect of imposed compressive residual stresses by shot peening is evaluated, usually only depth distribution of macroscopic (first kind) ones is taken into account [1, 9, 10]. The unique ability of X-ray diffraction methods to determine both the macroscopic residual stress and mean value of micro-strain and crystalline size in the irradiated volume is not commonly applied for complete characterization of degree of severe plastic deformation imposed into the surface layer affected by the shots′ stream. The diffraction line broadening caused by micro-strains and crystallite size reduction is taken into account by Prevéy in [11], where only the mutual effect (diffraction peak width) is considered and evaluated with connection with the cold work.
The aim of the paper is to present a complete X-ray diffraction study of the depth profile of real structure of shot-peened steel surfaces revealing courses of macroscopic stresses as well as of micro-strains. Comparative indentation study can enable to highlight advantages and lacks of method used. The broad range of selected samples includes common steels, tool steel and high alloyed steel as well.

2. Samples

The set of analysed samples (50 x 50 x 5 mm^3) was prepared from five steels: carbon steel to the heat treatment and surface quenching ČSN 12 050 (A), *Mn-Cr* steel to the cementation ČSN 14 220 (B), refractory *Cr-Mo-V* steel ČSN 17 134 (C), tool low-alloyed *Mn-Cr-V* steel for the cold working ČSN 19 313 (D), and high speed heavy duty *Mo-W-Cr* steel ČSN 19 852 (E). Summary of main alloying elements in steels is in Tab. 1.The samples were shot-peened by using two different intensities of blasting specified by using *Almen test* [12] as 0.2 mmA (samples signed by number 11) and 0.4 mmA (samples signed by number 13). Equally, the chosen intensity of the shot peening are minimum and maximum usually used in engineering practice. In order to analyse the stress gradients beneath the samples surface the layers of material was gradually removed by electrolytic polishing. A STRUERS device *Lectro-Pol-5* was used. The indentation measurements were performed on polished cross sections (10 x 5 mm^2), which were cut off from the central part of the samples. Prior to shot peening the samples were annealed (stress-relieved) in *Ar* at 550 °C for 2 hours.

Table 1. The chemical composition (only major alloying elements) of investigated steels.

Steel	Concentration, % weight							
	C	Mo	Cr	V	Co	W	Mn	Si
A	0.42-0.5	-	max 0.25	-	-	-	0.5-0.8	0.17-0.37
B	0.14-0.19	-	0.8-1.1	-	-	-	1.1-1.4	0.17-0.37
C	0.17-0.23	0.8-1.2	10-12.5	0.2-0.35	-	-	0.5-1	0.25-0.6
D	0.8-0.9	-	0.2-0.4	0.1-0.2	-	-	1.75-2.1	0.15-0.35
E	0.8-0.9	4.5-5.5	3.8-4.6	1.5-2.2	4.3-5.2	5.5-7	max 0.45	max 0.45

3. Used methods

3.1 X-ray diffraction

The measurement was performed on an ω-goniometer Siemens with Cr-Kα radiation. The line {211} of α-Fe phase was measured. Nine different tilts angles (ψ) from 0° to 63° were used. The $\sin^2\psi$ method was used for determination of macroscopic residual stress [13]. The X-ray elastic constants $\frac{1}{2}\, s_2 = 5.95 \cdot 10^{-6}$ MPa^{-1}, $-\, s_1 = 1.325 \cdot 10^{-6}$ MPa^{-1} were used in stress calculations. The single line Voigt function method was applied for corrections of instrumental broadening (Fe powder standard was measured for all nine inclination) and determination of microstrains and crystallite size (the data are not presented here) [14]. The microstress σ^{micro} was calculated from microstrains e using Hooke's law ($\sigma = e\, E$) [15] with the Young modulus E^{211} = 216 GPa to be comparable with macroscopic residual stress σ^{macro}. The Young modulus represents the average of two values from E^{211} computed in accordance with two extreme interaction models - Reuss and Voigt [13] for X-ray elastic constants calculation.

In order to obtain depth profiles of X-ray diffraction characteristics, the surface layers were removed by a *LectroPol–5* device for electrolytic polishing. The cooling system maintains the temperature in the range of 15 – 25 °C.

3.2 Indentation

The DSI (depth sensing indentation) curves were obtained by apparatus *NanoTest^{TM} NT600* equipped with diamond Berkovich indenter. Each sample was indented in several different points at each given distance from edge of treated surface. Maximal load was 50 mN; loading/unloading time was 20 s; dwell time at maximal load was 10 s. The improper curves (shifted, deformed, etc.) were excluded from calculations. Despite it each point in graphs is an average value from minimal three measurements. Indentation hardness was calculated according to the Oliver-Pharr method [16] which uses the unloading part of the DSI curve.

4. Results and discussions

4.1 X-ray diffraction

The samples were measured in two perpendicular directions. For reasons of clarity the average values for these perpendicular directions are used there. Ascertainment that shot peening is symmetrical treatment [12], give us the competence for the mentioned averaging. The single line Voigt function method [14] offer values of particle size D and microstrain e (or σ^{micro}) for each of the tilt angles ψ, i.e. for $\sin^2\psi$ = 0, 0.1, 0.2, … 0.8. The individual values depend on gradients, orientation of diffracting planes (angle ψ), and inaccuracy of used procedures. In order to compare these data with macroscopic residual stress, the values for all tilts and also for two perpendicular directions were averaged.

Figures 1 show the depth profiles of macro and micro stresses for all samples. The shot peening caused symmetrical both macroscopic and microscopic residual stresses. The macroscopic stresses are compressive.

The depth profile of microscopic stresses has nearly the same hyperbolic decreasing tendencies. The surface values differentiate both among steels and intensities of blasting. All microscopic stresses are decreasing to zero with increasing depth aside the refractory steel (C) which reaches the lowest value ca 100 MPa. This non-zero value is not evocated by shot peening and likely to have the origin in mass content of alloying elements. The depth profiles of macroscopic stresses show larger differences in the influenced depth than microscopic stresses.

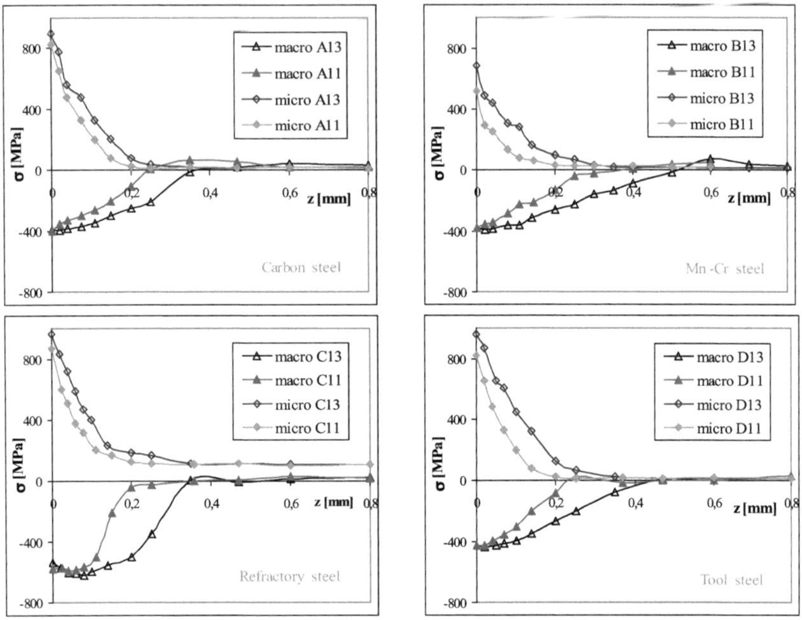

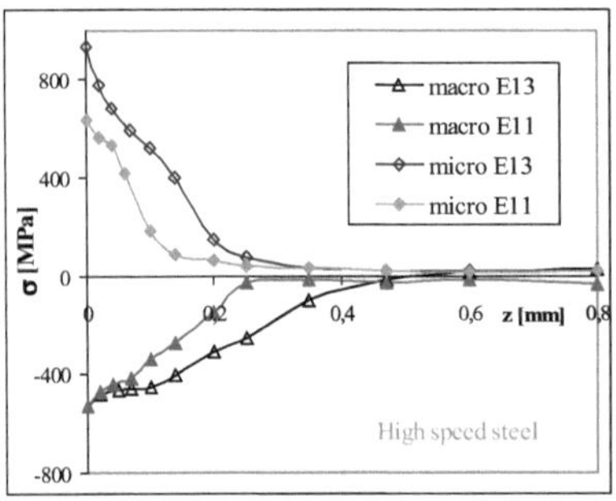

Figure 1. Depth profiles of macro and micro stresses for two intensity of blasting (11-low, 13-high) for all materials A, ..., E. The standard deviations are approximately ± 30 MPa.

The maximum values of macroscopic compressive stresses due to both the intensities (0.2 mmA – marked 11; 0.4 mmA - marked 13) applied are not different with respect to the experimental error. They reach the values of ~ 400 MPa for materials A, B, D and ~ 550 MPa for materials C, E. The last two materials have greater strength limits than materials A, B, D [17]. They have also higher hardness than remaining three steels. It could be as a conse-

quence of higher alloying, namely by chromium (Tab. 1). Thus, the magnitude of the compressive stress is primarily a function of the mechanical properties of the material.
The depth of the compressive layer is influenced primarily by intensity of shot peening, broadly speaking: approximately it is 0.2 mm for intensity of peening 0.2 mmA and 0.4 mm for intensity of peening 0.4 mmA. The widest affected zone has Mn-Cr steel (B) where the affected depths are approximately 0.3 and 0.5 respectively. The depth profiles of stresses are for all materials very similar. The most different is for the refractory steel (C) which has the widest zone of large macroscopic deformations near the surface but the depth of affected zone is smallest among all the observed materials. These two observations show to small deformation transfer ability. This finding is also in agreement with the lowest ductility found out in previous work [17]. Moreover the more intensively blasted sample (C13) has the maximum values of macroscopic stress in the depth of 0.08 mm (about 100 MPa larger compressive stress than the surface value). Presumably it is a result of the main role of plastic deformation.

4.2 Indentation

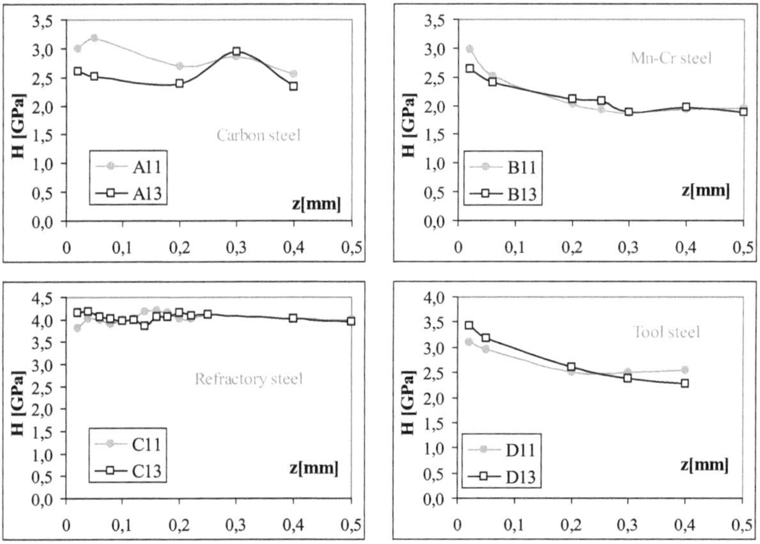

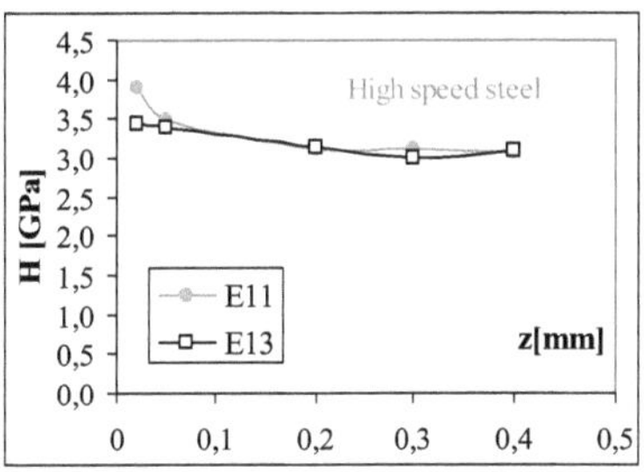

Figure 2. Depth profiles of hardness for two intensity of blasting (11-low, 13-high). The standard deviations are approximately ± 0.2 GPa.

Figures 2 show hardness depth profiles of corresponding samples investigated by XRD. The carbon steel (A) and refractory steel (C) do not show any distinct hardened zone. Some gradients of hardness can be observed only in material B, D, E. The peening process improved surface hardness of these steels on the average for about 0.8 GPa. The hardened zone is about 0.25 mm for both intensities of blasting. Parameters of the shot peening process have little effect on the magnitude of the hardness being primarily a function of the mechanical properties of the material. Nevertheless, hardness stays roughly directly proportional to macro stress.

The hardness was locally non-uniform on the studied samples, thus indentations applied on various places, although in the same distance from surface, lead to different results of hardness. The residual impressions in material after indentation were about 7 μm in diameter. The differences in hardness may by caused by inhomogeneties in the material and also by anisotropy of elastic constants. And so, the study of hardness gradients is difficult due to these inhomogeneties.

4.3 Relations between stresses and hardness

The figure 3 is parametrical plot between macro and micro stresses where on the axis x are plotted microscopic residual stresses σ^{micro} and on the axis y are plotted macroscopic residual stresses σ^{macro}. Since dependences for both intensities of blasting are for all materials very similar (whereas the curves for both intensities do not differ much from one another for the same material) and on the curves for different materials are shifted each other, it can be stated that the relations between macro and micro stress depend primarily on the material characteristics independently of intensity of peening. Thus, dependencies for only two materials (B, C) are plotted as limiting curves for the remaining materials (A, D, E).

The horizontal parts of the curves correspond to the zones where macroscopic stresses reach their maximum but microscopic stresses not yet. This occurred close to the samples surfaces. The vertical parts of the curves correspond to the zones where macroscopic stresses change rapidly as microscopic change slowly. This happened in the depths where the effect of blasting vanishes (0.1 – 0.5 mm). Compare these statements with figures 1.

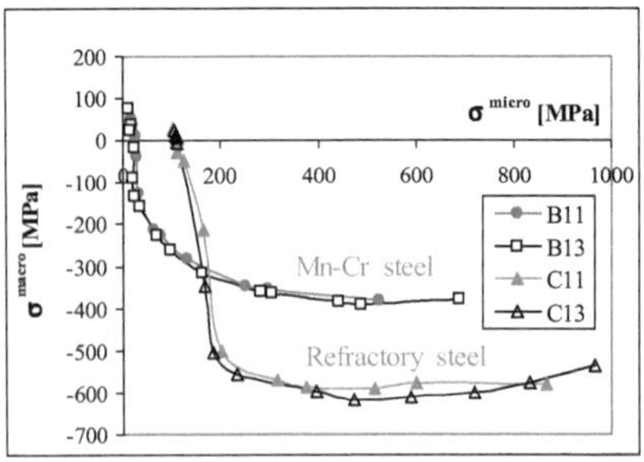

Figure 3. Parametric plot of macrostress vs. microstress for materials B and C.

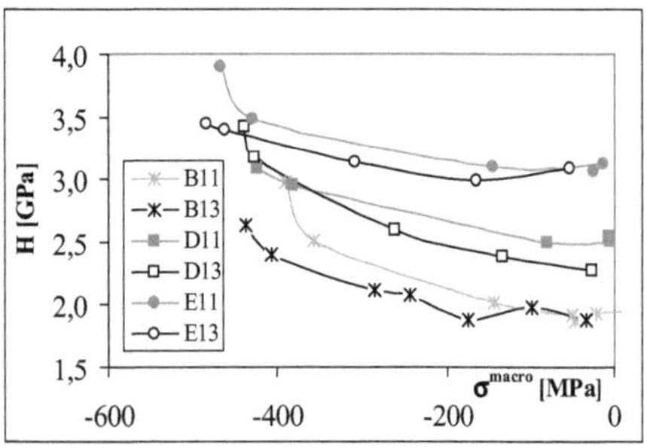

Figure 4. Parametric plot of macrostress vs. hardness for materials B, D, and E.

These experiments imply that relationships between macroscopic and microscopic stresses are independent on intensity of blasting for the particular material. Hence, knowledge of this dependence for a given material would enable to evaluate macro stresses from micro stresses and vice versa.
As mentioned afore, only materials B, D, and E show some distinct gradients of hardness. Figure 4 shows parametric plots of hardness H and macrostresses for these three materials. The values of macrostresses were extrapolated to the depth where hardness was measured. The curves are like linear lines where greater hardness corresponds to greater compressive residual macroscopic stresses. Similar relationship we presented elsewhere [18]. It is shown that material characteristics are more dominant than the intensity of shot peening - curves for different intensity of blasting are closer one to another than for different materials. In addition, the figure also shows influence of compressive stress on increase of hardness as reported for sputtered films as well [19].

5. Conclusions

The shot peening caused symmetrical both macroscopic and microscopic residual stress depth distributions. The macroscopic stresses are compressive and they reach the maximum surface value between - 400 and - 550 MPa. Depth profiles of particular types of stresses for all five investigated steels are similar. The former shows larger differences in course, the latter in surface value depending on both, material and intensity of blasting. The relationship between macroscopic and microscopic stresses is independent on intensity of blasting for the particular material. Hence, knowledge of this dependence for a given material would enable to evaluate macro stresses from micro stresses and vice versa.
Parameters of the shot peening process have only a little effect on the magnitude of the compressive macrostress induced which is primarily a function of the mechanical properties of the material. Subsurface range of this stress depends on intensity of the process; broadly speaking: it is approximately 0.2 mm for intensity of peening 0.2 mmA and 0.4 mm for intensity of peening 0.4 mmA.
Surface hardening by shot peening was observed namely for *Mn-Cr* steel (B), tool low-alloyed Mn-Cr-V steel (D), and high speed heavy duty *Mo-W-Cr* steel (E); on the average for about 0.8 GPa. Small hardening showed remaining two steels: carbon steel (A) and refractory *Cr-Mo-V* steel (C). Like linear dependence was found between macrostresses and hardness in subsurface region.
The indentation is a less sensitive method than X-ray diffraction. Surface roughness influences measurements for small indents. Deviations of results of the indentation are also greater because of a large influence of the surroundings of the indentation place (various grains, grain boundary, precipitate). On the other hand, namely the micro hardness in a larger depth indicates, that the investigated steels are considerable different. This is not clearly manifested by the residual stresses estimated by X-ray diffraction. The intensity of the shot peening has a less influence on the micro hardness, than on the results of X-ray diffraction measurements.

References

1. Wagner L., (editor), *Shot Peening*, 2005, (New York: Wiley - VCH), pp. 584.
2. Kobayashi, M., Matsui, T. & Murakami, Y., 1998, *Int. J. Fatigue*, **20**, 351.
3. Torres, M.A.S. & Voorwald, H.J.C., 2002, *Int. J. Fatigue*, **24**, 877.
4. Gao, Y.K., Yao, M., Shao, P.G. & Zhao, Y.H., 2003, *J. Mat. Eng. Perform.* **12**, 507.
5. Rodopoulos, C.A., Curtis, S.A., de los Rios, E.R. & Solis Romero J., 2004, *Int. J. Fatigue*, **26**, 849.
6. Harada, Y. & Mori, K., 2005, *J. Mater. Proc. Techn.*, **162–163,** 498.
7. Liu, G., Lu, J. & Lu K., 2000, *Mater. Sci. Eng. A*, **286**, 91.
8. Liu, G., Wang, S.C., Lou, X.F., Lu, J. & Lu, K., 2001, *Scr. Mater.* **44**, 1791.
9. Fontanaril, V., Frendo, F., Bortolamedi, Th. & Scardi, P., 2005, *J. Strain Anal. Eng. Design*, **40**, 199.
10. Sakakibara, T. & Sato, Y., 2005, *Powder Diffr.*, **20**, 117.
11. Prevéy, P.S., 2001, *Shot Peener*, **15**, 1.
12. Ganev, N., Čerňanský, M., Drahokoupil, J. & Kolařík, K., 2005, *CD-ROM of Proceedings EAN 2005*, 43rd International Conference "Experimental Stress Analysis 2005", June 7 – 9, 2005, Skalský Dvůr, Czech Republic.
13. Welzel, U., Ligot, J., Lamparter, P., Vermeulen, A.C., & Mittemeijer, E.J., 2005, *J. Appl. Cryst.*, **3**, 1.
14. De Keijser, Th.H., Langford, J.I., Mittemeijer, E.J. & Vogels, B.P., 1982, *J. Appl. Cryst.*, **15**, 308.
15. Warren, B.E. & Averbach, B.L., 1952, in *Imperfections in Nearly Perfect Crystals,* edited by W. Shockley, J.H. Hollomon, R. Mauer & F. Seitz (New York: John Wiley & Sons, Inc., London: Chapman & Hall), pp. 152-172.
16. Oliver, W.C. & Pharr, G.M., 1992, *J. Mater. Res.* **7,** 1564.
17. Ganev, N., Čerňanský, M., Boháč, P., Novák, V., Drahokouopil, J., Čtvrtlík, R. & Stranyánek, M., 2006, *Acta Mechanica Slovaca,* **10**, 137.Ganev, N., Čerňanský, M., Boháč, P., Drahokoupil, J., Čtvrtlík, R., Stranyánek M., 2005, 22nd Danubia – Adria Symposium on Experimental Methods in Solid Mechanics, September 28 – October 1, 2005, Monticelli Terme – Parma, Italy, Extended Abstracts, pp. 38-39.
19. Kulikovsky, V., Vorlíček, V., Boháč, P., Kurdyumov, A., Deyneka,A., Jastrabík, L., 2003, *Diamond Relat. Mater.*, **12**, 1378.

Acknowledgements. The research was supported by the Project No 106/07/0805 of the Czech Science Foundation, by the Project MSM 6840770021 of the Ministry of Education, Youth and Sports of the Czech Republic.

LINE BROADENING ANALYSIS; DETERMINATION OF CRYSTALLITE SIZE AND MICROSTRAIN

The "state of the art" of the diffraction analysis of crystallite size and lattice strain

E. J. Mittemeijer[1,*], U. Welzel[1]

[1] Max Planck Institute for Metals Research, Heisenbergstrasse 3, D-70569 Stuttgart, Germany
[*] e.j.mittemeijer@mf.mpg.de

Keywords: X-ray powder diffraction, line-profile analysis, crystallite size, microstrain, coherency of diffraction

Abstract. This paper addresses both old, but "renovated" methods and new methods for diffraction line-profile analysis. Classical and even extremely simple single-line methods for separating "size" and "strain" broadening effects have merit for characterization of the material imperfectness, but it is generally very difficult to interpret the data obtained in terms of microstructure parameters as used in materials science. Developments of recent years, focusing on distinct anisotropic line-broadening effects, as due to the type, orientation and distribution of dislocations and minute compositional variation, will be touched upon. The most promising development may be the synthesis of line profiles on the basis of a microstructure model and application of the (kinematical) diffraction theory without any further assumption, which contrasts with the other methods. This approach can in principle be applied in single-line and multiple-line variants and also in analyses of the whole diffraction pattern. The advantage is the direct evaluation of microstructure parameters as used in materials science. The challenge is to develop microstructure models which are flexible enough to be applicable in more than one case.....

Full paper published in Z. Kristallogr. **223** (2008) 552-560 / **DOI** 10.1524/zkri.2008.1213 (open access).

Z. Kristallogr. Suppl. **27** (2008) 101-111 / **DOI** 10.1524/zksu.2008.0014

Recent advancements in Whole Powder Pattern Modelling

P. Scardi[1,*]

[1] Department of Materials Engineering and Industrial Technologies, University of Trento, 38100 via Mesiano, 77, Trento, Italy
[*] Paolo.Scardi@unitn.it

Keywords: Line Profile Analysis, Whole Powder Pattern Modelling, dislocations, domain size distribution, powder diffraction

Abstract. Advantages of Whole Powder Pattern Modelling against conventional Line Profile Analysis methods are briefly reviewed, and a specific example is discussed on the possible ambiguity in the interpretation of the Williamson-Hall plot for polydisperse systems. Advancements in WPPM concerning dislocation line broadening are illustrated with examples taken from the recent literature. Reliability and limits in the application of WPPM to nanocrystalline systems are also discussed.

1. Introduction

In recent years, the growing interest in nanomaterials gave considerable momentum to diffraction Line Profile Analysis (LPA), recognised as one of the most used techniques to study crystalline domain shape and size distribution, as well as nature and amount of lattice defects [1]. As a consequence, LPA methods developed considerably: traditional methods based on diffraction peak integral breadths (e.g., Scherrer formula and Williamson-Hall (WH) plot [2,3]) or on the Fourier analysis of isolated peak profiles (e.g., the Warren-Averbach method [2,4]) have been paralleled by several new approaches with increasing tendency to deal with the full diffraction pattern [5,6]. It is useful to introduce a distinction between methods based on the use of some flexible but arbitrary profile function (e.g., Voigt, pseudo-Voigt, Pearson VII functions [7]), and methods that exclusively rely on physical models of the microstructure (e.g., describing coherent scattering effects from dispersed systems of crystalline domains, strain fields of lattice defects, etc.): we will refer to Whole Powder Pattern Fitting (WPPF) for the former and Whole Powder Pattern Modelling (WPPM) for the latter [8-10].
Profile fitting is almost invariably a need when dealing with X-ray Diffraction (XRD) patterns from finely dispersed and/or highly deformed systems – most cases of interest to LPA – to separate overlapping peak profiles and background. However, despite the simplicity in developing WPPF software [11] and the flexibility of this approach, WPPF is biased by the choice of the profile function which replaces, at some stage of the analysis, the experimental dataset, so that LPA is actually performed on the fitting parameters more than on the data. This can give results that do not match the original data [12].

A WPPM approach has a whole deal of advantages over WPPF and traditional LPA [8-14]. In synthesis: (i) possibly all effects contributing to the line profile can be considered (and compared) simultaneously; (ii) physical parameters, like domain size, density of defects, etc., are directly determined from the experimental data, without any intermediate stage of profile fitting; (iii) correct statistics, related to XRD data collection, is used. In addition to that, as a full pattern method, WPPM can properly consider overlapping among peaks and with the background, multiple phases, and suitably include instrumental effects, anisotropy and structural constraints. The present paper reviews some recent developments in WPPM, mostly related to the effects of dislocations. It is also shown how the presence of a distribution of domain sizes can affect results from the Williamson-Hall plot, one of the most popular integral breadth approaches. Finally, the limit of applicability of WPPM to nanodomains is explored and some preliminary results are shown.

2. Integral breadth analysis for polydisperse systems

A basic result of the diffraction theory is that the Scherrer formula, stating the inverse proportionality between domain size and integral breadth (β, ratio between peak area and peak maximum), is always valid, irrespective of domain shape and size distribution [15]. This result has motivated many researchers to use integral breadth methods with little regard to the presence of a distribution of domain sizes – a rule more than an exception in real cases of study. Considering for simplicity domain shapes described by one size parameter (D) dispersed according to a size distribution $g(D)$, then the Scherrer formula reads [9,12,16,]

$$\beta(d^*) = \langle L \rangle_V^{-1} = K_\beta M_3 / M_4 \,, \tag{1}$$

where d^* is the modulus of the scattering vector ($=2\sin\theta_B/\lambda$, with θ_B and λ as Bragg angle and X-ray wavelength, respectively), $\langle L\rangle_V$ is the volume-weighted mean column length (or Scherrer size), K_β is the *hkl*-dependent Scherrer (domain shape) constant and M_j are moments of the size distribution, defined as:

$$M_j = \int D^j g(D) dD \tag{2}$$

According to this definition, M_1 is the mean ($\overline{D}$) and M_2 - M_1^2 is the variance.
If we consider a lognormal distribution of spheres of diameter D,

$$g(D) = \exp\left[-(\ln D - \mu)^2\right)/2\sigma^2\Big]\Big/ D\sigma\sqrt{2\pi} \,, \tag{3}$$

mean size and Scherrer size are simply related to lognormal mean (μ) and variance (σ) as:

$$\overline{D} = exp\left(\mu + \sigma^2/2\right), \tag{4}$$

$$\langle L \rangle_V = 3/4\, exp\left(\mu + 7\sigma^2/2\right) \tag{5}$$

The distributions shown in figure 1a, all with the same Scherrer size of 10 nm, were calculated for increasing σ (from 0.01 to 0.8) and correspondingly decreasing μ (from 2.59 to 0.35) values, obtained by solving equation (5) for given $\langle L\rangle_V$ and σ. A portion of the simulated diffraction pattern for gold is shown in figure 1b for some of the distributions of figure 1a. The corresponding Williamson-Hall plot, shown in figure 1c, is the same for all patterns: a line parallel to the abscissa (size effect only) with intercept to β=1/10 nm^{-1}. Same WH plot and same Scherrer size correspond to quite different distributions, with the mean size varying

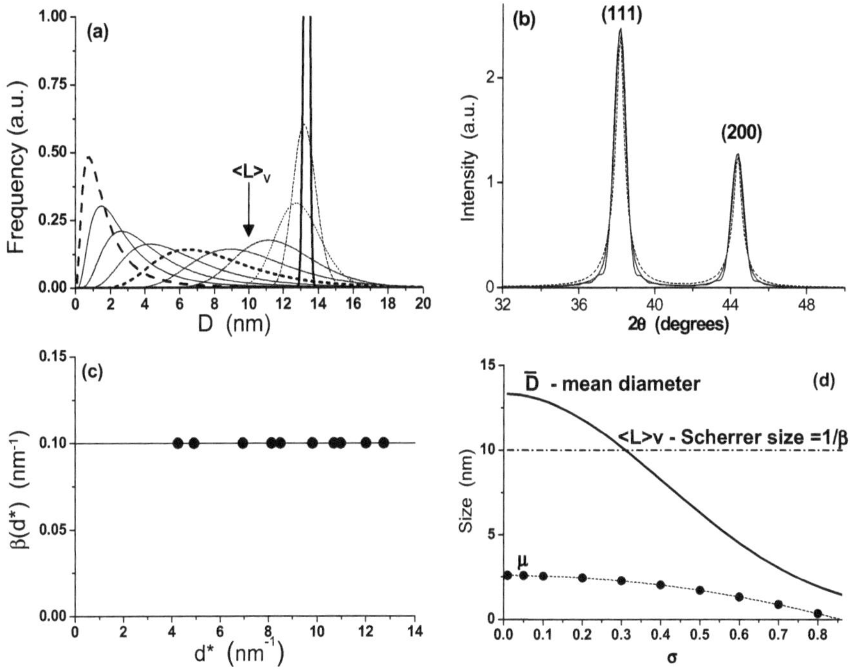

Figure 1. Lognormal distributions with constant Scherrer size (10 nm, see arrow) (a). Corresponding powder pattern of gold for the three distributions represented in (a) in bold: σ=0.01 (line), 0.4 (dot), 0.8 (dash) (b). Williamson-Hall plot for all simulated patterns (c). Mean diameter and lognormal mean (μ) for the various increasing values of lognormal variance (σ) used in the simulations (d).

from 13.3 nm (σ=0.01, μ=2.59) to about 2 nm (=0.8, μ=0.35) (see figure 1d).
This simple example suggests caution in interpreting integral breadth data. Results of Scherrer formula and WH plot can only be interpreted univocally if the studied system is monodisperse or if, for example in a series of samples, the distribution variance does not change significantly. Further ambiguity in the interpretation of the WH plot can be due to strain effects, generally related to the slope in the WH plot: the same slope could be produced by different density and arrangement of lattice defects.

3. Dislocation effects in Whole Powder Pattern Modelling

3.1 Extension to crystals of any symmetry

The WPPM philosophy is based on the simple observation that diffraction profiles result from a convolution of instrumental and sample-related effects. As such, the diffraction profile is conveniently described in terms of a Fourier transform [10,14]:

$$I_{\{hkl\}}\left(d^*_{hkl}\right) = k\left(d^*_{hkl}\right) \cdot \sum_{hkl} w_{hkl} \int_{-\infty}^{\infty} \mathbf{C}_{hkl}(L) \exp\left[2\pi i L\left(d^*_{hkl} - d^*_{\{hkl\}} - \delta_{hkl}\right)\right] dL \,, \quad (6)$$

where the sum is made on the *hkl* subcomponents of a given *{hkl}* Bragg reflection, with a weight w_{hkl} and shift δ_{hkl} from Bragg condition in reciprocal space (e.g., when $d^*_{hkl} = d^*_{\{hkl\}}$) caused by defects (e.g. faulting) [10,14]. *k* includes constants and known functions of d^*_{hkl} (square modulus of structure factor, Lorentz-polarization and absorption factor, etc.). The Fourier Transform (FT) of the peak profile, $\mathbf{C}_{hkl}(L)$, is the product of the FTs of peak profiles produced by the various instrumental and sample-related effects, as:

$$\mathbf{C}_{hkl}(L) = T_{pV}^{IP} \cdot A_{\{hkl\}}^{S} \cdot A_{\{hkl\}}^{D} \cdot (A_{hkl}^{F} + iB_{hkl}^{F}) \cdot A_{\{hkl\}}^{APB} \cdot \ldots, \quad (7)$$

where subscripts *IP*, *S*, *D*, *F*, *APB*, ... respectively stand for instrumental profile, size, dislocations, faulting, anti-phase domain boundaries, and any other possible effect contributing to the line profile [17].

Expressions for the FTs in equation (7) are available from the literature, so that the entire diffraction pattern can usually be modelled with relatively few physical parameters describing the microstructure, like the average dislocation density (ρ) and effective outer cut-off radius (R_e), deformation (α) and twin (β) faulting probabilities, anti-phase domain boundary probability (γ), etc.. Size effects can either be related to a given distribution of domain sizes (e.g. a lognormal, for which μ and σ can be refined) or to a general, free histogram distribution to be entirely determined by the WPPM [13].

Concerning dislocations, the FT is given by the Krivoglaz-Wilkens theory as [18]:

$$A_{\{hkl\}}^{D}(L) = \exp\left[-\frac{1}{2}\pi b^2 \overline{C}_{\{hkl\}} \rho d_{\{hkl\}}^{*2} \cdot L^2 f^*\left(L/R_e\right)\right], \quad (8)$$

where b is the modulus of the Burgers vector, f^* is a known function of L (Wilkens function [18]), and $\overline{C}_{\{hkl\}}$ is the average dislocation contrast factor, accounting for the anisotropy of the dislocation strain field and of the elastic medium [18] (see Appendix).

So far, one of the major limitations of WPPM and LPA methods based on equation (8) was the calculation of $\overline{C}_{\{hkl\}}$. Contrast factors were only known for a few slip systems of high-symmetry phases (cubic or hexagonal). In a recent work [19], following the basic indication of Popa [20] that the mean square strain $\langle\varepsilon^2_{\{hkl\}}\rangle$ (proportional to the average contrast factor) is a 4-th order invariant form of the Miller indices related to the Laue group of the studied phase, it was proposed that, in the most general (triclinic) form

$$d_{\{hkl\}}^4 \overline{C}_{\{hkl\}} = E_1 h^4 + E_2 k^4 + E_3 l^4 + 2\left(E_4 h^2 k^2 + E_5 k^2 l^2 + E_6 h^2 l^2\right) + 4\left(E_7 h^3 k + E_8 h^3 l + E_9 k^3 h + E_{10} k^3 l + E_{11} l^3 h + E_{12} l^3 k\right) + 4\left(E_{13} h^2 kl + E_{14} k^2 hl + E_{15} l^2 hk\right) \quad (9)$$

Considering the average contrast factor as an invariant ($\overline{C}_{\{hkl\}} = Inv\left(E_1 \ldots E_{15}, h, k, l\right)$), one can proceed according to two possible approaches:

i. provided that slip system, Burgers vector and elastic constants are known, $E_1, E_2 \ldots E_{15}$ can be calculated (the number of non-zero E_i, depends on the Laue group), so that equation (8) can be used in WPPM to obtain physical information on the dislocations in terms of ρ, R_e and dislocation character (e.g. an effective edge/screw fraction, f_E), in addition to other contributing effects (see equation (7)) among which the size distribution;

ii. if no information is available other than the cell symmetry, the $E_1, E_2 ... E_{15}$ constants can be treated as free modelling parameters.

Even if (ii) cannot directly provide quantitative information on the dislocation system, it can be used to guess possible slip systems and in any case to optimize in a consistent way the modelling of the powder pattern, so that other information (e.g., the size distribution or the peak area/position of interest in structural studies) can be better evaluated.

Concerning (i), it is worth noting that $E_1, E_2 ... E_{15}$ can now be calculated for any Laue group [19]. As an example, figure 2 shows a simulation of the effect of adding edge dislocation line broadening to the powder pattern of cassiterite (tetragonal SnO_2, Laue group 4/mmm) [19]. The powder is described by a lognormal system of spherical domains, with mean size 50 nm and standard deviation 15.3 nm. The inclusion of a dislocation broadening component adds a d*-dependent (strain) effect, that, as expected, progressively broadens line profiles with increasing angle. The increment in line broadening is clearly not monotonous with the diffraction angle, owing to the anisotropy effect introduced by the dislocation strain field and elastic medium (e.g., see (110) and (200) line in figure 2). To better appreciate this feature the average contrast factor is also shown in the figure (right axis).

In the cited reference [19] it was shown how WPPM can be used to investigate the characteristics of dislocations introduced in cassiterite by high energy grinding. Line broadening effects for different slip systems were compared and a best fit of the data gave information on the dislocations as well as on the domain size distribution.

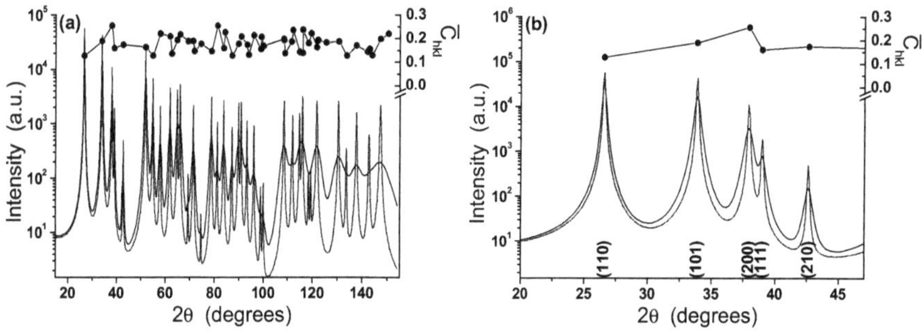

Figure 2. Simulated pattern (for CuKα₁ radiation) of SnO₂ (cassiterite) for a powder of spherical domains with (line) or without (dash-dot) contribution from <10-1>{101} edge dislocations ($\rho=5x10^{15}$ m^{-2}, R_e=10 nm). On the right, detail of the low angle region with Miller indices. Right axes refer to the average contrast factor (full symbol).

3.2 Parametric expressions of the average contrast factor

In the previous example $\overline{C}_{\{hkl\}}$ was calculated as an average over all the equivalent slip systems, supposed to be equally populated by dislocations. Each value was obtained by a numerical procedure [19], as it has always been the case even for the more symmetrical cubic and hexagonal systems. However, recent studies show that in some cases contrast factors can be calculated analytically [21,22], so that $\overline{C}_{\{hkl\}}$ can be written as a function of Miller indices, unit cell parameters and elastic constants.

So far a direct numerical calculation could only be avoided by using, in a rather limited number of cases concerning a few slip systems in cubic and hexagonal materials, some parametric expressions [23,24]. Those expressions were obtained by a pragmatic approach: contrast factors were numerically calculated for various combinations of the elastic constants (for cubic materials, anisotropy (Zener) ratio $A_i = 2c_{44}/(c_{11}-c_{12})$ and Poisson ratio $\nu = c_{12}/(c_{11}+c_{12})$), and then the coefficients of an empirical expression were optimized to represent the calculated contrast factors. In particular, for the cubic case the average contrast factor of equations (8) and (9) can be written as:

$$\overline{C}_{\{hkl\}} = \overline{C}_{h00}\left(1-q\cdot H\right), \tag{10}$$

where $H = \left(h^2k^2+k^2l^2+l^2h^2\right)/\left(h^2+k^2+l^2\right)^2$, while $\overline{C}_{h00}$ and q depend on the specific slip system and elastic constants (A_i, ν). Actually equation (10) should better be written as

$$\overline{C}_{\{hkl\}} = A + B\cdot H, \tag{11}$$

with $\overline{C}_{h00} = A$ and $q = -B/A$. In fact, equation (10) is indefinite for A=0 (as in the case of screw dislocations for the <001>{100} primary slip-system in the cubic Cu_2O phase [21]), whereas equation (11) can always be used.

The parameterization proposed in [23,24] for $\overline{C}_{h00}$ and q is:

$$\overline{C}_{h00} = a'\left[1-\exp\left(-A_i/b'\right)\right]+c'A_i+d', \tag{12}$$

$$q = a''\left[1-\exp\left(-A_i/b''\right)\right]+c''A_i+d'', \tag{13}$$

where the values of primed and double-primed a, b, c, d vary for $\overline{C}_{h00}$ and q and for different slip systems and range of Zener ratio: for screw dislocations in the primary *fcc* slip system (<-101>{111}), a'=0.1740, b'=1.9522, c'=0.0293, d'=0.0662, a''=5.4252, b''=0.7196, c''=0.0690, d''=-3.1970, for A_i>0.5 [23], and a'=0.0454, b'=0.1704, c'=0.1056, d'=0.0221, a''=48.5946, b''=0.0713, c''=10.3165, d''=-54.6536, for $A_i \leq 0.5$ [24].

Following the results of refs. [21,22] the average contrast factor for screw dislocations in the <-101>{111} slip system can be calculated explicitly, as a simple function of the Zener ratio [22]:

$$\overline{C}_{h00} = \sqrt{A_i}/6, \tag{14}$$

$$q = 3-2/A_i \tag{15}$$

It is worth noting that both expressions are independent of the Poisson ratio, and involve a functional dependence on the Zener ratio which is different from that in the parametric equations (12) and (13). Figure 3 shows the trends of the parameterizations (equations (12) and (13)) and of the exact analytical expression (equations (14) and (15)) as a function of the Zener ratio. As can be seen, the parameterization is a reasonably good approximation of the correct analytical value, even though a discrepancy exists and the difference diverges both for high and for low Zener ratios. To better appreciate this point, the difference (Δq) between correct analytical result and parametric value for q is plotted as a function of A_i for the two ranges where the parameterizations are valid, A_i>0.5 [23] and $A_i \leq 0.5$ [24], respectively for primed and double-primed a, b, c, d coefficients.

In conclusion, the parametric expressions, although of some practical utility in providing an acceptable approximation of the average contrast factors in some ranges of elastic constant

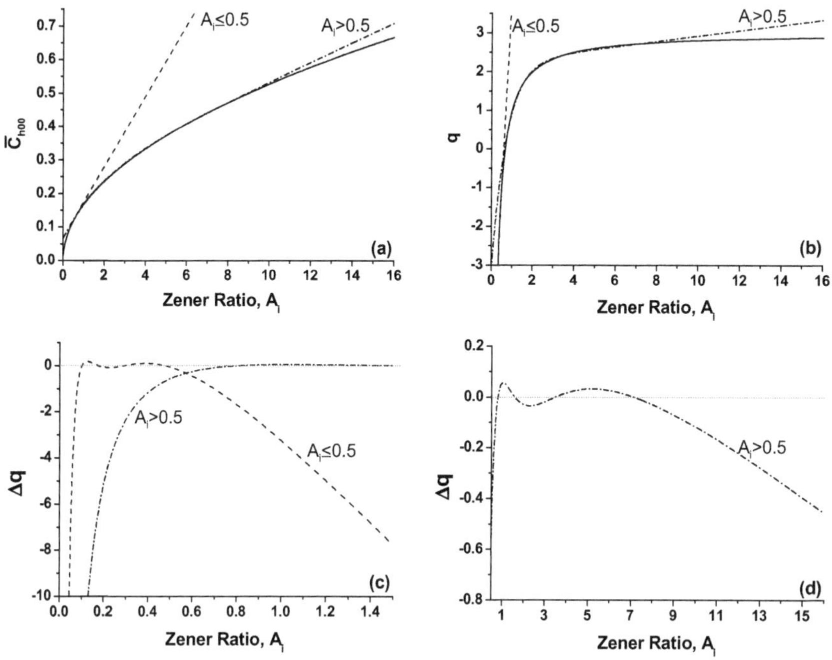

Figure 3. $\overline{C}_{h00} = A$ *(a) and q=-B/A (b) for screw dislocations in the <-101>{111} fcc slip system as a function of the Zener ratio, according to equation (14) and (15) (line) and to the parametric expressions (12) and (13) for Ai≤0.5 (dash) and Ai>0 (dash-dot). For the q parameter, the difference between correct analytical result and parametric value(Δq) is shown in (c) and in (d), respectively for low and high Zener ratio.*

values, may be misleading as they suggest a wrong functional dependence on the Zener ratio. On the contrary, besides providing exact $\overline{C}_{\{hkl\}}$ values, equations (14) and (15) and analogous expressions for other slip systems [21,22] are a real step forward in our understanding of materials properties: dislocation line broadening effects can now be studied directly in terms of the specific type and arrangement of dislocations and elastic properties of the crystalline medium.

4. Limits in WPPM application to finely dispersed powders

An open question on LPA, and WPPM as a consequence, concerns the reliability of the results when very small domain sizes are considered. It is expected that the "ghost" concept introduced by Wilson [15, 25], which is underlying the WPPM algorithm, should become imprecise as the domain size decreases and the atomic, discrete nature of matter starts to be important. Wilson's approach implicitly assumes continuous crystal shapes, ideally perfect solids that can just be an approximation of real nanocrystals.

An intrinsically more appropriate approach to describe the real shape of very small crystals is based on the use of the Debye equation [4,15]. Limits in the use of the Debye equation have been so far solely determined by the length of the calculations, a problem partly solved by smart algorithms and by the ever increasing computation power of modern computers [26]. To test the reliability of the WPPM when applied to nanometre-scale domain sizes, we recently started a systematic comparison between patterns synthesised by the Debye approach and WPPM. Detailed results are to be communicated soon [27]. So far it is interesting to note that the WPPM gives very good results, even for systems made of domains of a few nanometres. Figure 4 shows the pattern of a system of lognormally distributed spherical domains, with mean diameter of 3.36 nm and standard deviation of 0.51 nm, as obtained by means of the Debye equation (dot) [27]. The WPPM (line) gives a remarkably good result, both in terms of nearly perfectly flat residual (i.e. difference between pattern simulated by the Debye equation and WPPM result) and for the size distribution, that matches very well the one used to synthesize the pattern. The result is even beyond the most favourable expectations: in fact, while the Debye approach used in generating the pattern assumes a discrete distribution of sphere diameters, the WPPM adopts a continuous distribution, so that a difference could have been expected. This positive result will require further studies to better explore the limits in the applicability of WPPM, and in a more general sense of Wilson's approach, to the study of crystalline systems at the nanoscale.

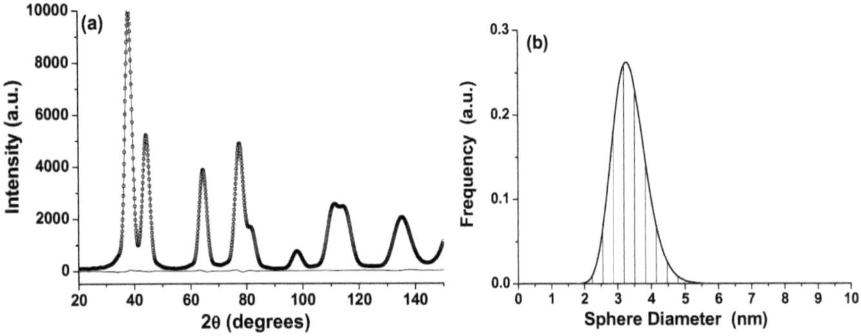

Figure 4. WPPM result (line) for a pattern simulated by the Debye equation approach [26,27] (dot) for a system of spherical nanodomains with lognormally distributed diameters (a). The discrete lognormal distribution used in the Debye equation (bar) is shown in (b) together with the distribution refined by WPPM (line).

5. Concluding remarks

Line profile analysis is definitely evolving toward full pattern methods. This progress has only partly been motivated by the ever increasing computation power of modern PCs: indeed, a more sensitive driving force is the need for reliable results, to be obtained by credible physical models of the material microstructure. Whole Powder Pattern Modelling is well representative of this new paradigm. In this work some recent advancements were discussed, underlying the extension of the WPPM study of dislocation effects to materials with any symmetry. Limits of the WPPM, and of the Wilson's approach to LPA in general, are being tested and preliminary results of a comparison between Debye equation methods and WPPM are really encouraging. The present review was also useful to point out some limitations of integral breadth methods, as well as the limited validity of analytical approximations of the Wilkens theory for the effects of dislocations on the diffraction peak profiles.

Appendix

Equation (8), introducing $k = \pi b^2 \overline{C}_{\{hkl\}} \rho d^{*2}_{\{hkl\}} / 2$ for simplicity, has a leading term [18]

$$\exp\left[-kL^2 Ln\left(R_e / L\right)\right], \tag{16}$$

which is directly related to the logarithmic term in the expression of the elastically stored energy for a crystal containing a dislocation [18]. As a direct consequence of this functional form, the peak profile corresponding to equation (8) is *not* Voigtian (a Voigtian profile is a convolution of Gaussian and Lorentzian profiles [28]). In fact, the FT of a Voigtian profile with integral breadths β_C and β_G, respectively for the Gaussian and Lorentzian components, is [28]:

$$T_V(L) = \exp\left(-2\beta_C |L| - \pi^2 \beta_G^2 L^2\right), \tag{17}$$

which is quite evidently different from equation (8) or (16). This is why Voigtian profiles are intrinsically not correct to describe dislocation line broadening. Similar expressions have been proposed as approximations of the Wilkens model. For example [29],

$$T_{AH}\left(L\right) = \exp\left(-\left|cL\right|^p\right), \tag{18}$$

with $1 \leq p \leq 2$ and c as a coefficient to be related to scattering vector, contrast factor, dislocation density and Burgers vector.

It is interesting to verify to what extent equations (17) or (18) can be used as approximations of the Wilkens expression (equation (8)). Figure 5a shows the FT, calculated from equation (8), for an edge dislocation system with ρ=5x10^{15} m^{-2} and R_e=10 nm. The peak profile considered is the (111) of gold, for which the Burgers vector modulus is $a_0/\sqrt{2}$ (a_0=0.408 nm) in the <-101>{111} slip system. The corresponding average contrast factor $\overline{C}_{\{111\}}$=0.159779, is given by equation (11) with A=0.330506 and B=-0.512181, calculated using gold elastic constants from [30].

If equations (17) or (18) are used to model the data of figure 5a by least squares, the results, shown in the same figure, only apparently match well the data simulated with equation (8). One should be very careful in judging discrepancies in a FT, as it is not so immediate to see to what extent those discrepancies affect the diffraction profile. Figure 5b shows the difference between the FTs of figure 5a. The difference between equation (8) and (18) is smaller than between equation (8) and (17), but in both cases the agreement is unsatisfactory, with systematic deviations. Figure 5c shows the peak profiles corresponding to the FTs of figure 5a: differences are quite evident, even more so when looking at the detail of figure 5d.

As a final remark on the general problem of modelling strain contributions to line broadening, it is worth underlying that different sources can indeed produce similar line broadening results. The basic reason is in the functional form of equation (9), the same for any strain source (i.e., $<\varepsilon^2_{\{hkl\}}> = Inv\left(E_1 ... E_{15}, h, k, l\right)$): what is different is the value of $E_1, E_2 ... E_{15}$,

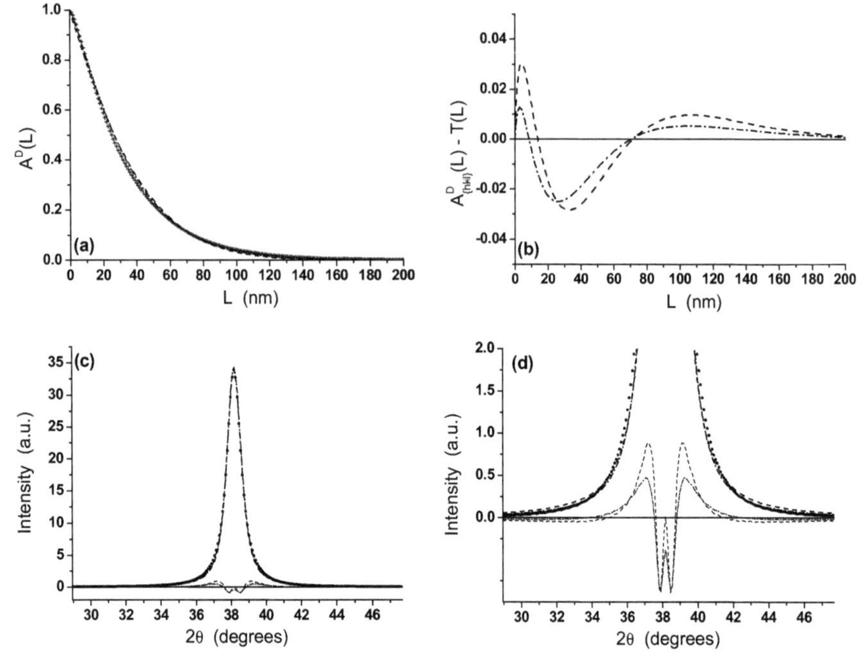

Figure 5. Fourier Transform of the (111) gold peak profile produced by dislocations according to equation (8) (dot), (17)(dash) and (18) (dash-dot) (a); difference between FTs from equation (8) - (17) (dash), and (8) - (18) (dash-dot) (b); corresponding peak profiles from Wilkens model (equation (8), dot) and from equation (17) (dash) and (18) (dash-dot), with corresponding residuals (c); a detail of (c) is shown in (d).

some of which can be zero owing to the symmetry of the material and/or its properties, and the exact functional dependence on L of the FT of the peak profile. It is therefore important that: (i) data quality be the highest, so to allow one to distinguish and single-out fine details in the profile shape; (ii) physically sound models be developed to properly account for line broadening sources. From this last point of view, approximate parameterizations and FT expressions should possibly be avoided.

References

1. Suryanarayana, C., 2001, *Prog. Mat. Sci.*, **46**, 1.
2. Klug, H.P. & Alexander, L.E., 1974, *X-ray diffraction procedures for polycrystalline and amorphous materials,* 2nd ed. (New York: Wiley).
3. Williamson, G.K. & Hall, W.H., 1953, *Acta Metall.* **1**, 22.
4. Warren, B.E., 1990, *X-ray diffraction, 2nd unabridged ed.* (New York: Dover).

5. Snyder, R.L., Fiala, J. & Bunge, H.J. (editors), 1999, *Defects and microstructure analysis by diffraction* (New York: Oxford University Press)
6. Mittemeijer E.J. & Scardi, P. (editors), 2004, *Diffraction analysis of the microstructure of materials* (Berlin: Springer)
7. Young, R.A. & Wiles D.B., 1982, *J. Appl. Cryst.*, **15**, 430.
8. Scardi, P., Dong, Y.H. & Leoni, M., 2001, *Mat. Sci. Forum*, **378-381**, 132.
9. Scardi, P. & Leoni, M., 2001, *Acta Cryst. A*, **57**, 604.
10. Scardi, P. & Leoni, M., 2002, *Acta Cryst. A*, **58**, 190.
11. Dong, Y.H. & Scardi, P., 2000, *J. Appl. Cryst.*, **33**, 184.
12. Scardi, P., Leoni, M. & Delhez, R., 2004, *J. Appl. Cryst.*, **37**, 381.
13. Leoni, M. & Scardi, P., 2004, *J. Appl. Cryst.*, **37**, 629.
14. Scardi, P. & Leoni, M., 2006, *J. Appl. Cryst.*, **39**, 24.
15. Guinier, A., 1994, *X-ray diffraction: in crystals, imperfect crystals, and amorphous bodies* (New York: Dover).
16. Langford, J.I. & Wilson, A.J.C., 1978, *J. Appl. Cryst.*, **11**, 102.
17. Scardi, P., 2008, in *Powder Diffraction Theory and Practice*, edited by R.E. Dinnebier & S.J.L. Billinge (Cambridge, UK: Royal Society of Chemistry), pp. 378-416.
18. Wilkens, M., 1970, in *Fundamental Aspects of Dislocation Theory*, NBS Special Publication No. 317, Vol. II, edited by Simmons, J.A., de Wit, R. & Bullough, R. (Washington, DC: National Bureau of Standards), pp. 1195-1221.
19. Leoni, M., Martinez-Garcia, J. & Scardi, P., 2007, *J. Appl. Cryst.*, **40**, 719.
20. Popa, N.C., 1998, *J. Appl. Cryst.*, **31**, 176.
21. Martinez-Garcia, J., Leoni, M. & Scardi, P., 2007, *Phys. Rev. B.*, **76**. In press.
22. Martinez-Garcia, J., Leoni, M. & Scardi, P., 2008, *Phys. Rev. Letters*. Submitted.
23. Ungár, T., Dragomir, I., Révész, A. & Borbély, A., 1999, *J. Appl. Cryst.*, **32**, 992.
24. Dragomir, I.C. & Ungár, 2002, *Powder Diffr.*, **17**, 104.
25. Wilson, A.J.C., 1962, *X-ray optics,* 2nd ed. (London: Methuen).
26. Cervellino, A., Giannini, C. & Guagliardi, A., 2006, *J. Comp. Chem.*, **27**, 995
27. Beyerlein, K., Cervellino, A., Leoni, M. & Scardi, P., 2008. In preparation
28. Langford, J.I., 1978, *J. Appl. Cryst.*, **11**, 10
29. Adler, T. & Houska, C.R., 1979, *J. Appl. Phys.*, **50**, 3282; **50**, 3288.
30. Every, A.G. & McCurdy, A.K., 1992, in *Landolt–Börnstein: Crystal and solid state physics new series*, group III, 29, edited by D.F. Nelson (Berlin: Springer).

Acknowledgements. I wish to thank staff and students of my research group for constant and qualified support, and in particular M. Leoni, M. D'Incau and J. Martinez-Garcia.

On the simulation of the anisotropic peak broadening in powder diffraction

V. B. Zlokazov[1,*]

[1] JINR, FLNPh, Dubna, Russia
[*] zlokazov@nf.jinr.ru

Keywords: powder diffraction, conference, anisotropic broadening, Rietveld analysis.

Abstract. Distortions of powders is formally a phenomenon more general than strains of monocrystals, including not only the distortion of crystallites but also different types of powder imperfection - heterogeneousness, non-isotropy, finite size and the specific shape of the crystallites.
The article derives formulae to describe measures for the deformation of peaks in the diffractograms - the structure factors distortions and the anisotropic broadening. On the basis of these formulae the diffraction profiles can be simulated and compared with the experimental data.

1. Introduction

Both distortion of the powder and its imperfection heterogeneousness, non-isotropy, finite size and the specific shape of the crystallites) lead to the following physical model: the coherent scattering is carried out on a set of powder crystallites each of which has its own unit cell [1].
Supposing that the ideal unit cell parameters and atomic positions are denoted by x, we can describe the different types of the crystallites by s - the array of all relevant local fluctuations of x, so that $s = 0$ represents the ideal crystallite (no distortions); a profile from such an imperfect or distorted powder is a superposition of diffraction maxima from all such crystallites with different weights $p(s)$ and is described by the following formula

$$f(x) = \int p(s)\, y(x,s)\, ds\,, \qquad (1)$$

where $y(x,s)$ is the diffraction pattern due to crystallites characterized by distortions of sth type.
The function $p(s)$ can be considered either as a deterministic weight function, or the density of a statistical distribution.
In the reciprocal space the corresponding lattices will be superimposed on each other with the same weights.
Therefore, depending on $p(s)$ the diffraction scattering on such a system of lattices can be considered as scattering on a crystal, where unit cell and corresponding reflection positions,

and atomic coordinates fluctuate, so that the resulting reciprocal density will be smeared in comparison with the profile from an ideal sample; in such a scattering each reflection with indices H=h,k,l from ideal sample will be matched by a set of lattice points from some region around H, and the structure factor will be sum of scatterings on atoms with coordinates in a "smeared" unit cell, different from coordinates in the ideal sample.

Here we can pose the two mostly typical problems of the analysis of the diffractograms from such powders:

* extract the information about the ideal polycrystals (undeformed or / and perfect);
* extract the information about the factors of distortion or characteristics of imperfection.

In the general case (1) is ill-posed, so that we have very little chances to get a direct solution. More attractive is the parametrical way. The parametrization of (1) is necessitated to balance the two extremes, contradicting each other:

* take the minimal possible number of parameters to keep the solution of (1) the most stable;
* take the maximum possible number of parameters to provide for an acceptable accuracy of the solution.

It is very difficult to find an appropriate compromise between the both; so along with the solving (1) by parametrical methods the simulation of $f(x)$ (using known or tested weight function p(s) and crystallite unit cells) and the subsequent analysis of the data obtained is strictly necessary.

Let us take the diffracted part of the profile and consider the case of a single phase. The profile obtained from the ideal sample will be considered as basic and having the perfect unit cell. This basic profile corresponds to $p(0)$, i.e. at s=0.

The formal purpose of the analysis of such data is the evaluation of $y(x,s)$ and $p(s)$.

At any s the diffractional part of the profile is described by the formula (omitting the overall scale factor, that of polarization, etc.)

$$y(x) = \sum_H m_H \left|F_H\right|^2 L_H P(z) \qquad (2)$$

where x = channel (ToF) or 2θ),
H = vector of Miller indices h,k,l of the diffraction peak,
F_H = structure factor, and m_H = its multiplicity,
L_H = Lorentz factor,
d = interplanar distance (d=$1/q$),
$P(z)$ = diffraction peak profile function,
$z = (x - K)/W$,
K = channel of the peak center and W = its halfwidth,
the position of the Hth reflection d_H is

$$d_H = 1 / \sqrt{(H,H)}, \qquad (3)$$

where the scalar product $(H,H) = Q$ and is written in terms of the reciprocal space as follows:

$$Q = a^{*2}h^2 + b^{*2}k^2 + c^{*2}l^2 + 2a^*b^*\cos(\gamma^*)hk + 2a^*c^*\cos(\beta^*)hl + $$
$$+ 2b^*c^*\cos(\alpha^*)kl;$$

The formula for the structure factor for a reflection $H = (h,k,l)$ at the atoms X_j , j=1,2,..,na reads

$$F_H = \sum_j b_j N_j \exp(2\pi i(H,X_j))\exp(-(B_j H,H)) \qquad (4)$$

Here b_j, N_j, X_j, B_j - scattering factor, occupation factor, coordinates and thermal tensor of the jth atom, respectively.

The profile (1) will be characterized by the following features

* peak broadening, both uniform and anisotropic;

* change of the structure factor, its leakage into reflections, which emerged due to smearing of the lattice, and due to change of fluctuations of the atomic positions in the smeared direct cell;

* possible change of the symmetry type; thus reflection equivalences normally won't be preserved.

In the literature instances for the analytical solution of the equation (1) are given. They are based on rather arbitrary assumptions about quantities entering (1).

For many cases it is assumed that in fact $y(x,s)$ depends only on the difference x-s, so that we can write $y(x)$, where x is a simple function variable. This transforms the integral in (1) to a convolution one.

Let us consider some particular cases of functions $y(x)$ and $p(s)$.

* Gaussian probability density and the Gaussian peak model

$$p(s) = \frac{1}{\sqrt{2\pi}\sigma}\exp\left(-\frac{s^2}{2\sigma^2}\right), y(x) = \frac{1}{W}\exp\left(-\frac{x^2}{W^2}\right). \qquad (5)$$

We have

$$f(x) = \int \frac{A(s)}{W} y\left(\frac{x-K\pm s}{W}\right) p(s)\,ds, \qquad (6)$$

where s has the same meaning as in (1).

Usually for the sake of simplicity K=0, and $A(s) = A$. However, this is wrong. The amplitude, i.e. squared absolute value of the structure factor, will be constant, if the distortion did not change the atomic coordinates in the cell (e.g., in case of the simplest textures). In the general case, A(s) will not be constant.

Omitting the details, with the help of the Fourier transform we get

$$f(x) = \frac{A}{\sqrt{W^2+\sigma^2}}\exp\left(-\frac{x^2}{W^2+\sigma^2}\right). \qquad (7)$$

- Lorentz function and the Cauchy distribution

$$y(x) = \frac{A}{W\left(1+x^2/W^2\right)};\; p(s) = \frac{1}{\pi\sigma}\frac{1}{1+s^2/\sigma^2}. \qquad (8)$$

The Cauchy distribution has neither the expectation, nor the variance (the corresponding integrals don't converge). But in a convolution we can use it. Similarly to (5),(6) we get

$$f(x) = \frac{A}{(W+\sigma)\left(\left[x/(W+\sigma)\right]^2+1\right)}.$$

- Let us consider the case of asymmetric Gaussian probability density, i.e. with different left and right σ s and the Gaussian peak model. Evidently, the solution will be

$$f(x) = \begin{cases} \frac{A}{\sqrt{W^2+\sigma_l^2}} \exp\left(-\frac{x^2}{W^2+\sigma_l^2}\right) & \text{if } x<0; \\ \frac{A}{\sqrt{W^2+\sigma_r^2}} \exp\left(-\frac{x^2}{W^2+\sigma_r^2}\right) & \text{if } x>0 \end{cases} \qquad (9)$$

At x=0 the function has a break, but a more realistic model of asymmetric weight function is that with a continuous transition of σ_l to σ_r what practically leads to a function coinciding with (9) everywhere apart from a small neighborhood of zero, where it will have a maximum shifted from zero.

The implication of this is: in the exactly symmetric case of distortion its effects are peak width increase and peak amplitude decrease; in the asymmetric case a shift of the peak center and peak asymmetry appear.

2. The general case of the diffraction from distorted crystallites

To find the solution of (1) in the general case the following assumptions will be made:

* $p(s)$ is normalized, i.e. $\int p(s)ds = 1$, and has a finite expectation and a variance.

* Since at least the approximate periodical structure of the sample is preserved, the distortions are small and symmetric, i.e., the parameter deviations from zero are equally possible:

$$p(s) = p(-s).$$

A small distortion means, that $p(s)$ decreases very fast if s grows. Its symmetry in the expansion of $y(x,s)$ in s allows us to neglect the odd (in particular, first) derivatives, since in the convolution with $p(s)$ they annul each other.

Then for an arbitrary partition of the region, where $p(s)$ is not equal to zero, we can write:

$$f(x) = \int p(s)y(x,s)ds \approx \int p(s)\left[y(x,0)+0.5\left(y''_{ss}(x,0)s,s\right)+o(\|s\|)\right]ds =$$
$$y(x,0)\int p(s) + 0.5\sum_s y''_{ss}(x,0)\int p(s)(s,s)ds + \varepsilon.$$

From this we have

$$f(x) = y(x,0) + 0.5\sum_s C(s)y''_{ss}(x,0) + \varepsilon. \qquad (10)$$

The differentiation is carried out with respect to s, which describe the deviations of the unit cell and those of the atomic parameters;

s is a vector, and y''_{ss} is a matrix.

$C(s) = \int p(s)(s,s)\,ds$; due to the finite variance of $p(s)$. $C(s)$ will be finite.

$\varepsilon = \int p(s)\,o(|s|)\,ds$ is the integral accuracy of (1), which will be the smaller, the faster $p(s)$ tends to zero while $|s|$ tends to the infinity.

Using (10) and (2), we see, that approximately (with ε accuracy) the diffraction parts of the smeared profile and the basic one will differ by the quantity

$$y''_{ss}(x,0) = \sum_H \frac{m_H}{L_H}\left[\left(|F_H|^2\right)''_{XX} C_X P(z) + |F_H|^2 P''_{AA}(z) C_A\right], \qquad (11)$$

where X and A denote, respectively, atomic coordinates and cell parameters (direct or inverse), and the constants C_X and C_A are distortion measures of X and A.

We don't take the Lorentz factors L_H into account. Multiplicities can be taken constant, because sometimes it is appropriate to use the lowest (triclinic) symmetry, since within each group of equivalent reflections in case of symmetries higher than orthorhombic indices h,k,l can substitute each other.

3. Parametrization of the distorted profiles

The purpose of Rietveld analyse of diffraction data from distorted crystallites are not only the usual crystallographic parameters but also additional parameter vectors C_X and C_A, and, in the general case, the function $p(s)$.

Formulae, obtained by the straightforward differentiation of (11), are too bulky and cumbersome and lead only to ill-conditioned least squares matrices in the Rietveld procedure. Therefore, 3 groups of principal parameters, additional to the crystallographic ones, were built

* **Peak shape**; As such the following expression was taken

$$P_n(z) = P(z) + (A_0 + A_1 x) P''_{XX}(z), \qquad (12)$$

which describes the line, slowly and linearly changing along the axis d (or 2θ), and $z = (x-K)/(W_0 + W_n)$,

where W_0 will be the standard halfwidth, and W_n will describe the broadening due to the distortion or imperfection of the powder.

* **The change of the peak halfwidth**. The most obvious broadening of the peaks is caused by the fluctuation of the reflections.

As a measure for such broadening for any H we can take the mean deviation of $d(H)$ from the basic one $d_0(H)$, which will be equal to

$$0.5\frac{\partial^2 d}{\partial A_i \partial A_j}|\delta A_i||\delta A_j| \qquad (13)$$

Here $A_i = a, b, c, \alpha, \beta, \gamma =$ parameters of the direct (or the inverse) unit cell, ΔA_i - norm of the ith average parameter deviation.

Note that (13) is only the most probable average fluctuation of the peak halfwidth, and its connection with the characteristics of the region of fluctuating points, which correspond to H on the lattices of the reciprocal space is completely unclear (usually this region is supposed to be an ellipsoid, though without any sufficient reasons [2]).

Further we shall use as parameters not the angles, but their cosines, because it gives us simpler expressions.
Let us consider the 2nd derivative of d with respect to a^* (similarly to b^*, c^*). From (3) we have

$$\frac{\partial^2}{\partial a^{*2}} = \frac{-2h^2Q + 6R^2}{Q^{2.5}}$$

where $R = \partial Q/\partial a = 2\left(a^* h^2 + b^* \cos\gamma^* hk + c^* \cos\alpha^* kl\right)$.

The numerator in this expression will contain the products of the 2nd and 4th power of h, and, respectively, d^3 and d^5. It is obvious, that a quantity, which to the maximum extent is responsible for the selective peak h anisotropy, is the product h^4d^5; it will be responsible for the large halfwidth deviations depending on h in peaks, close to each other by d. Next, the numerator of the 2nd derivative of d with respect to $\cos\alpha^*$ (similarly to $\cos\beta^*$, $\cos\gamma^*$ will contain the products of k^2l^2 and d^5. This product $k^2l^2d^5$ will be also responsible for the large selective halfwidth deviations depending on k^2l^2.
Fine effects, described by products of lesser powers of the indices, can be also parametrized [2], but, it seems, that they may be included into the general (uniform) peak broadening.
Thus, we have for the halfwidth $W_f = W_0 + W_1u(d) + W_a$.
The parameter W_1 provides for the general peak broadening depending on the function $u(d)$.
The member W_a contains items like $h^4\ A_a/Q^{2.5} = d^5\ h^4\ A_a$, or $d^5k^2l^2A_\alpha$ where A_a, A_α are parameters describing the integral cell deformation along the a-link and α-angle.
Similarly, the parameters are derived for other links and angles.
Finally, the parameter dependence of the peak halfwidth is

$$W_f = W_0 + u\left(d(x)\right)W_1 + d^5(x)\begin{pmatrix} h_n^4A_a + k_n^4A_b + l_n^4A_c \\ +k_n^2l_n^2A_\alpha + h_n^2l_n^2A_\beta + h_n^2k_n^2A_\gamma \end{pmatrix}$$

h_n, k_n, l_n = Miller indices, normalized by the maximum values, and $d(x)$ are either proportional to x (time-of-flight case), or inversely proportional to x (for the angular distributions).
* **Change of peak amplitudes**, i.e., structure factors;
The 2nd derivative with respect to X-coordinate (similarly to Y and Z) with account of (4) is

$$\frac{\partial^2 F_H}{\partial X^{*2}} = F_H 4\pi^2 h^2 .$$

Therefore, for the mean fluctuation of $\left|F_H^*\right|^2$ we have

$$\delta F_H = \left|F_H^*\right|^2 d^5(x)\left(h_n^4F_X + k_n^4F_Y + l_n^4F_Z\right) = \left|F_H^*\right|^2 \delta F$$

where F_X, F_Y, F_Z are parameters, which characterize the average deviations of atomic positions in the distorted cell.
The final parametrical description of the distorted sample is:

$$y(x) = \sum_H m_H \left|F_H\right|^2 \left(1 + \delta F_H\right)/L_H\left(P(z) + \left(A_0 + A_1x\right)P''_{XX}(z)\right), \qquad (14)$$

where $z = \{x-K\}/W_f$.
In case of symmetries higher than orthorhombic it is more appropriate to use the Miller indices formed as if the symmetry is the lowest - triclinic. It allows us to account better for the non-uniform role of equivalent indices for the total sum.

4. Simulation

On the basis of (1) we can simulate data which imitate those obtained in real experiments with distorted samples and compare the both to see whether our model of the latter is correct. If the assumption about the symmetry of the function p(s) is correct, then the above-given method can be applied to the real data and we can obtain the estimates of parameters A_0, A_1, F_X, F_Y, F_Z, A_a, A_b, A_c, A_α, A_β, A_γ. The analysis of the simulated data can shed light on the meaning of these parameters and provide for a better understanding of them.
In principle, having an estimate of $y(x,s)$ in (1) we can obtain the estimate of $p(s)$. But solving (1) is a known ill-conditioned problem and chances to solve it are small even for an exactly known $y(x,s)$. Therefore, at this stage of treating the anisotropy problem the method of simulation is more reliable.

5. Conclusion

The simulation according to the above-mentioned algorithm produces data which qualitatively repeat the effects observed in real distorted data, and is an excellent means to illustrate the influence of the weight function $p(s)$ features as imitators of real physical factors on the data quality.
A parametrical description of the profile (14), which takes both the general and the selective or anisotropic (depending on h,k,l indices) data distortion into account provides not only for a better fitting of the distorted profile, but also allows the Rietveld analysis to obtain estimates of the basic (non-distorted) crystal parameters and at the same time of measures and character of powder distortion imperfection) with acceptable accuracy.

References

1. Balzar, D., Audebrand, N., Diamond, M.R., Fitch, A., Hewat, A., Langford, J.L., LeBail, A., Louer, D., Masson, O., McCowan, C.N., Popa, N.C., Stephens P.W. & Toby, B.H., 2004, *J. Appl. Cryst.*, **37**, 911.
2. Popa, N.C., 1998, *J. Appl. Cryst.*, **31**, 176.

Z. Kristallogr. Suppl. **27** (2008) 121-126 / **DOI** 10.1524/zksu.2008.0016

XRD and FTIR characterization of nanocrystalline YVO_4: Eu derived by co-precipitation process

E. Borca[1,*], M. Bercu[1], S. Georgescu[2], S. Hodorogea[1,3], E. Cotoi[2]

[1] Faculty of Physics, University of Bucharest, Romania
[2] National Institute for Laser, Plasma & Radiation Physics, Bucharest-Magurele, Romania
[3] INCDIE ICPE-CA, Bucharest, Romania
[*] lia_borca@yahoo.com

Keywords: XRD, line broadening, crystallite size, FTIR, nanophosphors

Abstract. Nanosized YVO_4 powders doped with Eu^{3+} were synthesized through a co-precipitation process. The influence of annealing temperature on the microstructural features was investigated by X-ray powder diffraction (XRD) and Fourier Transform Infrared spectroscopy (FTIR). Nanocrystalline YVO_4 phase is obtained just after the precipitation process without thermal treatment. The integral breadth approach based on Williamson-Hall plots was used for analysis of XRD line broadening. By this simple method, a dramatic increase of apparent the size of coherently diffracting domains is evidenced when the temperature of thermal treatment is increased from about 500 to 600°C. XRD data were compared with the annealing temperature behaviour of FTIR absorption bands corresponding to the VO_4^- group.

1. Introduction

Yttrium orthovanadate (YVO_4) is known as an attractive phosphor-producing host lattice for several rare earth (RE) ions [1]. These luminescent materials are of interest in the field of developing optoelectronic devices and systems as new displays, X-ray imaging or up-conversion phosphors [2]. For practical applications of phosphors it is desirable to have a fine particle size and chemical purity. Yttrium orthovanadates are conventionally synthesized by solid-state reactions which requires high temperatures (>1600°C) and ball-milling. In such conditions, it is difficult to control the structural characteristics and the purity of the synthesized powders. Alternatively, much attention was focussed on the wet chemistry methods used to prepare well-dispersed nanopowders with controlled structural characteristics ([3-5] and references therein).

In this paper, we use a direct precipitation method [3] to prepare rare earth (Eu and Tb) doped YVO_4 nanopowders. The evolution of the crystalline phase fraction and the changes of microstructural characteristics induced by the thermal treatments were characterized by means of XRD analysis and FTIR spectroscopy.

The nanocrystallite size of annealed samples at various temperatures was estimated by the analysis of X-ray diffraction peak broadening, using the classic integral breadth approach [6] based on the Williamson–Hall (W-H) plot [7]. Beside the peculiarities of this method, discussed and reviewed extensively [8-9], the simple approach allows us to observe the evolution of nanoparticle crystallization by changing the thermal treatment conditions.

2. Experimental

Nanopowders of $YVO_4:RE^{3+}$ (RE=Eu,Tb) have been obtained by a direct precipitation reaction [3]. A solution of NH_4VO_3 was adjusted to pH 12.5 with NaOH. Then solutions of rare earth nitrates ($Y(NO_3)_3$ and $RE(NO_3)_3$) were mixed in a molar ratio according to the formula $Y_{1-x}RE_xVO_4$; the mixture was added dropwise into the NH_4VO_3 solution; a yellow opalescent colloid was obtained. After heating 1 h at 60°C with magnetic stirring, a milky colloid containing nanoparticles $YVO_4:RE^{3+}$ has been obtained. The colloid was separated from the solution by filtering, washed with distilled water and dried at 60°C in air.

The following compositions, corresponding to YVO_4:Eu (5at.%) and YVO_4:Tb (1at.%) were independently synthesized. Samples of the obtained powders were heated in air for 2 h at different temperatures in the interval 200-1200°C.

XRD measurements were carried out at room temperature, using CuKα radiation (40 kV and 40m A) on a diffractometer Bruker-AXS type D8-ADVANCE. The data were collected in the range 2θ between 15° and 80° with a scan step of 0.04°.

Infrared absorption spectra were recorded on a SCIMITAR type FT-IR spectrometer (model 2001, Digilab) in the range of 4000-400 cm^{-1}.The samples were prepared using a KBr pellet technique.

3. Results and discussion

3.1 XRD analysis

YVO_4 crystal has a zircon type structure and belongs to the tetragonal space group $I4_1/amd$ with cell dimensions: a = 7.119 Å, c = 6.289 Å [10].

XRD patterns for YVO_4:RE (RE = Eu, Tb) powders samples annealed at different temperatures between 200 and 1200°C were recorded. Regardless the dopant, the diffraction patterns showed the same dependence of microstructure of YVO_4 powders of the thermal treatment temperature.

Figure 1 presents the XRD patterns recorded on as prepared YVO_4:Eu powder (colloid dried at 60°C) and on powder specimens annealed at 300°C, 400°C, 500°C, 600°C and 900°C, respectively. Despite the broadening caused by the size and distorsion of small particles, all patterns are in agreement with the zircon-type structure of bulk YVO_4 (JCPDS #17-0341). It is obvious that nanocrystalline YVO_4 phase was obtained just after the precipitation process without thermal treatment. Moreover, diffraction patterns registered on samples annealed at 500°C or lower temperature shows only small differences between specimens (i.e. broadened, low intensity peaks), which indicate a weak dependence of microstructure characteristics on annealing temperature in this range. Thermal treatment at 600°C results in a dramatic sharpening of the diffraction maxima. Further, XRD patterns indicate small changes for samples heat-treated at higher temperatures.

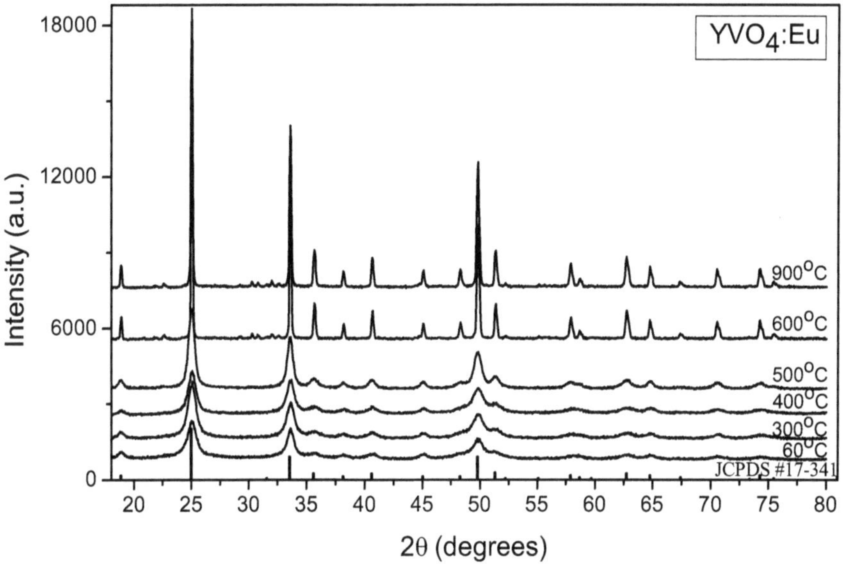

Figure 1. XRD patterns for just prepared YVO4:Eu nanopowder (dried at 600°C) and for YVO_4:Eu specimens annealed at various temperatures.

The full width at half maxima (FWHM) is determined by profile fitting procedure, using a Lorentzian peak shape. A correction for instrumental peak broadening was introduced using a silicon powder as standard sample. The FWHM (w) and the integral breadth (β) of Lorentzian function are related by:

$$\beta = \frac{\pi w}{2} \tag{1}$$

The integral breadths of diffraction peaks are used in a W-H analysis based on the well known expression [7]:

$$\beta = \beta_s + \beta_d = \frac{1}{< D >_V} + 2\varepsilon d^* \tag{2}$$

where $<D>_V$ is the apparent volume-weighted size of crystallites (coherently scattering domains), β_s, β_d are the integral breadths of size and strain respectively components, ε is the upper limit for the lattice strain and d^* is the reciprocal-space variable.

Figure 2 presents examples of W-H plots obtained for two samples of YVO_4:Eu, one annealed at 500°C, the other annealed at 600°C (*left side*) and two samples of YVO_4:Tb annealed at 200°C and 1200°C, respectively (*right side*). Regardless of the dopant, the W-H plots of all samples annealed at $T \leq 500$°C show a large spread of data. On the other hand, in the case of each specimen annealed at 600°C or higher temperatures, all reflexions are well fitted by a linear dependence with non-zero slope, revealing the presence of both size and strain effects on peak broadening. The apparent crystallite size ($<D>_V$) was calculated as the inverse of intercept for all samples.

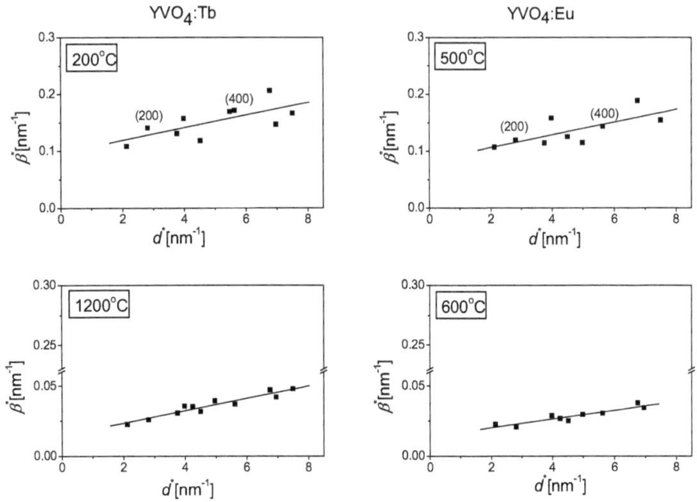

Figure 2. W-H plots for selected specimens with different compositions, annealed at different temperatures.

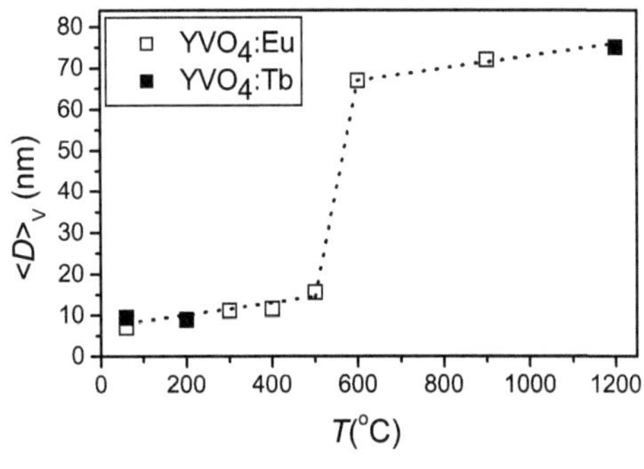

Figure 3. Effect of annealing temperature on apparent volume weighted crystallite size for two series of YVO_4:RE powders (RE=Eu, Tb). The dotted line is only guide for the eyes.

Even if this is a crude approximation, the same linear dependence (Eq. 2) was used to estimate the apparent size for samples annealed at 500°C or lower temperatures. At this stage of analysis, the apparent size parameter is used to check the influence of the thermal treatment on the nanocrystalline particle growth.

In figure 3 the plot of the apparent size $<D>_V$ against the annealing temperatures (T) for specimens of two independently prepared samples is shown: YVO_4:Eu (5at.%) and YVO_4:Tb (1at.%). As expected, an increase of apparent crystallite size with the temperature of thermal treatment was obtained. Unexpected is the dramatic increase of size parameter obtained by changing the temperature of thermal treatment from about 500°C to 600°C observed for YVO_4:Eu (5at.%) samples. The data obtained for YVO_4:Eu (5at.%) and YVO_4:Tb (1at.%) at 60°C and 200°C are in good agreement. Also, the value of the apparent crystallite size for YVO_4:Tb sample annealed at 1200°C continues the linear dependence of the YVO_4:Eu ones for $T \geq 600$°C. From this it can be concluded that it is a reproducible effect determined by important microstructural changes induced by thermal treatment at about 600°C. The refinement of XRD analysis correlated with TEM measurements will provide a further insight into the particle microstructure.

3.2 FTIR spectroscopy

The FTIR measurements are in agreement with the XRD results in respect of the increase of the crystalline phase fraction as a result of thermal treatment.

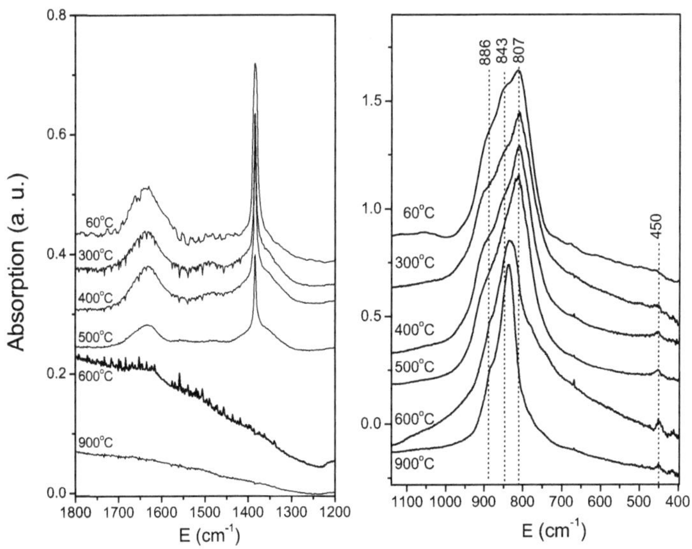

Figure 4. FTIR spectra of YVO_4:Eu specimens annealed at different temperatures. Left: absorption peaks due to the impurities adsorbed on the large surface of nanoparticles (NO_3^-, OH^-); right: characteristic absorption peaks of VO_4^- group (~843 cm^{-1}) and Y-O or Eu-O bond (450 cm^{-1}).

Figure 4 shows the FTIR spectra of YVO_4:Eu specimens annealed at different temperatures. Two spectral ranges are illustrated, one from 1100 to 400 cm^{-1} and the other from 1800 to 1200 cm^{-1}. Inside the first spectral range a strong peak at ~850 cm^{-1} can be observed, a superposition of absorptions at 807 cm^{-1}, 843 cm^{-1}, 858 cm^{-1} associated with characteristic vibrations of the VO_4^- group and a weak absorption peak of the Y-O bond (~450 cm^{-1}) [2,

11]. The shape and the intensity of the envelope of VO_4^- absorptions (around 850 cm^{-1}) are dependent of the annealing temperature. Increasing the temperature, the size of the crystalline grains increases and the absorption peak becomes narrower. A significant change is observed for the sample annealed at 600°C compared with that annealed at 500°C, in good agreement with the XRD results. This result is in accordance with the analysis of optical spectra measured on the same samples which also reveals a sharp increase of the crystalline quality for the powders annealed at 600°C (these results will be published elsewhere).
The absorptions observed at 1384 cm^{-1} and 1636 cm^{-1} indicate the presence of impurities corresponding to NO_3^- and OH^- groups. These absorption bands disappear in the case of samples annealed at $T > 600$°C as a result of thermally stimulated purification of the nanoparticles. This process enhances the growth of the crystalline phase volume from the core of the particle up to its surface. The increase of the order/disorder ratio in nanoparticles is related to the decrease of the frequency dispersion of VO_4^- vibrators revealed by the sharpness of the absorption peak at 843 cm^{-1} after annealing above 500°C. This behaviour of the IR absorption band narrowing is expected when an amorphous/crystalline phase transition appears.

4. Concluding remarks

Nanocrystalline YVO_4 powders doped with Eu or Tb were synthesized through a simple precipitation process.
The influence of thermal treatment temperature on nanoparticle microstructure was revealed by XRD and FTIR measurements. Simplified integral breadth approach (W-H plot) analysis of X-ray diffraction data indicate a sharp increase of apparent crystallite size obtained by annealing the samples in between 500°C and 600°C.
FTIR spectrometry data are in good agreement with XRD results indicating a clear increase of order around VO_4- vibrators in the case of YVO_4 nanoparticles annealed above 500°C.

References

1. Levine, A.K. & Palilla, E.C., 1964, *Appl. Phys. Lett.* **5**, 118.
2. Sun, Y., Liu, H., Wang, X., Kong, X. & Zhang, H., 2006, *Chem. Mater*. **18**, 2726.
3. Yanhong, L.& Guangyan, H., 2005, *J. Sol. State Chem.* **178**, 645.
4. Ma, J., Wu, Q.& Ding, Y., 2007, *Mat. Lett.*, **61**, 3616.
5. Althues, H., Simon, P. & Kaskel, S., 2007, *J. Mater. Chem.*, **17**, 758.
6. Klug, H.P. & Alexander, L.P., 1974, *X-ray Diffraction Procedures for Polycrystalline and Amorphous Materials*, 2nd edition, New York, Wiley.
7. Williamson, G.K. & Hall, W.H., 1953, *Acta Metall.*, **1**, 22.
8. Scardi, P., Leoni, M.& Delhez, R., 2004, *J. Appl. Cryst.*, **37**, 381.
9. Lucks, I., Lamparter, P. & Mittemeijer, E.J., 2004, *J. Appl. Cryst.*, **37**, 300.
10. Baglio, J.A. & Gashurov, G., 1968, *Acta Cryst.* **B24**, 292.
11. Huignard, A., Buissette,V., Laurent,G., Gacoin,T. & Boilot, J–P., 2002, *Chem. Mater.*, **14**, 2264.

Acknowledgements. This work was supported was supported by the Romanian Education and Research Ministry (Contract CEEX – D11-53/2005).

Z. Kristallogr. Suppl. **27** (2008) 127-133 / **DOI** 10.1524/zksu.2008.0017

Cumulants and moments in the line profile analysis

M. Čerňanský[1,*]

[1] Institute of Physics of the ASCR, v. v. i., Na Slovance 2, 182 21 Prague 8, Czech Republic
[*] cernan@fzu.cz

Keywords: cumulants, moments, diffraction profile, single line method

Abstract. An analogy between diffraction profiles and probability distributions of random variables is a basis for the utilization of the mathematical formalism of the theory of probability in the line profile analysis. Namely, the first and the second moments were used to the describing of a position and a breadth of diffraction profiles. Higher moments were introduced also in the line profile analysis, especially the fourth moments for the estimation of the particle size and the microstrain from a single line. However, the formulas for these higher moments are more complicated. Besides the moments, cumulants of random variables were introduced in the theory of probability. The purpose of the contribution is to show, that the cumulants can be used very advantageously in the line profile analysis, especially instead of the higher moments of the diffraction profiles and in the case when number of convoluted functions is a higher as two.

1. Moments

A random variable x is completely described in the probability theory by the probability function f (the probability density, the probability distribution). The shape of the probability function can be characterized in a more condensed way by the raw moments (taken about 0 – also called "crude moments"), which are defined by the equations [1]

$$m_n = \int_{-\infty}^{\infty} x^n f(x)dx, \qquad n = 0, 1,... \tag{1}$$

and by the central moments, taken about the centroid m_1

$$\mu_n = \int_{-\infty}^{\infty} (x - m_1)^n f(x)dx, \quad n = 0, 1,... \tag{2}$$

$m_0 = \mu_0 = 1$ from the definion of the probability function f and evidently $\mu_1 = 0$. The other central moments can be expressed by the raw moments using the binomial expansion in (2)

$$\mu_n = \sum_{k=0}^{n} \binom{n}{k} (-1)^{n-k} m_k m_1^{n-k} \tag{3}$$

and the raw moments can be expressed by the central moments analogously

$$m_n = \sum_{k=0}^{n} \binom{n}{k} \mu_k m_1^{n-k} \tag{4}$$

The first raw moment m_1 is the centroid C, which describes the position of the function f. The second central moment is the variance W characterizes a width of the function f. Its asymmetry or skewness is associated with the third central moment, $\gamma_1 = \mu_3/\sigma^3$, where σ is the standard deviation, i.e. the square root of the variance W. The flattening or excess is associated with the fourth central moment $\beta_2 = \mu_4/\sigma^4$, $\gamma_2 = \beta_2 - 3$.
The function f can be recovered from its moments. The first possibility is the formula of Edgeworth [1]

$$f(x) = \left[\varphi(x) - \frac{1}{3!}\frac{\mu_3}{\sigma^3}\varphi^{(3)}(x) + \frac{1}{4!}\left(\frac{\mu_4}{\sigma^4} - 3\right)\varphi^{(4)}(x) + \frac{10}{6!}\left(\frac{\mu_3}{\sigma^3}\right)^2 \varphi^{(6)}(x) - \cdots \right], \tag{5}$$

where φ is the Gaussian function with its derivatives $\varphi^{(i)}$. The moments can be substituted by above defined coefficients γ_1 and γ_2 of the asymmetry and the excess respectively [1]. This formula was alredy considered as possible profile function [2].
The second way is the Gram-Charlier A series [3]

$$f(x) = \alpha(x)\left\{1 + \frac{1}{6}\mu_3 H_3 + \frac{1}{24}(\mu_4 - 3)H_4 + \ldots\right\}, \tag{6}$$

where μ_i ($i = 3, 4$) are the central moments,

$$\alpha(x) = \frac{1}{\sqrt{2\pi}} \exp(-\frac{1}{2}x^2) \tag{7}$$

and

$$H_3 = x^3 - 3x, \quad H_4 = x^4 - 6x^2 + 3 \tag{8}$$

are Hermite polynomials.

The third possibility is to define the inverse Fourier transform of the function f

$$\varphi(t) = \int_{-\infty}^{+\infty} f(x) \exp(itx) dx , \tag{9}$$

so that

$$f(x) = \int_{-\infty}^{+\infty} \varphi(t) \exp(-itx) dt \tag{10}$$

and the expansion of the exp(*itx*) in (9) gives

$$\varphi(t) = \sum_{k=0}^{\infty} (it / k!)^k m_k , \tag{11}$$

where m_k are raw moments of the function *f*, which can be reconstructed from (10) and (11). Further relations between moments and the Fourier transform are e.g. in [4].

2. Moments and convolution

If the function *h* is the convolution of the functions *f* and *g*, i.e.

$$h(x) = \int f(y) g(x-y) dy , \tag{12}$$

the multiplication by x^n in (12) and its integration gives [5]

$$m_{n,h} = \sum_{k=0}^{n} \binom{n}{k} m_{n-k,f} m_{k,g} \tag{13}$$

for the raws moments of functions *f, g, h*. Especially, it means

$$m_{1,h} = m_{1,f} + m_{1,g} \tag{14}$$

$$\mu_{2,h} = \mu_{2,f} + \mu_{2,g} \tag{15}$$

$$\mu_{3,h} = \mu_{3,f} + \mu_{3,g} \tag{16}$$

$$\mu_{4,h} = \mu_{4,f} + 6\mu_{2,f}\mu_{2,g} + \mu_{4,g} \tag{17}$$

The eqs. (14) and (15) indicate the aditivity of the centroids and the variances of the convoluted functions, i.e. $C_h = C_f + C_g$, $W_h = W_f + W_g$. These simple relations were probably a motivation for the application of the moments in the X-ray diffraction with the purpose to consider instrumental effects [6]. Subsequently it was showed that the variance has physical significance [7,8].

More complicated relations hold for the fourth and higher moments. The fourth moments were used for an estimation of the particle size, microstrain and its gradient [9] and for the estimation of the particle size and microstrain from the single line [10,11].

3. Generating functions and cumulants

Eq. (9) is the special case of a definition of the generating function $G(t) = G(t;f)$ to the probability function $f(x)$ in the statistical mathematics. Generally for continuous variables [12]

$$G(t) = \int f(x) t^x dx \tag{18}$$

Putting $t = e^a$ in (18), we obtain the moment generating function

$$G(e^a) = \int f(x) e^{ax} dx = 1 + m_1 a + m_2 a^2 / 2! + m_3 a^3 / 3! + \ldots = M(a) \tag{19}$$

and the expansion of the logarithm of the moment generating function $M(a)$ gives the cumulant generating function

$$K(a) = \ln M(a) = \left(\kappa_1 a + \kappa_2 a^2 / 2! + \kappa_3 a^3 / 3! + \ldots\right), \tag{20}$$

where κ_n are cumulants and fulfil the recursive relation with the raw moments

$$\kappa_n = m_n - \sum_{k=1}^{n-1} \binom{n-1}{k-1} \kappa_k m_{n-k}. \tag{21}$$

Especially $\kappa_1 = m_1$. Number n is the order of moments and of cumulants.
In terms of the central moments:

$$\kappa_2 = \mu_2 \tag{22}$$

$$\kappa_3 = \mu_3 \tag{23}$$

$$\kappa_4 = \mu_4 - 3\mu_2^2 \tag{24}$$

The main property of cumulants is their additivity for the convoluted functions in all orders [12,13]. The second advantage of cumulants is their simple additivity in the case of more than two convoluted functions, also in the all orders [13]. This property of cumulants was proposed to the analysis of a combination of geometrical abberations [2] and it can be very useful in the line profile fitting or modelling, when a relatively large number of functions are convoluted [14]. Cumulants and Edgeworth series expansion was used in crystallography formerly to a generalized treatment for thermal-motion effects [15].
Moments and cumulants make possible to find a physical profile f from the measured profile h and the instrumental profile g. A number of members can be small in a formula for moment expansion, e. g. an expansion to the third or to the fourth moment is enough in the Edgeworth series expansion [1].

4. Particle size and microstrain from the single line

The particle size and microstrain can be estimated from the second and the fourth central moments of the physical profile of a single line [10]. Newerheless, the mathematical expressions are rather complicated, although approximations were used for the computing of the moments in [10]. These approximations were removed in [11], but a graphical method was proposed for the estimation of the particle size and microstrain.

Purpose of this part is to show, that cumulants and moments enable to estimate the mentioned quantities analytically. The second and the fourth central moments of the physical profile f can be easy estimated from the corresponding moments of the measured profile h and the instrumental profile g according the eqs. (15) and (17)

$$\mu_{2,f} = \mu_{2,h} - \mu_{2,g} \quad \text{and} \quad \mu_{4,f} = \mu_{4,h} - \mu_{4,g} - 6(\mu_{2,h} - \mu_{2,g})\mu_{2,g} \tag{25}$$

The second and the fourth cumulants of the physical profile f are given by eqs. (22) and (24). It is assumed, that profiles f^D and f^ε refer to particle size and microstrain respectively and that the physical profile f is the convolution of the profiles f^D, f^ε. The second and the fourth cumulants of the physical profile f are then connected by relations

$$\kappa_2 = \kappa_2{}^D + \kappa_2{}^\varepsilon \tag{26}$$

and

$$\kappa_4 = \kappa_4{}^D + \kappa_4{}^\varepsilon \tag{27}$$

where $\kappa_2{}^D$ and $\kappa_4{}^D$ are the second and the fourth cumulant of the profile f^D, analogously $\kappa_2{}^\varepsilon$ and $\kappa_4{}^\varepsilon$ are the second and the fourth cumulant of the profile f^ε. It follows from eqs. (22) and (24)

$$\kappa_2{}^D = \mu_2{}^D \quad \text{and} \quad \kappa_4{}^D = \mu_4{}^D - 3(\mu_2{}^D)^2 \tag{28}$$

and

$$\kappa_2{}^\varepsilon = \mu_2{}^\varepsilon \quad \text{and} \quad \kappa_4{}^\varepsilon = \mu_4{}^\varepsilon - 3(\mu_2{}^\varepsilon)^2 \tag{29}$$

It is generally assumed, that the profile f^ε, corresponding to the microstrain, is Gaussian [16]. It means $f^\varepsilon = A \exp(-r^2 s^2)$, where normalization factor A can be evidently omitted, r is connected with the microstrain [16] and s is the angle of diffraction expressed as a distance in reciprocal space, measured from the point *hkl* along the radius vector. The zeroth central (and raw) moment – integral intensity – of f^ε is [17]

$$\mu_0^\varepsilon = \int_{-\infty}^{+\infty} \exp(-r^2 s^2) ds = \frac{\sqrt{\pi}}{r} \tag{30}$$

the second and the fourth central moment of f^ε are [17]

$$\mu_2^\varepsilon = \frac{1}{\mu_0^\varepsilon} \int_{-\infty}^{+\infty} s^2 \exp\left(-r^2 s^2\right) ds = \frac{1}{2r^2}, \quad \mu_4^\varepsilon = \frac{1}{\mu_0^\varepsilon} \int_{-\infty}^{+\infty} s^4 \exp\left(-r^2 s^2\right) ds = \frac{3}{4r^4} \tag{31}$$

so that $\mu_4^\varepsilon = 3\,(\mu_2^\varepsilon)^2$ and $\kappa_4^\varepsilon = 0$. The fourth cumulant of Gaussian function is equal to zero. This fact simplyfies substantially the estimation of the particle size and the microstrain from the single line. The particle size can be estimated only from eq. (27), because $\kappa_4 = \kappa_4^D$ and the microstrain can be estimated then from eq. (26). The particle size estimation is easy if, in addition, μ_4^D is a multiple of the square of μ_2^D . Eqs. (26) and (27) are generally a system of two equations for two unknown quantities – particle size D and microstrain ε.
These unknown quantities are inserted into system of eqs. (26) and (27) by means of moments of functions, which are assumed as forms of the profiles f^D and f^ε. Cauchy function is often considered as the form of the profile f^D [16]. It is necessary to cut of Cauchy profile in calculations of its moments, because corresponding integrals cannot be calculated in infinite intervals. The mentioned ways of the moment expansion is also problematic for Cauchy profiles from this reason. Moments were used also in the interpretation of physical profiles [18-20].

Conclusions

Basic formulae were summarized for moments and cumulants, which are convenient in the line profile analysis. They make possible to find physical profiles from measured and instrumental profiles. Cumulants of diffraction profiles were proposed to a modification of the method for estimation of the particle size and the microstrain from the single line and this method was discussed. It was shown, that the fourth cumulant of Gaussian profile is zero.

References

1. Cramér, M., 1946, *Mathematical Methods of Statistics* (Princeton: Princeton University Press).
2. Wilson, A.J.C., 1973, *J. Appl. Cryst.* **6**, 230.
3. Kendall, M.G. & Stuart, A., 1969, *The Advanced Theory of Statistics*, Vol. **1** (London: Charles Griffin).
4. Bracewell, R.N., 1986, *The Fourier Transform and Its Applications* (New York: McGraw-Hill, Inc., 2nd ed.).
5. Fingerland, A., 1960, *Czech. J. Phys. B*, **10**, 233.
6. Spencer, R.C. 1931, *Phys. Rev.* **38**, 618.
7. Tournarie M., 1956, *C. R. Acad. Sci. Paris* **242,** 2016, 2161.
8. Wilson, A.J.C., 1963, *Mathematical Theory of X-Ray Powder Diffractometry* (Eindhoven: Philips Technical Library).
9. Mitra, G.B., 1964, *Brit. J. Appl. Phys.* **15**, 917.
10. Kagan, A.S.& Snovidov V.M., 1965, *Fiz. Met. Metalloved.* **19**, 191.

11. Snovidov, V.M., Kagan, A.S. & Kovalskii, A.E., 1968, *Zav. Lab.* **34,** 1086.
12. Aitken, A.C., 1947, *Statistical Mathematics* (Edinburgh and London: Oliver and Boyd).
13. Janossy, L., 1965, *Theory and Practice of the Evaluation of Measurements* (Oxford: Oxford University Press).
14. Scardi, P. & Leoni, M., 2006, *J. Appl. Cryst.* **39**, 24.
15. Johnson, C.K., 1969, *Acta Cryst. A*, **25**, 187.
16. Warren, B.E., 1959, *Progress in Metal Physics* **8**, 147.
17. Dwight, H.B., 1961, *Tables of Integrals and Other Mathematical Data* (New York: The Macmillan Company, 4th ed.).
18. Harrison, J.W., 1966, *Acta Cryst.* **20**, 390.
19. Harrison, J.W., 1967, *Acta Cryst.* **23**, 175.
20. Ungár, T.H., 1967, *Acta Cryst.* **23**, 174.

Acknowledgements. The research was supported by the Project No 106/07/0805 of the Czech Science Foundation.

Z. Kristallogr. Suppl. **27** (2008) 135-141/ **DOI** 10.1524/zksu.2008.0018

X-ray line-broadening analysis of dislocations in a single crystal of Zr

M. Griffiths[1,*]

[1] Atomic Energy of Canada Limited, Chalk River Laboratories, Chalk River, Ontario, K0J 1J0, Canada
[*] griffitm@aecl.ca

Keywords: Dislocations, strain, irradiation, line-broadening

Abstract. The X-ray diffraction line broadening due to micro-strains near dislocations is complicated by the effect of grain size and inter-granular strain distributions on the line profile for deformed polycrystalline materials. Use of irradiated single crystals to study the effect of dislocation strain on diffraction line profiles eliminates this problem. The diffraction line profiles of single crystals of Zr were measured before and after irradiation in a nuclear reactor. The irradiation produces damage in the material in the form of dislocation loops. Using the original material as a standard the deconvoluted diffraction peak represents the broadened peak profile that is due to the dislocation loops only. Fourier analysis of the deconvoluted peak using the method of Warren and Averbach shows that the magnitude of the micro-strain component is consistent with the dislocation density measured by transmission electron microscopy with some allowance for reduction of the lattice strain due to loop alignment (polygonisation). The Fourier analysis also gives a value for the coherent diffraction domain size that is consistent with the mean spacing between the dislocations. These results are discussed in terms of the effect of the lattice strain induced by dislocation loops on the diffraction line profile.

1. Introduction

In deformed polycrystalline materials with a hexagonal close-packed structure such as Zr, X-ray diffraction line profiles are dependent on a number of different factors: (i) instrumental; (ii) inter-granular strain distributions; (iii) coherent diffracting domain size; and (iv) lattice micro-strains (local lattice distortion due to dislocations), [1]. The size and strain effects can be characterized by the Fourier analysis method of Warren and Averbach [2] and Warren [3], provided that the only contributions to the line profile arise from these latter two factors, (iii) and (iv). Other methods of analysing the diffraction line profile rely on knowing the individual effect of the different factors (domain size and strain) on the shape of the line profile [4,5]. The instrumental broadening effect, (i), can be extracted from the line profile using conventional deconvolution methods with an appropriate standard [3]; the line-broadening can then be analysed relative to that standard. It is not just the instrumental effect that is nullified by using a standard; the standard also represents the original state of the material against which any subsequent changes in state can be measured. For example, if one takes an annealed material that is subsequently cold-worked one can measure the effect of the cold-working directly by using the original annealed material as a standard. Often annealed single crystals are used as standards against which many different metallurgical states can be measured. In these cases, all analyses represent the material state relative to the single crystals.

The presence of inter-granular strain distributions, (ii), introduces an error into the determination of dislocation densities because the strain distributions are dependent on the extent of deformation and their effect on the diffraction line profile cannot be readily separated from the effect of dislocations. However, if one introduces dislocations into a material without generating inter-granular strains, then it would be possible to study the direct effect of dislocations on the line profile. Unfortunately it is extremely difficult to deform single crystals of Zr to produce dislocations without introducing twins. Twinning would then introduce smaller domains (grains) and inter-granular strains that confound the analysis of strain due to dislocations. Introduction of dislocations into a Zr single crystal can be achieved by irradiation to produce dislocation loops. The irradiation does not produce any grain boundaries and there will be no broadening due to inter-granular strains or small crystallites because the material is still a single crystal after irradiation.

Having obtained a diffraction line profile that has no contribution from inter-granular strain distributions, one can separate the strain and domain size components [2]. The coherent diffracting domain size and dislocation micro-strains are convoluted but can be separated because the domain size broadening is independent of the order of the diffraction line whereas the strain broadening is not. In the Warren-Averbach treatment the diffraction line profile is transformed into its Fourier components and processed to separate the domain size and micro-strain effects (after correction for instrumental broadening). This method has been adopted for the determination of dislocation densities in deformed Zr and Zr-alloys [1,6,7,8,9,10] but there is a difficulty in validating the dislocation density conversion from the line-broadening data because of the effects of inter-granular strain distributions [1]. Inter-granular strains are particularly important in hexagonal materials because of the deformation anisotropy [11].

The aim of this work is to study the effect of dislocations on the X-ray diffraction line-profile of Zr (a hexagonal close-packed metal) from measurements on single crystals before and after neutron irradiation.

2. Experimental

X-ray diffraction specimens (about 100 mm^2 area x 0.5 mm thick) were prepared by cutting slices from Zr single crystals that had been grown by multi-pass zone melting and refinement. Each specimen was then chemically polished using a solution of 5% HF, 45% HNO_3 and 50% H_2O to remove at least 0.025 mm - the depth of damage from the low speed diamond wheel used to cut the specimen. X-ray diffraction peak profiles for $\{10\bar{1}0\}$ planes were then obtained using a Rigaku X-ray diffractometer with a 12 kW (rotating anode) generator with Ni-filtered, CuK_α radiation. The single crystals were then irradiated in the NRU nuclear reactor to a fast neutron fluence of approximately 3×10^{24} $n.m^{-2}$ (E > 1 MeV) producing radiation damage of about 0.5 displacements per atom and then re-measured on the same X-ray diffractometer with the same conditions described above.

Transmission electron microscopy (TEM) specimens were prepared after irradiation and measurement with X-rays by further chemically thinning the samples to 0.1 mm thickness and punching out 3-mm-diameter discs. Foils were then made by electro-polishing with a solution of 10% perchloric acid in methanol at about -40°C with a current density of about 0.01 $amps.mm^{-2}$ using a Materials Science Northwest twin-jet apparatus. TEM analyses were performed using a Philips CM 30 electron microscope operating at 300 kV.

3. Results

The effect of irradiation on Zr single crystals was investigated for specimens cut parallel with the $\{10\bar{1}0\}$ plane using the original unirradiated material as a standard. Figure 1(a) shows an electron micrograph showing dislocation loops that are produced by irradiation. Figure 1(b) shows the same area imaged with an 0002 diffracting vector; the loops do not exhibit diffraction contrast with this diffracting vector indicating that they have $<11\bar{2}0>$ type Burgers' vectors. In relation to the hexagonal lattice the loops have <a> type Burgers' vectors that are parallel with the basal plane (0001) and have habit planes that are close to $\{10\bar{1}0\}$ prism planes.

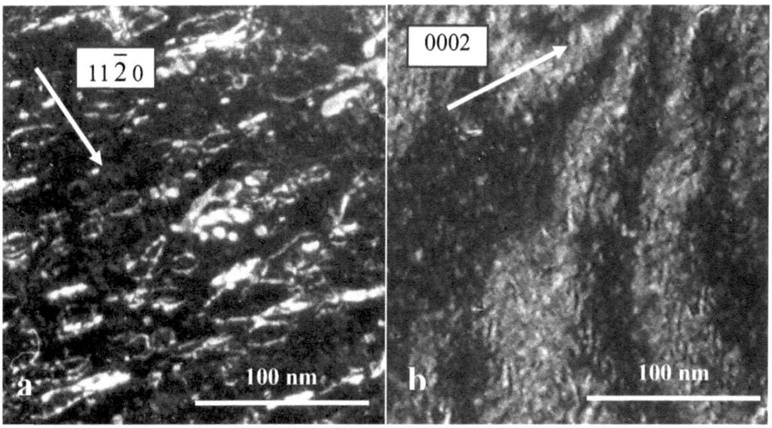

Figure 1. Micrograph showing dislocation loops in neutron irradiated single-crystal Zr following neutron irradiation at 573 K to a dose of about 0.5 displacements per atom: (a) Diffracting vector $11\bar{2}0$ *showing <a>-type dislocation loops; (b) diffracting vector* 0002 *showing zero contrast from the <a>-type dislocation loops.*

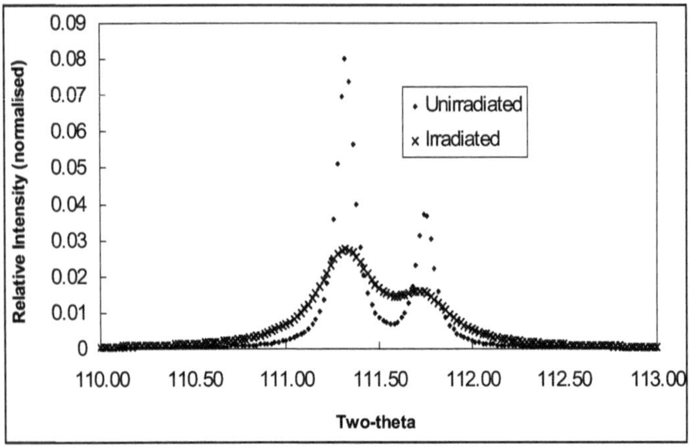

Figure 2. Diffraction line profiles for the $\{30\bar{3}0\}$ *prism planes of single-crystal Zr before and after irradiation showing the effect of the increase in dislocation density due to irradiation.*

The effect of irradiation on the X-ray diffraction line profiles is illustrated in Figure 2 for the $\{30\bar{3}0\}$ line profile. The broadening due to irradiation can be extracted by separating the effects of instrumental broadening from the line profile of the irradiated sample by the Stokes method, [12], using the original unirradiated material as a standard. In this way all instrumental and any other unknown effects of the material on the line broadening are extracted from the line profile. After applying deconvolution in angular space (2 θ) the resulting line profile represents the effect of the irradiation damage only and is shown in Figure 3. The deconvoluted peak has a shape that is close to Lorentzian in nature.

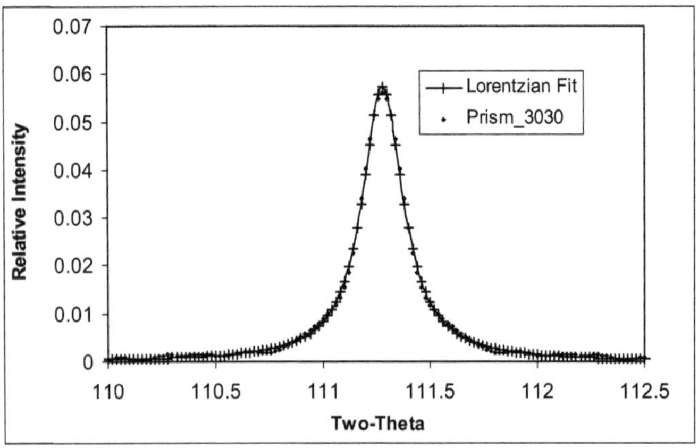

Figure 3. Deconvoluted $\{30\bar{3}0\}$ *prism plane diffraction line profile (angular-space).*

The line shape is consistent with a Lorentzian distribution.

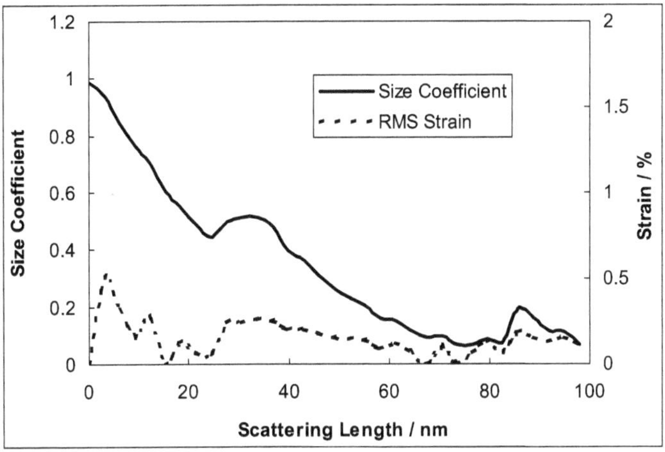

Figure 4. Size coefficients and root-mean-squared lattice strains derived from Fourier analysis using the method of Warren and Averbach [2].

In order to separate out the strain and domain size effects, both the $\{20\bar{2}0\}$ and $\{30\bar{3}0\}$ line profiles for the unirradiated standard and the irradiated samples were processed similarly, first converting to wave-space (2sinθ/λ) then correcting for instrumental effects following the Fourier transform method of Stokes [11], and finally processing the Fourier coefficients by the method of Warren and Averbach [2]. The resulting size coefficients and root-mean-squared strains (ε) obtained by combining the data from the two orders of reflection are illustrated in Figure 4. The results show that the root-mean squared strain for scattering lengths in the range of 20-70 nm is about 0.15%. The domain size (D), determined by taking the x-intercept of the tangent to the curve in a plot of the size coefficient against scattering length, is about 70 nm. The dislocation density was determined from the domain size and the root-mean-squared strain (ε) using the method of Williamson and Smallman [5]. In this method the dislocation density (ρ) is given by:

$$\rho = \frac{K < \varepsilon >^2}{F\, b^2} \tag{1}$$

K is a factor dependent on the shape of the strain distribution (taken as 20 [6, 7]) and F is a factor that adjusts for the strain energy of the dislocation as the outer cut-off radius varies. This latter factor takes into account the fact that polygonisation will reduce the total strain energy due to dislocations and is given by [5]:

$$F = \frac{\ln(D/2r_0)}{\ln(r/r_0)} \tag{2}$$

Where D is the domain size derived from the Fourier size coefficients, r_0 is the inner cut-off radius and r is taken as 50nm [7, 13]. For heavily deformed material with a relatively uniform distribution of dislocations, i.e. no dislocation polygonisation or pile-up, it has been assumed in some treatments that the coherent diffracting domain size is comparable with the average distance between dislocations [6]. This case corresponds with the condition where the dislocation density determined from analysis of particle (domain) size broadening is equivalent to that determined from the analysis of lattice strain broadening [5]. The outer cut-off radius for a uniform distribution of dislocations is then defined by the domain size and can be used in a simple derivation of dislocation densities based on analysis of the lattice strain [7].

Assuming that the coherent diffracting domain size corresponds with the mean spacing between dislocations [6] there are three different methods for estimating the dislocation density (direct measurement by transmission electron microscopy, from the root-mean-squared strain (ε) and from the domain size (D)). These are compared in Table 1. The dislocation densities determined by the three methods differ by a factor of three. These differences can be reconciled by the fact that the dislocations are in the form of loops that are not uniformly distributed in the material.

Table 1. Domain size(nm) and Dislocation densities (10^{14} m^{-2}) measured from TEM micrographs compared with those determined from line-broadening data for $\{10\bar{1}0\}$ *planes in Zr single crystals irradiated at 573 K to a dose of about 0.5 displacements per atom.*

	Domain Size	Dislocation Density TEM	Dislocation Density $\frac{1}{\sqrt{D}}$	Dislocation Density $\frac{K.<\varepsilon>^2}{F.b^2}$
Unirradiated	>1000	<0.01	<0.01	<0.01
Irradiated	63	7.1	2.5	4.22

4. Discussion

The results show that: (i) the dislocation line profile due to dislocation loops in irradiated, single crystal Zr has a Lorentzian shape; (ii) the coherent diffracting domain size determined from a Fourier analysis of the diffraction peak corresponds with the approximate spacing between dislocations.

In the method of Warren and Averbach [2] the Fourier coefficients of the diffracted line profiles are equated to the coherent diffracting domain and the distortion or micro-strain within this domain. The results show that the domain size derived from the Fourier analysis changes from a value >1000 nm to 63 nm simply by introducing dislocations. This supports the notion that one can treat the domain size as being equal to the average distance between dislocations because there is no other factor that can affect the domain size other than dislocation loop formation.

One can speculate that the micro-strains cancel on average mid-way between dislocations of opposite sign. Therefore, for a uniform array of dislocations coherency will exist only within the region close to the dislocation where the interference from other surrounding dislocations is minimal. Although a uniform array of straight dislocations is not present in this case the dislocation density derived assuming that the domain size is equal to the mean spacing between dislocations is comparable with the value derived from the mean micro-strain. The fact that TEM shows higher values than that derived from the X-ray diffraction line profile simple reflects the fact that the dislocation loops are not straight dislocations and are not uniformly distributed.

5. Conclusions

The diffraction line profile due to dislocation loops in irradiated Zr has a Lorentzian shape. Fourier analysis shows that the magnitude of the micro-strain component is consistent with the dislocation density measured by transmission electron microscopy with some allowance for reduction of the lattice strain due to loop clustering. The domain size that arises from the analysis by the method of Warren and Averbach [2] has a value that is consistent with the mean spacing between the dislocations.

References

1. Griffiths, M., Sage, D., Holt, R.A. & Tome, C.N., 2002, *Proc. TMS 2001, Mater. Metall. Trans. A*, **33**, 859.
2. Warren, B.E. & Averbach, B.L., 1950, *J. Appl. Phys.*, **21**, 595.
3. Warren, B.E., 1969, *X-ray Diffraction* (Reading, MA: Addison-Wesley Publ.).
4. Williamson G.K. & Hall, W.H., 1953, *Acta Metall.*, **1**, 22.
5. Williamson G.K. & Smallman, R.E., 1956, *Phil. Mag.*, **1**, 3.
6. Aqua E.N. & Owens, C.M., 1967, *Trans. Met. Soc. AIME*, **239**, 155.
7. Griffiths, M., Winegar, J.E., Mecke J.F. & Holt, R.A., 1992, *Adv. X-ray Anal.*, **35**, 593.
8. Klimanek, P. & Kuzel, R., 1988, *J. Appl. Cryst.*, **21**, 59.
9. Kuzel, R. & Klimanek, P., 1988, *J. Appl. Cryst.*, **21**, 363.
10. Klimanek, P. & Kuzel, R., 1989, *J. Appl. Cryst.*, **22**, 299.
11. MacEwen, S.R., Christodoulou, N. & Salinas-Rodriguez, A., 1990, *Mat. Trans. A*, **21A**, 1083.
12. Stokes, A.R., 1948, *Proc. Phys. Soc. London*, **61**, 382.

Z. Kristallogr. Suppl. **27** (2008) 143-150 / **DOI** 10.1524/zksu.2008.0019 143

XRD line profile analysis of calcite powders produced by high energy milling

H. Pesenti[1], M. Leoni[1], P. Scardi[1,*]

[1] Department of Materials Engineering and Industrial Technologies, University of Trento, 38100 via Mesiano, 77, Trento, Italy
[*] Paolo.Scardi@unitn.it

Keywords: diffraction, line profile analysis, carbonates, ball milling, WPPM

Abstract. Natural calcite crystals were ground to powder by high energy mechanical milling. Depending on the grinding conditions, powders can be pure calcite or calcite-aragonite mixtures, with a complex equilibrium between the two polymorphs of calcium carbonate. In any case, the mechanical milling is effective in reducing domain size to the nanoscale, introducing large microstrains whose exact nature is not yet totally clear. Domain size distribution and possible sources of microstrain were investigated by XRD line profile analysis, with the support of electron microscopy and BET surface area analysis.

1. Introduction

Even if calcium carbonate is one of the most known and diffuse minerals on the Earth crust, several features are still object of active research. For example, the crystallographic aspects of the transformation between the two most frequent polymorphs, calcite and aragonite, is well known [1], but the mechanisms are still debated [2-4]. In particular, the role of lattice defects is still to be clarified.

In a more general sense, most mineral properties change by mechanical activation. In Earth science, fine dispersion of crystalline domains and incorporation of lattice defects have an influence on surface reactivity [5], uptake of contaminant species and in particular on dissolution kinetics [6,7] and transport properties [8]. In technological applications involving mineral powders, domain size and defectivity influence nearly any forming process and many properties of ceramic materials [9].

Mechanical grinding is one of the most effective and simple means to comminute and activate mineral powders: grinding can lead to extended solubility, phase transformations, mechano-chemical reactions, amorphization and formation of metastable compounds [10]. X-ray Diffraction (XRD) is a powerful tool to study mechanically activated minerals: besides following structural evolutions (as in polymorphic phase transformations), XRD can provide a detailed picture of the microstructure by means of Line Profile Analysis (LPA).

In the present study we propose some preliminary results on calcite powders produced by high energy grinding. Nature and extent of calcite-aragonite transformation and lattice defect incorporation are discussed.

2. Experimental

Mineral calcite was ground by dry milling using two Fritsch mills: planetary P4 (milling at 20°C) and vibrating cup P9 (milling at -20°C). Steel equipment was used for the milling. In particular, for the planetary mill, 25 balls of 12 mm diameter were employed, with a powder/ball weight ratio of 3/175. Jars and main disk were counter rotated at 200 rpm. To limit heating of the system, a pause of 60 min was enforced every 20 min of milling.
Milling in the P9 was done with a 667 g cylinder and 8.22g of powder in a cup oscillating under the impulses of a motor rotating at 1100 rpm. As the system dissipates much more heat than a traditional planetary mill, cycles of 3 minute milling followed by 5 minute pause were employed.
A Rigaku IIID-max diffractometer, operated with a Cu tube (40kV, 30mA) and equipped with a secondary graphite monochromator was employed for the X-ray data collection. Patterns were recorded in the 18–90° 2θ range with a step size of 0.05° and a fixed counting time of 45s per step.
A FEI XL30 Environmental Scanning Electron Microscope (ESEM) was employed for the microscopy analysis: specimens were observed at low pressure (ca. 4 Torr) without the need for any metallization.

3. Results and discussion

The grinding efficiency of the two mills used in this study is markedly different. As shown in figure 1, ball milling (in P4 mill) induces at some stage a partial transformation of calcite to aragonite. Since the average temperature was kept under control by the long pauses between relatively short grinding cycles, the polymorphic transformation is likely induced by the high energy transfer typical of the impulsive interactions occurring during ball impacts. Grinding is so energetic that it also causes, as a downside, powder contamination from the ferritic iron of the balls, identified from a weak diffraction line around 44.6°.
A quantitative phase analysis by the Rietveld method (using Brukers' *TOPAS* v. 3.00), including the small contamination (always below 1%) and the 'natural' quartz contaminant (also present in the same amount in the pristine mineral) shows that the transformed fraction of aragonite reaches about 40%. In agreement with other studies, it seems therefore not possible to completely transform calcite to aragonite by grinding [3].
Grinding in the vibrating cup should give faster effects: the P9 mill is indeed known as a 'fast' mill, able to rapidly homogenize powders, even when starting from millimetre-size particles. However, as shown in the XRD patterns of figure 2, the grinding conditions are such to progressively broaden the peak profiles - pointing out the progressive grain comminution and incorporation of defects - without inducing a transformation of calcite to aragonite in visible amounts. It is also worth noting that no contamination from the vial and cylindrical rotating body was introduced. It is therefore concluded that the P9 mill operated at -20°C is more appropriate than the P4 mill to introduce controlled amounts of defects and progressive grain comminution. Grinding effects apparently saturate above 60 minutes of active milling, and the peak broadening in the calcite phase seems slightly less pronounced than in the same phase after extensive milling in P4.

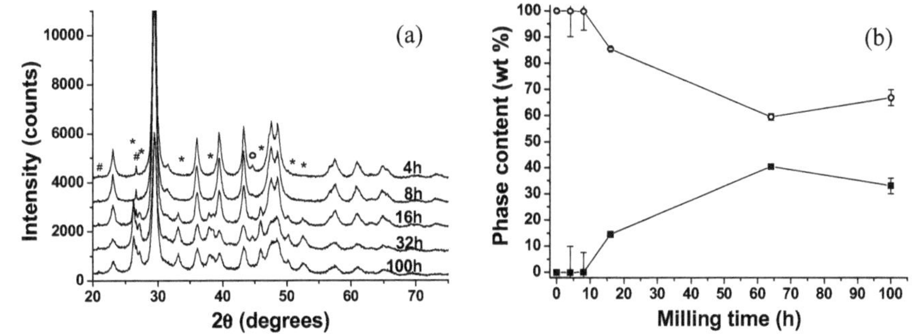

Figure 1. XRD patterns of calcite ground in the P4 mill for increasing time with indication of main peaks of aragonite (*), quartz (#) and ferrite (○) (a-iron) contaminant) (a); corresponding phase percentages for calcite (circle) and aragonite (square) are shown in (b).

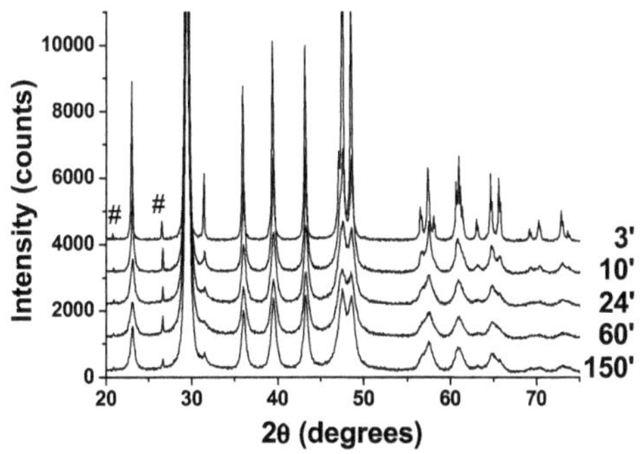

Figure 2. XRD patterns of calcite ground in the P9 mill for increasing time. The main reflections of the quartz impurity are marked by (#).

Just for the sake of a *preliminary* analysis of the line broadening effects, the XRD patterns of calcite ground in P9 mill were analysed by *MarqX*, a software for whole powder pattern fitting [11]. *MarqX* allows the separation of diffraction peaks (pattern decomposition) and background, also taking into account the instrumental profile component. Results can then be used in a Williamson-Hall plot (integral breadth versus inverse of interplanar distance, d^* (scattering vector modulus)), shown in figure 3a, from which both the increase in intercept and in slope with milling time are quite evident.

Even though the exact nature of the lattice defects is not known at this stage, inverse of intercepts in the WH plot were used to evaluate an effective size, as equivalent sphere diameter, given by volume-weighted mean column length times the Scherrer constant for a sphere, 4/3 [12,13]. The trend, shown in figure 3b as a function of the grinding time, apparently suggests a fast decrease of the effective size to a saturation value lower than 50 nm.

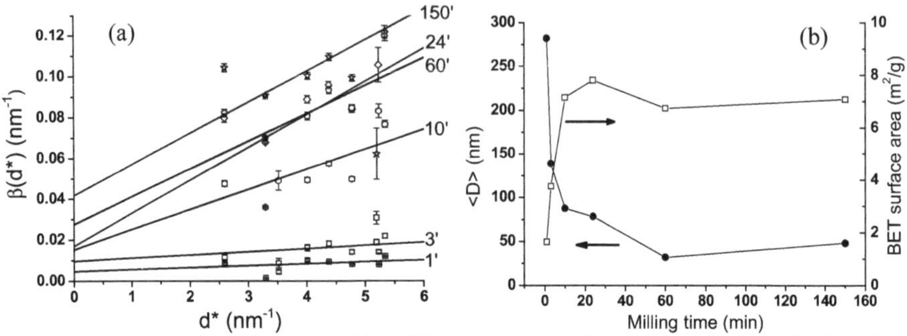

Figure 3. Williamson-Hall plot (integral breadth versus inverse of interplanar distance) for the calcite powder ground in P9 for increasing time (a); effective domain size, estimated from the plot in (a), together with surface area from BET (b);

Electron microscopy shows the evolution of powder morphology after progressively longer grinding (figure 4). The grain size apparently decreases in the first minutes of grinding, up to 24', when coalescence effects lead to the formation of agglomerates made of finely divided grains. As a consequence, the surface area measured by BET, shown in figure 3, goes through a maximum at about 24', and tends to saturate to a somewhat smaller value for longer grinding time.

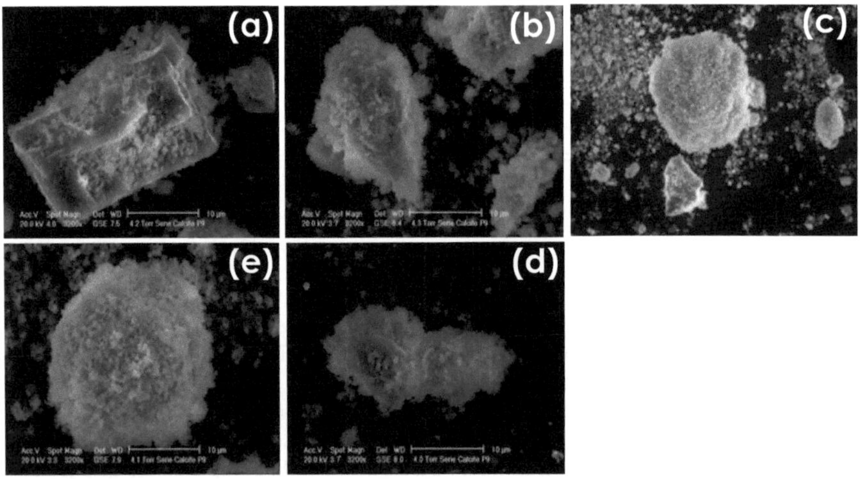

Figure 4. ESEM pictures of calcite powders ground in P9 mill for 1' (a), 3' (b), 10' (c), 24' (d), 60' (e).

The above results underline the importance of a clear distinction between domain size - as measured by XRD methods - and grain (or agglomerate) size visible by microscopy. BET seems more affected by the latter than by the former dimension, even though effects of domain size and lattice defects on dissolution kinetics of nanocrystalline mineral powders have been demonstrated [7].

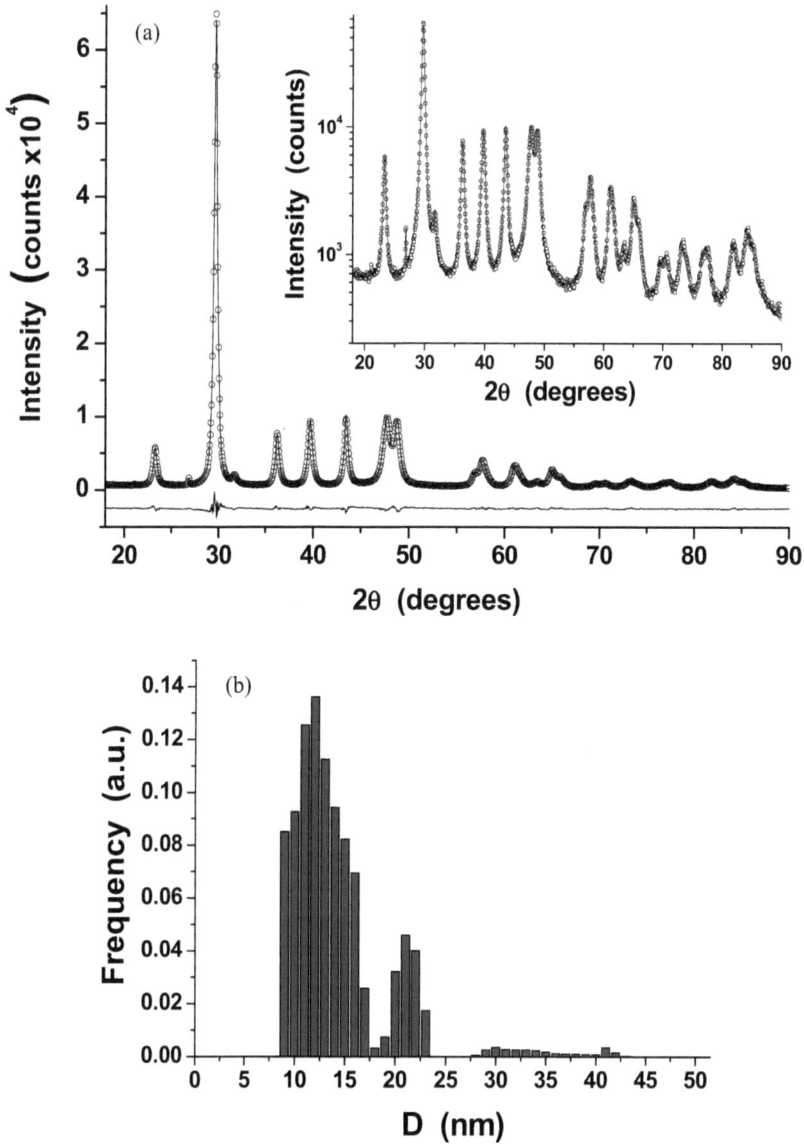

Figure 5. WPPM results for the calcite powder ground for 150′ in the P9 mill. Experimental data (dot), modelled pattern (line) and difference between the two (line below). The result is also shown in Log scale in the inset (a). Corresponding domain diameter distribution (b);

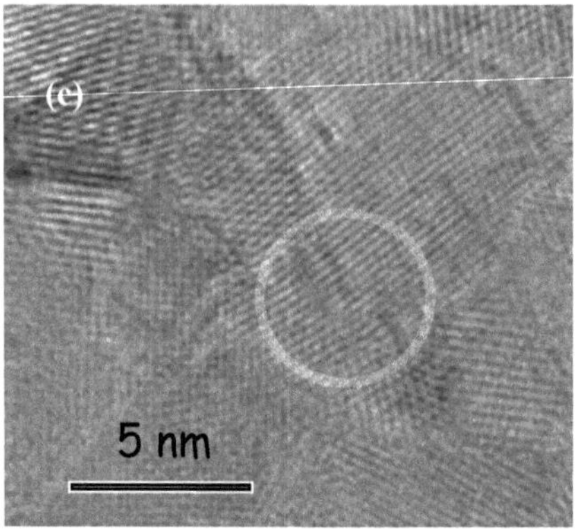

Figure 5 (continued). High resolution TEM micrograph, showing a dislocation (see ring) in calcite nanodomains (courtesy of M. Gemmi and G. Artioli) (c).

The literature on calcite deformation is quite abundant (see [1,14] and references therein), but mostly refers to observations on single crystal or polycrystalline carbonates with micrometer-size or larger grains. The basic mechanism, especially for low temperature, is deformation twinning, but dislocation glide is also significant at higher temperatures and high deformation rates [1,15].

Much less is known on the deformation of small, nanoscale domains, especially when considering the peculiarity of mechanical grinding. Milling mostly involves impulsive forces between vial, grinding bodies and powder. Even if the average temperature can be kept constant (as in the P9 mill) the energy released locally during the fast impacts is considerably high. Under these conditions it could be put forward that dislocations glide should be favoured with respect to twinning as the active plasticity mechanism.

Transmission Electron Microscopy (TEM) on the ground powders is in progress. So far we were able to observe several dislocations in very finely divided crystalline domains, as in the picture shown in figure 5. More extensive studies will be necessary to determine the prevailing slip system(s) and other possible mechanisms of deformation induced by grinding.

A detailed LPA cannot be performed with the present state of knowledge on these powders. However, the XRD pattern of the calcite ground for 150' in the P9 mill was studied by Whole Powder Pattern Modelling (WPPM), using the approach recently devised for systems were details on the dislocation slip systems are not known [16].

Details on the WPPM can be found elsewhere [13,16-18]. Here it is worth underlying that crystalline domains, which in this case give the main contribution to line broadening, were assumed equiaxial and modelled as spheres whose diameters are distributed according a free histogram distribution, to be determined by WPPM. As to the strain broadening component, without specific assumptions on the exact mechanism and on specific dislocation slip systems, the anisotropic peak broadening was modelled by means of an effective contrast factor, described in rather general terms as a 4th order invariant form of the Miller indices [16]. The specific invariant, used in the Fourier Transform of the peak profile component due to lattice strain, is that for the calcite Laue group (3m1):

$$\overline{C}_{hkl} = \frac{E_1\left(h^2+k^2+hk\right)^2+2E_2l^2\left(h^2+k^2+hk\right)+E_3l^4+\left(4/3\right)E_4l\left(2h^3-2k^3+3h^2k-3hk^2\right)}{d_{hkl}^4}, \quad (1)$$

where E_1, E_2, E_3, E_4 are to be determined by WPPM. Instrumental profile components, lattice parameters, background and other factors and correction terms were included as described elsewhere [13,16-18].

The WPPM result for the calcite powder ground for 150' in the P9 mill is shown in figure 5. Even if the modelling is not completely satisfactory (the residual is not totally featureless), the convergence is sufficiently robust to give a reliable result on the domain size distribution. In particular we can see that the domain size is actually in the nanometre scale (in qualitative agreement with the preliminary analysis made by the WH plot), quite far from the agglomerate size. Further studies are in progress: (i) to find out the lattice defects introduced by the grinding; (ii) to assess the role of lattice defects in the polymorphic transformations of calcium carbonates during grinding in the P4 mill; and (iii) to correlate morphology, BET and LPA with the dissolution kinetics and, more generally, with the surface reactivity of the carbonates.

4. Concluding remarks

Calcite grinding is strongly affected by the milling conditions: a highly energetic grinding in planetary ball mill induces a partial transformation of calcite to aragonite, whereas grinding in a vibrating cup mill (under controlled bowl temperature of -20°C) keeps calcite stable.

Grinding has in any case the effect of reducing carbonate domains to the nanoscale, steeply increasing the defect density. Even if at the present state of knowledge there is no definitive evidence on the prevailing lattice defects in ground calcite, a reliable domain size distribution could be obtained by Whole Powder Pattern Modelling for the calcite ground in the vibrating cup mill. Preliminary TEM analysis shows dislocations in milled calcite powders, and a range of domain size entirely compatible with the WPPM result.

As expected, XRD results do not match evidence from BET surface area analysis. The latter is in fact related to the grain size, where 'grains' can either be single crystals or, as it seems the case here, aggregates of fine crystals, whose size changes in a non monotonous way with the length of the milling process. ESEM micrographs clearly confirm this conclusion

References

1. Wenk, H.-R. (editor), 1985, *Preferred orientation in deformed metals and rocks: an introduction to modern texture analysis* (Orlando: Academic Press).
2. Burns, J.H. & Bredig, M.A., 1956, *J. Chem. Phys.*, **25**, 1281.
3. Dandurand, J., Gout, R. & Schott J., 1982, *Tectonophysics*, **83**, 365.
4. Martinelli, G. & Plescia, P., 2004, *Phys. Earth Planet Inter.*, **142**, 205.
5. Vaughan, D.J. & Pattrick, R.A.D. (editors.), 1995, *Mineral Surfaces*, The Mineralogical Society Series, **Vol. 5**. (Chapman & Hall).

6. Schott, J., Brantley, S., Crerar, D., Guy, M. & Willaime, C., 1989, *Geoch. Cosmoch. Acta*, **53**, 373.
7. De Giudici, G., Biddau, R., D'Incau, M., Leoni, M. & Scardi, P., 2005, *Geoch. Cosmoch. Acta*, **69**, 4073..
8. Eberl, D.D., 2004, *American Mineralogist*, **89**, 1784.
9. Kingery, W.D., Bowen, H.K. & Uhlmann, D.R., 1976, *Introduction to Ceramics*, 2nd edition (New York: Wiley).
10. Suryanarayana, C., 2001, *Progr. Mater. Sci.*, **46**, 1-184
11. Dong, Y.H. & Scardi, P., 2000, *J. Appl. Cryst.*, **33**, 184.
12. Klug, H.P. & Alexander, L.E., 1974, *X-ray diffraction procedures for polycrystalline and amorphous materials* (New York: Wiley).
13. Scardi, P. & Leoni, M., 2001, *Acta Cryst. A*, **57**, 604.
14. De Bresser, J.H.P., Spiers, C.J., 1997, *Tectonophysics*, **272**, 1.
15. Langenhorst, F., 2002, *Bull. Czech Geol. Survey*, **77**, 265.
16. Leoni, M. Martinez-Garcia, J. & Scardi, P., 2007, *J. Appl. Cryst.*, **40**, 719.
17. Scardi, P. & Leoni, M., 2002, *Acta Cryst. A*, **58**, 190.
18. Leoni, M. & Scardi, P., 2004, *J. Appl. Cryst.*, **37**, 629.

Acknowledgements. Authors wish to thank M. Gemmi (Univ. of Milan) and G. Artioli (Univ. of Padua) for preliminary TEM results on ground calcite. The work was partly funded by the PRIN fund of the Italian Ministry for University and Research

Z. Kristallogr. Suppl. **27** (2008) 151-158 / **DOI** 10.1524/zksu.2008.0020 151

Refining real structure parameters of disordered layer structures within the Rietveld method

K. Ufer[1,*], R. Kleeberg[1], J. Bergmann[2], H. Curtius[3], R. Dohrmann[4]

[1] Institute of Mineralogy, TU Bergakademie Freiberg, Brennhausgasse 14, 09596 Freiberg, Germany
[2] Ludwig-Renn-Allee 14, 01217 Dresden, Germany
[3] Institute of Energy Research (IEF-6: Safety Research and Reactor Technology), Forschungszentrum Jülich, Germany
[4] BGR/LBEG, Stilleweg 2, 30655 Hannover, Germany
[*] kristian.ufer@bgr.de.

Keywords: DIFFaX, BGMN, Rietveld refinement, layered structures, planar disorder, layered double hydroxide (LDH)

Abstract. A new approach is proposed for the refinement of disordered layered structures within the Rietveld method. It is shown that it is possible to introduce the existing recursive treatments of stacking disorder in a Rietveld refinement by using a supercell approach in combination with a complex manipulation of the structure factor calculation. The applicability of this approach is demonstrated by the refinement of disordering parameters of data simulated by the established program DIFFaX [1]. In addition, structural parameters and contents of disordered layered double hydroxides in mixtures with a reference material were successfully refined from measured laboratory diffraction data.

1. Introduction

The physicochemical properties of materials are often directly related to their structural characteristics. X-ray powder diffractometry in combination with the Rietveld method [2] proved to be a suitable method for the determination of microstructural parameters. Unfortunately, this approach fails in the presence of materials which show strong structural defects like stacking disorder in layered materials. In the reciprocal space, structural disorder leads to intensity fractions continuously distributed along reciprocal rods. Therefore it is not treatable by the classical Rietveld method. To overcome this problem in the case of turbostratic disorder, a partly filled supercell was introduced [3]. Further treatments are necessary for stacking disorder consisting of translations/rotations with distinctive amounts and probabilities or mixed layer stacking. The literature reports several attempts for the treatment of stacking faults, e.g. the explicit calculation of finite stacks [4] or a recursive treatment [1]. Numerous studies proved that the simulation of powder patterns is an appropriate way to obtain disorder parameters of faulted layer structures. Publicly available programs containing automatic refinement procedures of the recursive method do exist [5,6]. However, the refinement of mixtures containing at least one strongly disordered layer structure has not yet been implemented. In this study, we demonstrate an alternative way to combine the recursive treatment of stacking disorder with a least-squares refinement procedure.

DIFFaX

One of the frequently used programs to treat diffraction phenomena of layer structures with coherent planar faults is DIFFaX [1]. It describes diffraction in a recursive way. The comparison of experimental and simulated powder patterns was successfully used to describe planar faults of layered structures, e.g. LDHs [7].
For the application of DIFFaX it is necessary to define one or more crystallographic layers. These layers can be regarded as an infinite repetition of unit cells in two dimensions (layer), analogous to the three-dimensionally repeating unit cell of “classic” structure factor calculation. For interconnections between layers it is necessary to define stacking vectors and the probability of the occurrence of these vectors.

Rietveld program code BGMN

The program code BGMN [8] was chosen for the implementation of the recursive treatment in a Rietveld program. The required equations as given in [1] are formulated in the BGMN built-in structure description language. As is common with computer languages, BGMN can handle looping, calls to other program codes, etc. To simulate powder patterns of disordered layered materials it is necessary to calculate intensities along reciprocal rods. Therefore the standard unit cell is enlarged perpendicular to the layer to n times the original repeating distance of the layers. In this study, enlargements of up to 100 times the original layer thickness were used. This supercell, on the one hand, is still compatible with the traditional periodic calculation of the Rietveld method. On the other hand, it generates additional hkl positions on the reciprocal rods to reproduce a continuous distribution. For the case of turbostratically disordered structures, the supercell was only filled with one layer to destroy the periodicity [3].
In order to adapt BGMN to the Treacy algorithm [1] some modifications of the BGMN code were made:

- The BGMN kernel was enabled for computing partial structure factors for only parts of the defined atomic positions, their complex values are manipulable within the structure language.
- The structure description language may define a total structure factor F.
- The structure description language was expanded by sub-programs for solving linear equations (even complex).

These sub-programs are coded in the BGMN kernel to be much faster compared to the structure language interpretation. Unlike turbostratic disorder, there is in principle no need for any correction factor to get the correct weight content. However to obtain the correct X-ray-

density (which will then be used for microabsorption correction), both the density and the scaling factor parameter have to be multiplied by the enlargement factor of the supercell.
Even symmetric peaks which are unaffected by the stacking disorder are built up of overlapping discrete hkl reflections. To produce a continuous diffraction line without artificial ripples, hkl dependent broadening factors are introduced.

Example: layered double hydroxides (LDHs)

From investigations [9] of corrosion products of research reactor fuel elements in salt brines it is known that layered double hydroxides (LDHs) are secondary corrosion products. Further experiments [10] showed that these LDHs have the ability to incorporate repository-relevant radionuclides. In order to examine this incorporation behavior, several LDHs containing trivalent lanthanides were synthesized. One of these materials, a samarium bearing Mg-Al-Cl LDH, will be addressed in this study.
LDHs consist of brucite-like sheets in which some divalent cations M^{II} have been substituted by trivalent ions M^{III} to form positively charged sheets. The cationic charges created in the sheets are compensated by the presence of anions A^{n-} in the interlayer. In addition, water molecules are also present in the interlayer. The structure units as the octahedral sheet and interlayer together form one layer. A general formula for LDHs is $[M^{II}_{1-x}M^{III}_{x}(OH)_2]^{x+}[A^{n-}]_{x/n}$ yH_2O.
The planar nature of the LDHs leads to a wide variety of polytypes [11]. In natural and synthetic CO_3^{2-} intercalated LDH phases the trigonal $3R_1$ polytype and the hexagonal $2H_1$ polytype predominantly exist. Both polytypes form trigonal prismatic interlayer polyhedra, which are occupied by the interlayer anion complexes. In the $3R_1$ structure adjacent layers are translated by a stacking vector (2/3,1/3,1/3) relative to each other. The $2H_1$ polytype is formed by an alternate stacking of layers, which are rotated around 180° parallel to the adjacent layers. Both stacking forms can occur in the same crystallite, so that the two polytypes are end members of various sequences of the two stacking arrangements [7].

2. Experimental

Synthesis of the Mg-Al-Sm LDH
Deionised water was boiled and stored under an argon atmosphere. This water was used for all experiments. Chemicals were purchased from Merck and Aldrich and used without further treatment. All experiments were performed under an argon atmosphere. The Mg-Al-Sm-Cl LDH was obtained according to the following co-precipitation process: 250 ml of water was placed in a three-necked glass flask, stirred and a pH of 10 was adjusted using 2 M NaOH solution. 250 ml of a mixed aqueous solution of MgCl2·6H2O (0.3 M) and AlCl3·6H2O (0.09 M) and, secondly, 10 ml of Sm2O3 (0.005 M) dissolved in 0.01 M HCl solution were added. This addition was performed over a period of five hours while the temperature was set to 70°C and the pH was maintained at 9.5 by addition of 2 M NaOH solution. Afterwards the temperature was raised to 90 °C and stirring was continued for 24 hours. After switching off the heater and cooling down to room temperature, the precipitated material was filtered and then dialysed at 60 °C. For the dialysis, a dialysis hose was filled with the substance and placed in 2-L vessel containing deionised water. The water was repeatedly

analysed for chloride using the cuvette test LCK 311, DR. LANGE. It was changed each time until it was chloride-free. The dialysed precipitate was filtered and dried in a desiccator.

Powder XRD measurements

The XRD data were collected with a Bruker D8 Advance™ diffractometer with Bragg-Brentano geometry. The diffractometer was equipped with a Cu tube, an automatic divergence slit (irradiating 6 mm sample length) and a position-sensitive detector (Vantec™) with an active length of 3°. The patterns were recorded from 8° to 130° 2Θ with a step width of 0.007° 2Θ and a total measuring time of 16 hours for the measurement of the pure LDH sample and 5 hours for the measurement of the mixture with 9.09 wt.% ZnO (1 portion of ZnO plus 10 portions of sample).

3. DIFFaX simulations

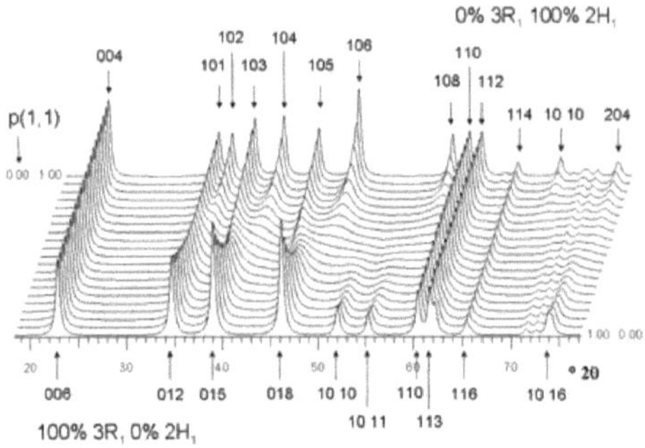

Figure 1. DIFFaX simulations of LDH powder patterns with different stacking probabilities p(1,1). Calculated for λ = 1.540598 Å.

The DIFFaX simulations were carried out with program version 8.07. The X-ray powder patterns were calculated from 5° to 130° with a step width of 0.04° 2Θ using a wavelength of 1.540598 Å. To introduce instrumental broadening a constant Lorentzian-shaped broadening of 0.5° FWHM was used. The brucitic sheets were built up of undisturbed octahedra. The hydroxyls at the corners of the octahedra were only considered as oxygen. The centres of the octahedra were occupied by 2/3 Mg and 1/3 Al. Along the z direction, the CO_3^{2-} anion was placed in the centre of the interlayer parallel to the brucitic sheet. The oxygen ions point in the direction of the octahedral hydroxyls and are at a distance of 1.28 Å from the carbon in the centre. To calculate the diffraction patterns of the $2H_1$ polytype it was necessary to introduce an additional layer. This second layer was generated from the original layer by a 180° rotation around the vertical axis. $3R_1$ stacking was implemented by a stacking vector (2/3,1/3,1) for layer 1 to layer 1 and (1/3,2/3,1) for layer 2 to layer 2. For the $2H_1$ stacking (layer 1 to layer 2 and layer 2 to layer 1) a vector (0,0,1) was used. The chosen lattice parameters were a = 3.06 Å and c = 7.80 Å. The layer width and the number of layers in the stack are regarded as infinite. The probabilities of the stacking transitions depend on each other. This means that the whole stacking can be described statistically by just one parameter, since p(1,1) is equal to p(2,2) and p(1,1) + p(1,2) = 1 and p(2,1) + p(2,2) = 1, defining

p(n,m) as the stacking probability of layer n to layer m. These structural parameters were successfully used for the simulation of Mg-Al-Eu LDHs [12] with an alternative algorithm [13].
One dataset was created for different stacking probabilities p(1,1). p(1,1) was varied in steps of 0.05 from 0 (pure $2H_1$ stacking) to 1 (pure $3R_1$ stacking). The occupancy of the interlayer anion complex was chosen as 1/12 for these simulations. All atoms were defined without Debye-Waller factors. The calculated patterns are shown in figure 1. Some of the peaks are indexed according to the lattice of the pure polytypes. It can be seen that symmetrical peaks only occur for the two end members.
A second dataset was created for different occupancy factors of the interlayer anion complex. The occupancy was varied from 0 to 0.2 in steps of 0.01. This reflects the chemically reasonable range for interlayer carbonate occupancy in LDHs. As stacking probability p(1,1) = 0.5 was chosen for these calculations.

4. Refinement of simulated data

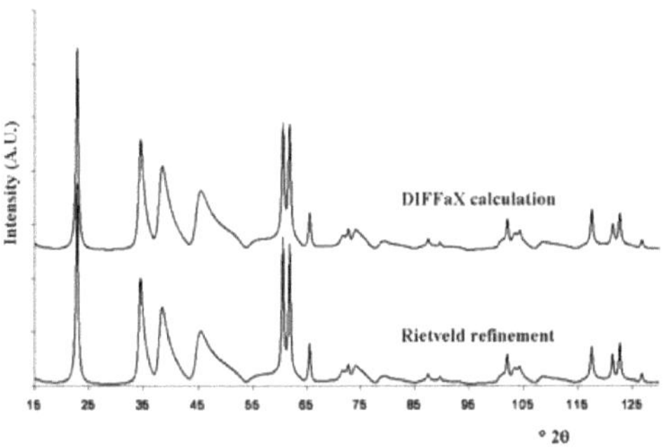

Figure 2. Comparison of a DIFFaX simulation and a Rietveld refinement of an LDH with a stacking probability of p(1,1) = 0.5. Calculated for λ = 1.540598 Å.

The same structural parameters, peak shape and wavelength are chosen as for the DIFFaX simulations for the refinement of the simulated datasets. The size of the supercell for the continuous calculation of the hk rods was enlarged to 100 times the original layer thickness. For all refinements a scaling factor and a constant peak broadening parameter were refined. In addition, for the dataset with varying stacking parameters, p(1,1) was refined with 0.5 as a starting value. $p(CO_3)$ was set to 1/12 for these calculations. In figure 2, the Rietveld refinement of the simulation for p(1,1) = 0.5 is shown.
For the refinement of $p(CO_3)$ the stacking probability was set to 0.5 (highest degree of disorder) and $p(CO_3) = 0$ was chosen as the starting parameter.

5. Refinement of observed data

The structure for the refinement of the Mg-Al-Sm-Cl LDH was derived from [14]. In this work, the structure of a Zn-Al-Cl LDH with pure $3R_1$ stacking is given. Since disordered stacking is examined here, this 3-layer polytype was reduced to a single layer structure with P3m1 (space group no 156). The occupancy for Mg^{2+} p(Mg) was chosen to be refined between the constraints 0.667 and 1. For M^{III} the occupancy 1-p(Mg) was chosen. This M^{III} occupancy consists mainly of Al^{3+} with partial substitution by Sm^{3+}. This partial substitution p(Sm) was refined using constrains between 0 and 0.1, which is the highest possible value calculated by initial weight. The occupancies of Cl^- were coupled to the octahedral occupancies to compensate the positive charge. To introduce additional interlayer water the Cl^- position was also partly occupied by oxygen p(H_2O). In addition, the following parameters were refined: lattice parameters a and c, preferred orientation, broadening parameters and the distance of the Cl^- position from the threefold axis (distCL).
The sample contained a small amount of $Sm(OH)_3$. The structure model for the refinement of $Sm(OH)_3$ was taken from [15]. For the sample containing 9.09 wt.% ZnO as a known internal standard we used a structure model from the ICSD database (no 65120).
As additional parameters the zero point error, the sample displacement error and a polynomial function for background modelling were refined.

6. Discussion

The results of the refinements of the simulated datasets are graphically presented in figure 3. The estimated standard deviation is smaller than the graphical symbol for all points. It can be seen that the refined values for the stacking probability are slightly underestimated. The refinement of the interlayer carbonate occupancy led to slightly overestimated values. Although the misfit is larger than the estimated standard deviation, the refinements are in good agreement with the predefined values.

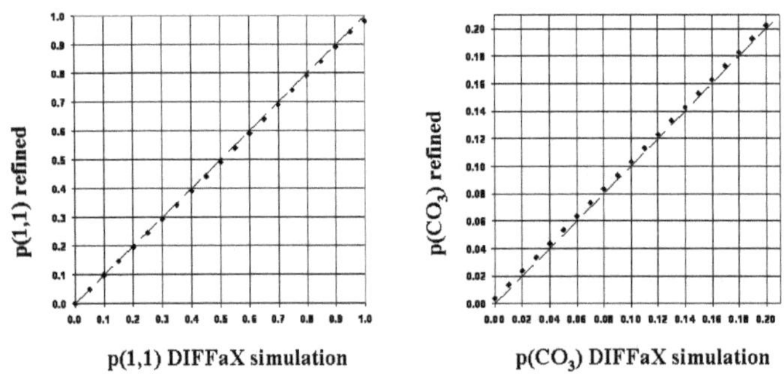

Figure 3. Comparison of parameters used for DIFFaX simulations and values obtained from the Rietveld refinements. Left: p(1,1). Right: p(CO_3).

In view with the fact that we are comparing simulated data of two calculation methods, the agreement of the refined values is satisfactory. With both methods, the DIFFaX calculations and the Rietveld method, structure factors are calculated by the same recursive treatment of diffraction, although the generation of the powder diffractograms is slightly different. Therefore small mismatches are unavoidable.

Table 1. Refinement results of Mg-Al-Sm-Cl LDH. ESD: estimated standard deviation.

	LDH + $Sm(OH)_3$		**LDH + $Sm(OH)_3$ + ZnO**	
	Refinement	**ESD**	**Refinement**	**ESD**
R_{wp} [%]	4.95		5.21	
R_{exp} [%]	1.32		2.35	
a [Å]	3.071511	0.000068	3.07218	0.00023
c [Å]	7.99770	0.00036	8.00664	0.00089
p(1,1)	0.5566	0.0013	0.5485	0.0015
p(Sm)	0.1[1]		0.1[1]	
p(Mg)	0.8366	0.0099	0.813	0.012
p(H_2O)	0.1110	0.0012	0.1251	0.0012
dispCL	0.20013	0.00052	0.19576	0.00061
LDH [wt.%]	95.190	0.028	95.291[2]	0.029
$Sm(OH)_3$ [wt.%]	4.810	0.028	4.709[2]	0.029
ZnO [wt.%]	-	-	10.122	0.051

[1]constraint value reached [2]calculated as free of standard admixture

Figure 4 shows the refined powder pattern of the Mg-Al-Sm-Cl LDH. The refined parameters are listed in table 1. The pure LDH pattern is shown separately. 95.2 wt.% LDH and 4.8 wt.% $Sm(OH)_3$ were found. The recovery of the added internal standard led to 10.1 wt.%.

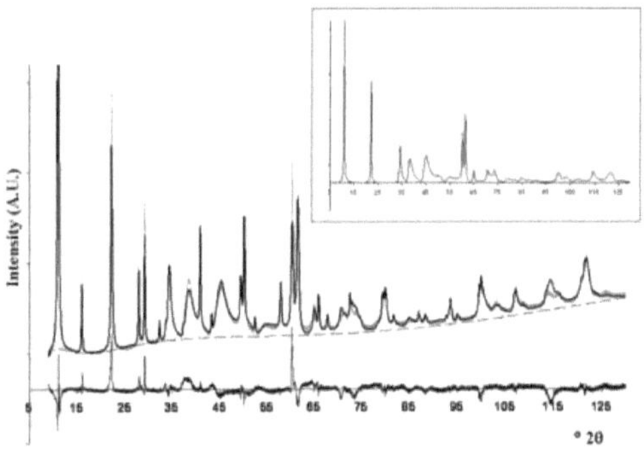

Figure 4. Refinement of Mg-Al-Sm-Cl LDH sample. In the box the pure LDH pattern from the refinement is shown. Cu $k\alpha_{1,2}$ radiation.

The structural parameters of the two refinements are in good agreement. The refined value of the stacking probability p(1,1) = 0.55 indicates that this Mg-Al-Sm-Cl LDH shows a high degree of disordering slightly favouring of $3R_1$ stacking. A model of just two stacking vectors in an infinite crystal certainly is a simplification. The refinement could be improved by introducing additional stacking vectors, turbostratic stacking, or size effects.

From the initial weight, the Sm occupancy was calculated as 0.1, regarding the material as pure LDH. Considering 4.8 wt.% $Sm(OH)_3$ impurity, the Sm occupancy is reduced by about 70 %. This is clearly in disagreement with the refined occupancy of 0.1. The results from the refinement of simulated data showed that it is possible to refine occupancies in the presence of stacking disorder. However, the limits of the Rietveld method are clearly reached when three different occupancies on the same crystallographic position have to be refined.

7. Conclusion

The refinements of the simulated datasets show that the recursive treatment of diffraction of planar faulted stacking was successfully implemented in the Rietveld program BGMN. Disordering parameters and ideal structure parameters can now be refined simultaneously. The applicability was demonstrated by a refinement of, on the one hand, disordering parameters and, on the other hand, quantities of disordered layered double hydroxides in a mixture with a reference material.

References

1. Treacy, M.M.J., Newsam, J.M. & Deem, M.W., 1991, *Proc. R. Soc. London*, **A433,** 499.
2. Rietveld, H. M., 1967, *Acta Cryst.*, **22**, 151.
3. Ufer, K., Roth, G., Kleeberg, R., Stanjek, H., Dohrmann, R. & Bergmann, J., 2004, *Z. Kristallogr.*, **219,** 519.
4. Drits, V.A. & Tchoubar, C., 1990, *X-ray Diffraction by Disordered Lamellar Structures*. (New York: Springer), p. 371.
5. Leoni, M., Gualtieri, A.F. & Roveri, N., 2004, *J. Appl. Cryst.*, **37**, 166.
6. Casas-Cabanas, M., Rodríguez-Carvajal, J. & Palacín, M.R., 2006, *Z. Kristallogr. Suppl.*, **23,** 243.
7. Radha, A.V., Shivakumara, C. & Kamath, P.V., 2005, *Clays Clay Min.*, **53**, 520.
8. Bergmann, J., Friedel, P. & Kleeberg, R., 1998, *Commission of Powder Diffraction. International Union of Crystallography CPD Newsletter,* **20,** 5.
9. Mazeina, L., Curtius, H. & Fachinger, J., 2003, *Clay Min.*, **38** 35.
10. Stumpf, T., Curtius, H., Walther, C., Dardenne, K., Ufer, K. & Fanghänel, Th., 2007, *Environ. Sci. Tech.*, **41**, 3186.
11. Bookin, A.S. & Drits, V.A., 1993, *Clays Clay Min.*, **41**, 551.
12. Curtius, H. & Ufer, K., 2007, *Clays Clay Min.*, **55**, 354.
13. Plançon, A., 2002, *J. Appl. Cryst.*, **35**, 377.
14. Ennadi, A., Legrouri, A., De Roy, A. & Besse, J.P., 2000, *J. Solid State Chem.*, **152,** 568.
15. Mullica, D.F., Milligan, W.O. & Beall, G.W., 1979, *J. Inorg. Nucl. Chem.*, **41**, 525.

Acknowledgements. We are grateful to the Federal Institute for Geosciences and Natural Resources (BGR) for financial support.

Microstructural characterisation of Cr-Al-N nanocomposites deposited by cathodic arc evaporation

Ch. Wüstefeld[1,*], M. Dopita[1], V. Klemm[1], D. Heger[1], D. Rafaja[1]

[1] Institute of Materials Science, TU Bergakademie Freiberg, D-09599 Freiberg, Germany
[*] wuestefeld@ww.tu-freiberg.de

Keywords: powder diffraction, line broadening, nanocrystalline materials, microstructural defects, nanocomposites

Abstract. The Cr-Al-N nanocomposites with a variable [Cr]/([Cr]+[Al]) ratio were deposited by cathodic arc evaporation. The investigations using XRD and TEM were done to analyse the microstructure of the coatings. The XRD phase analysis showed that Cr-rich coatings up to [Cr]/([Cr]+[Al]) = 0.54 consisted of a single fcc phase. With increasing Al-content, wurtzitic AlN developed as a second phase. The Cr-Al-N nanocomposites are composed of nano-sized clusters that consist of partially coherent nanocrystallites. This microstructural model was obtained from the combination of the XRD line profile analysis and TEM. The mutual misorientation of the fcc nanocrystallites within the clusters was between 0.5° and approx. 1° as calculated from the onset of their partial coherence. Detailed TEM investigations performed on Cr-rich coatings revealed that in the single phase coatings the nano-sized clusters are separated by dislocation walls. The dislocations building the walls could be identified as screw dislocations.

1. Introduction

The hard and ultra-hard coatings based on the Cr-Al-N system are an appropriate alternative to the Ti-Al-N and Ti-Al-Si-N system. Due to their properties e.g. high hardness, good adhesion, temperature stability and oxidation resistance, Cr-Al-N coatings are used as wear protective coatings with high oxidation resistance e.g. for punching of perforated sheets [1] or high end spindle bearings [2]. Furthermore Cr-Al-N coatings are considered as a promising candidate for advanced high speed machining applications [3]. The nanocomposites, like Ti-Al-N and Cr-Al-N, are characterised by a decomposition process where Al segregates from the fcc-(Ti,Al)N or fcc-(Cr,Al)N phase respectively and forms wurtzitic w-AlN [4-6]. With the following investigations we want to find out to which extend the phase segregation is responsible for the formation of nanocrystallites in the Cr-Al-N nanocomposites and what is the driving force for the formation of nanocrystallites in single-phase Cr-Al-N coatings.

2. Experimental details

The Cr-Al-N coatings under study were deposited using the cathodic arc evaporation in nitrogen atmosphere from two laterally rotating arc cathodes; one was made of chromium, the second one of aluminium. The coating process was carried out in the coating chamber π-80 from PLATIT [7]. The base pressure was 5×10^{-3} Pa and the working pressure was 1.3 Pa. The ion current on the chromium cathode was 80 A and the ion current on the aluminium cathode was 120 A. The bias voltage was -75 V and the deposition temperature was approximately 450 °C. Seven polished plates of cemented carbide were mounted on the substrate holder having different distances and angles to the two cathodes as shown in figure 1. During the deposition, the substrates were not rotated. Hence seven samples with a variable [Cr]/([Cr]+[Al]) ratio as well as a variable coating thickness (see figure 2) were produced in one coating process and the preferred orientation of the crystallites was not superimposed by the sample rotation.

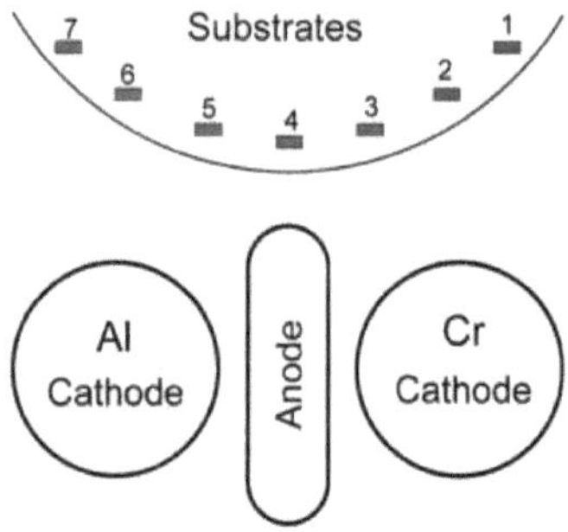

Figure 1. Illustration of the positions of the samples in the deposition chamber.

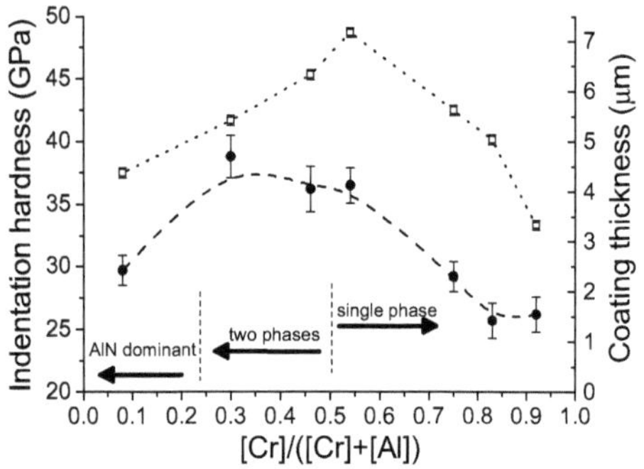

Figure 2. Hardness (●) and thickness (□) of the Cr-Al-N coatings.

The electron probe microanalysis with wavelength-dispersive spectroscopy (EPMA/WDS) was applied to determine the overall chemical content of Cr, Al, N and O in the coatings in 40 points across each sample. The nitrogen concentration was 50 at.% within the experimental accuracy in all samples. As the spectral signal from oxygen was below the (overall) background intensity, the maximum oxygen contents in the samples was estimated to be lower than 0.1 at.%. The chemical compositions of the coatings are given in figure 3.

The phase compositions of the coatings, the stress-free lattice parameters, the size of crystallites, the size of clusters of partially coherent crystallites and the mutual misorientation of the partially coherent crystallites in the clusters were determined by X-ray diffraction (XRD). For the XRD experiments, the glancing angle X-ray diffraction (GAXRD) geometry with a constant angle of incidence of 3° was chosen in order to reduce the penetration depth of the radiation and to keep the penetration depth nearly constant with increasing diffraction angle [8]. The GAXRD experiments were performed on a D8 diffractometer from Bruker AXS using copper radiation. The D8 diffractometer was equipped with a parabolic Göbel mirror in the primary beam. A Soller collimator with the acceptance of 0.12° and a flat LiF monochromator, which changed the $K_{\alpha 2}/K_{\alpha 1}$ intensity ratio to 0.08, were situated in the diffracted beam in front of the scintillation detector.

For the calculation of the stress-free lattice parameter, the position of the (200) line of the fcc phase was determined with the classical $\sin^2\psi$-method on a PTS 3000 X-ray diffractometer from Seifert equipped with an Eulerian cradle and with a mono-capillary in the primary beam. The texture measurements were also performed on this diffractometer. The pole figures (111), (100) and (110) were measured for the fcc-(Cr,Al)N phase and a background intensity correction was done [9].
HRTEM investigations were done on a 200kV analytical high-resolution TEM JEM 2010 FEF from Jeol equipped with ultra-high-resolution objective lens (C_s=0.5mm) and an in-column energy filter which selects only the elastic electrons for the image formation. HRTEM was used to confirm the crystallite and the cluster size. The analysis of the dislocation structures was done using the diffraction contrast in the bright field image. The samples for HRTEM, which were prepared in plane-view orientation, were mechanically pre-thinned and subsequently etched by ion beam. Finally, plasma cleaning was done on the specimens.
The hardness of the coatings was obtained from the indentation experiments done with a computer-controlled Nano Indenter XP from MTS. Continuous stiffness measurements [10] using the sinus mode, where a small oscillation force is added to a quasi-static force, were performed for the measurement of the hardness as a function of the penetration depth up to a maximal depth of 1200 nm. The force, the displacement amplitude and the phase angle between these two signals were measured at any time and thus the hardness could be obtained as a function of the penetration depth. Twenty indentations were done on a polished site; the indentation hardness given in figure 2 was calculated from the values obtained in indentation depths from 50 to 100 nm, where the hardness was affected neither by the surface roughness nor by the substrate.

3. Experimental results

3.1 Phase composition and the stress-free lattice parameters

The GAXRD patterns given in figure 3 show the development of the phase composition in all samples. In the coatings with $[Cr]/([Cr]+[Al]) \leq 0.46$, w-AlN develops as a second phase. The third phase which is obvious in the diffraction pattern is tungsten carbide from the substrate. The formation of a second phase containing Al can also be concluded from the dependence of the stress-free lattice parameter on [Cr]/([Cr]+[Al]). Low Al contents in the Cr-Al-N coatings lead to the reduction of the lattice parameter according to the Vegard rule. In our previous paper [11] a function describing the dependence of the stress-free lattice parameter of the fcc crystal on the stoichiometry ratio of Al and Si was presented. As shown in figure 4, for the Cr-rich samples the stress-free lattice parameter depends linearly on the [Cr]/([Cr]+[Al]) ratio which was measured in the whole coating by EPMA/WDS. At higher Al contents and respectively lower [Cr]/([Cr]+[Al]) ratios (≤ 0.46) the stress-free lattice parameter is higher than expected from the linear Vegard-like dependence of the stress-free lattice parameter on the Cr-content in the films. This departure indicates the formation of a second phase containing Al, because Al segregates from the host fcc structure and forms wurtzitic AlN. The stress-free lattice parameters presented in figure 4 were calculated using the $\sin^2\psi$ method for the 200 reflection as done in Ref. [12] for the Cr-Al-Si-N coatings. For uniaxial residual stress the lattice parameter of cubic materials depends linearly on

$\sin^2\psi$, see equation (1). Calculating $a_{\perp}^{hkl}$ and $a_{\parallel}^{hkl}$ for ψ=0° and ψ=90°, respectively, a_0 can be obtained using equation (2) with $\nu^{200} = 0.214$ taken from [13].

$$a_{\psi}^{hkl} = (a_{\parallel}^{hkl} - a_{\perp}^{hkl})\sin^2\psi + a_{\perp}^{hkl} \tag{1}$$

$$a_0^{200} = (a_{\parallel}^{200} - a_{\perp}^{200})\frac{2\upsilon^{200}}{\upsilon^{200}+1} + a_{\perp}^{200} \tag{2}$$

The stress-free lattice parameter of the coatings with the [Cr]/([Cr]+[Al]) ratio from 0.92 to 0.3 shown in figure 4 were calculated in this way, whereas the lattice parameter at [Cr]/([Cr]+[Al])=1 was taken from the ICSD database [14]. The previously described way of the calculation of the stress-free parameter was chosen due to elastic anisotropic behaviour of our coatings which is one topic in Ref. [15]. Nevertheless we had to disregard the dependence of the Poisson's ratio on the chemical composition because of unknown data.

3.2 Preferred orientation of crystallites

Texture analysis of the fcc phase in the coatings up to [Cr]/([Cr]+[Al]) $\geq$ 0.30 showed that the <111> crystallographic direction was nearly perpendicular to the sample surface. The maximum inclination from the sample normal was approx. 5°and appeared in $Cr_{0.30}Al_{0.70}N$. The (200) and (220) pole figures revealed that the fcc crystallites have also a preferred orien tation in the plane of the coating. This "three-dimensional" texture was also observed in the Cr-Al-Si-N [12] and Ti-Al-N [16] coatings. The preferred orientation of the Cr-Al-N and Cr-Al-Si-N nanocomposites is described in detail in Ref. [9].

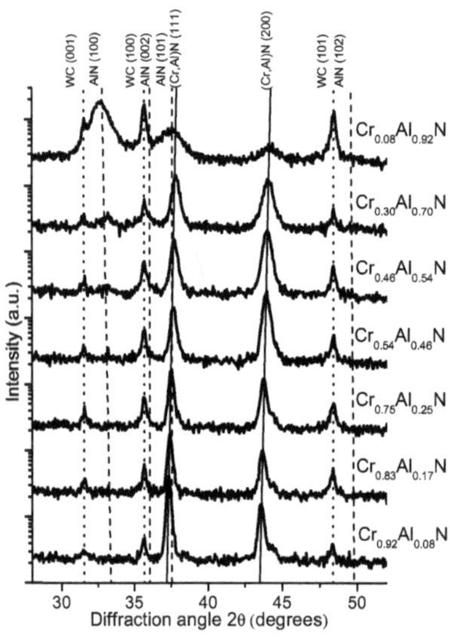

Figure 3. Parts of the GAXRD patterns of the Cr-Al-N coatings.

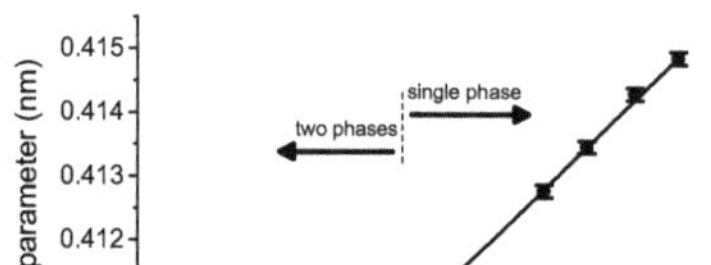

Figure 4. Dependence of the stress-free lattice parameters of the fcc crystal on the Cr-stoichiometry ratio. The solid line shows the calculated lattice parameter according to the Vegard-like dependence presented in [11].

3.3 The crystallite and cluster size

Plotting the XRD line broadening in the fcc phase vs. $\sin\theta$ showed a similar curve for all investigated samples: For small $\sin\theta$ the line broadening increased slightly up to $\sin\theta = 0.6...0.8$, then the line broadening increased steeply and saturated at high diffraction

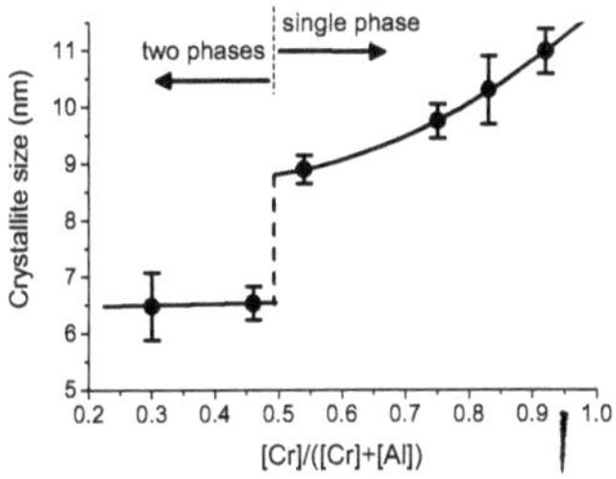

Figure 5. Dependence of the crystallite size of the fcc phase on the chromium stoichiometry ratio of the investigated coatings.

angles, e.g. as shown for the example of $Cr_{0.92}Al_{0.08}N$ in Ref. [17] in this issue. This behaviour of the line broadening was also observed for Ti-Al-N, Ti-Al-Si-N and Cr-Al-Si-N nanocomposites [12, 18] and could be explained by the partial coherence of slightly disoriented nanocrystallites [19]. Partial coherence was supported by the observed in-plane texture of the fcc crystallites. The existence of the partial coherence in the investigated coatings means that the neighbouring nanocrystallites in the Cr-Al-N nanocomposites under study are strongly preferentially oriented, which was explained by the fact that they build nano-sized clusters. This was proven by TEM and XRD. Considering the theory of partial coherence [17,19], the size of the nanocrystallites and the clusters as well as the mutual misorientation of the nanocrystallites can be calculated from the dependence of the line broadening on $\sin\theta$. The size of nanocrystallites that was calculated from the line broadening at high diffraction angles decreased with increasing Al content (see figure 5). For the Cr-Al-N coatings consisting of a single fcc phase, a slight reduction of the crystallite size was observed with increasing Al content. Whereas a steep decrease of the crystallite size appeared in the coatings containing fcc-(Cr,Al)N and w-AlN. A HRTEM micrograph of the nanocrystallites in the $Cr_{0.30}Al_{0.70}N$ coating is given in figure 6. The nanocrystallites form nano-sized clusters whose size could be calculated from the extrapolated XRD line width at $\sin\theta = 0$. An example of such a cluster having a size of ≈ 60 nm is shown in figure 7.

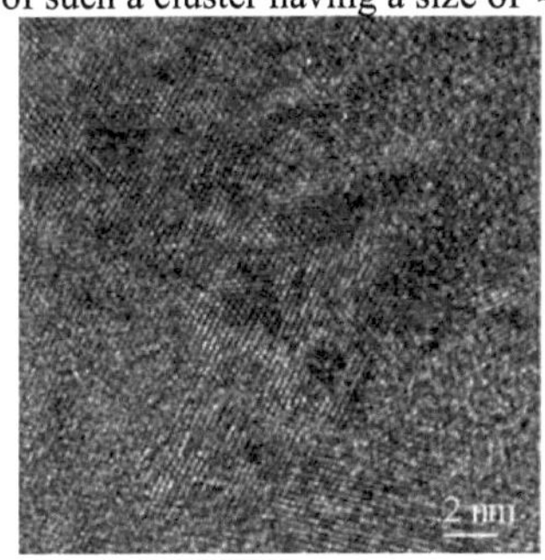

Figure 6. HRTEM image of the sample $Cr_{0.30}Al_{0.70}N$ showing partially coherent nanocrystallites.

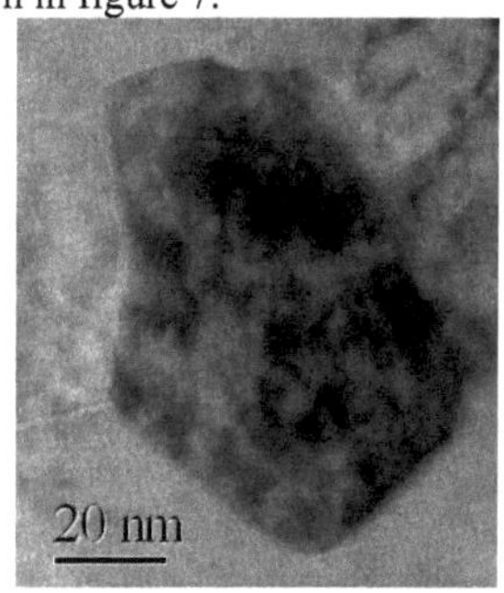

Figure 7. TEM picture of a nano-sized cluster observed in the $Cr_{0.75}Al_{0.25}N$ coating.

3.4 Hardness of the coatings

The hardness of the coatings (see figure 2) obtained by continuous stiffness measurements using the sinus mode are in good agreement with the hardness values received from the hardness measurement given in [11]. Comparing figure 5 and 2, the reduction of the crystallite size with increasing Al content can be assigned to the increase of the hardness. Due to this finding we can conclude that the hardness of the coatings can be enhanced by small crystallites as it was observed for other nanocomposites [20]. Although the coating thickness was not constant in all samples the hardness should not be influenced by the coating thickness because the highest hardness was not observed in the thickest coating and in addition the hardness was obtained from a small penetration depth (50 to 100 nm) in comparison to the total coating thickness where the coating hardness did not depend on the indentation depth. The highest hardness was observed for $Cr_{0.30}Al_{0.70}N$ which could be attributed to the inter-twining of the fcc and wurtzitic phase resulting in the formation of local intrinsic stresses at the crystallites boundaries as explained by the microstructure model in Ref. [18]. Due to the predominance of the softer w-AlN, the hardness for the $Cr_{0.08}Al_{0.92}N$ coating was lower than the coatings under study which contained two phases.

3.5 Dislocation structure in the single-phase $Cr_{0.92}Al_{0.08}N$ coating

The analysis of the XRD line broadening showed a slight disorientation of the nanocrystallites within the clusters, which was calculated from the position of the steep increase in the line broadening vs. sin θ plot. The mutual misorientation rose slightly from 0.5° up to ≈ 1° with increasing Al contents as presented in [11]. For single phase materials these small misorientations α are caused by dislocation walls which can be described by equation (3) [21]:

$$\vec{B} = 2\left(\vec{r} \times \vec{l}\right) \cdot \sin\frac{\alpha}{2} \tag{3}$$

where $\vec{B}$ represents the sum of the Burgers vectors of the dislocations which are intersected by a vector $\vec{r}$ lying in the dislocation wall. The unit vector $\vec{l}$ lies in the dislocation wall and is parallel to the rotation axis of the neighbouring crystallites. The presence of dislocations in our single-phase coatings could be proven by TEM investigations of the sample $Cr_{0.92}Al_{0.08}N$. The TEM micrographs in figure 8 were taken in the two-beam diffraction condition with $[\bar{1}\,\bar{1}\,1]$ as beam direction. Figure 8a shows three dislocations building a dislocation network with a distance of 22 nm. The TEM image from figure 8b shows a dislocation wall between area 1 and 2 which is oriented perpendicular to the diffraction vector $(2\bar{2}0)$. Nanodiffraction in area 1 and 2, respectively, showed that one direction remained the same in both diffraction patterns. This direction could be identified as $[2\bar{2}0]$ and represents the rotation axis causing misorientation between the area 1 and 2. The rotation axis and the diffraction vector are parallel to each other so that the diffraction contrast does not change in this area. The dislocation wall and the rotation axis are perpendicular, see diffraction pattern of area 1 in figure 8b. Therefore, it can be assumed that the dislocation wall consists of screw dislocations and hence the rotation axis is a twist axis. Regarding the TEM micrograph from figure 8a and assuming screw dislocations, it is possible to calculate the misorientation produced by this dislocation wall using equation (3) rewritten in the following form: $\left|\vec{b}\right| \approx 2\left|\vec{r} \times \vec{l}\right| \cdot \sin\frac{\alpha}{2}$ (4)

where $\vec{b}$ is the Burgers vector which is $a/2<110>$ for fcc materials supported by TEM investigations and a is the lattice parameter which is 0.4145 nm as shown in figure 4. The term $|\vec{r}\times\vec{l}|$ represents the projection of the distance between dislocations into the imaging plane of TEM which is 22 nm for our example (see figure 8a). The misorientation α calculated by equation (4) is approx. $0.75°$ and is very close to the misorientation obtained from the analysis of the XRD line broadening, which was 0.54° [11].

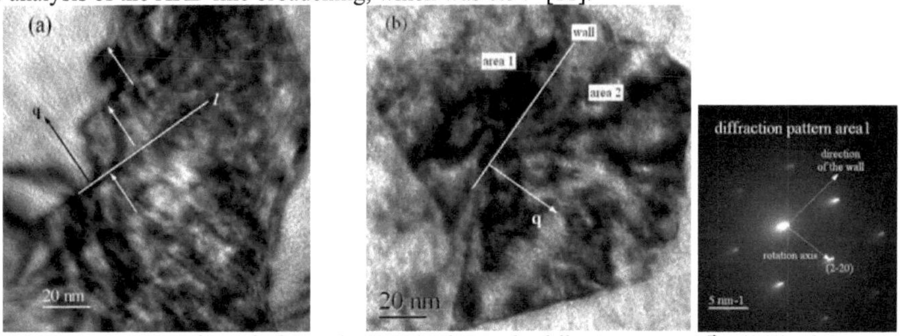

Figure 8. TEM images of sample $Cr_{0.92}Al_{0.08}N$ (a) taken at the diffraction vector $\vec{q} = (242)$ showing three dislocations (white arrows) with unit vector $\vec{l} = 1/\sqrt{2}\,(101)$. (b) showing a dislocation wall which is perpendicular to the diffraction vector $\vec{q} = (2\bar{2}0)$.

4. Summary and discussion

XRD phase analysis gave the evidence of the formation of fcc-(Cr,Al)N in all samples and w-AlN at $[Cr]/([Cr]+[Al]) \leq 0.46$. The phase segregation was also confirmed by a departure of the stress-free lattice parameter from the linear Vegard-like dependence that was expected for fcc-$Cr_{1-x}Al_xN$. Partial coherence of fcc crystallites in the coatings was observed by the analysis of the XRD line broadening. These partially coherent nanocrystallites formed nano-sized clusters. The size of the nanocrystallites decreased with increasing Al content in the coatings. The presence of dislocation walls separating the nanocrystallites as observed by TEM for the single-phase $Cr_{0.92}Al_{0.08}N$ coating could be the result of plastic deformation of the clusters probably during the deposition which leads to the decrease of the crystallite size in single-phase coatings. The crystallite size is substantially smaller in the coatings containing fcc-(Cr,Al)N and w-AlN than in the single phase coatings containing only the fcc phase. We assume that this is a result of the phase segregation during the deposition process, during which w-AlN develops at the boundaries of the fcc crystallites.

References

1. Spain, E., Avelar-Batista, J.C., Letch, M., Houdsen, J. & Lerga, B., 2005, in *Surf. Coat. Technol.*, **200**, p.1507.
2. Brecher, C., Spachtholz, G., Bobzin, K., Lugscheider, E., Knotek, O. & Maes, M., 2005, in *Surf. Coat. Technol.*, **200**, p.1738.
3. Ding, X-Z., Zeng, X.T., Liu, Y.C., Fang, F.Z. & Lim, G.C., et. al., 20078, in *Thin Solid Films*,. doi:10.1016/j.tsf.2007.05.019.**516**, 1710.

4. Sugishima, A., Kajioka, H. & Makino, Y., 1997, in *Surf. Coat. Technol.*, **97**, p.590.
5. Vepřek, S., Männling, H.-D., Jílek, M. & Holubář, P., 2004, in *Mater. Sci. Eng.*, **A366**, p.202.
6. Rafaja, D., Šíma, M., Klemm, V., Schreiber, G., Heger, D., Havela, L. & Kužel, R., 2004, in *J. Alloys Compd..*, **378**, p.1107.
7. Holubář, P., Jílek, M. & Šíma, M., 2000, in *Surf. Coat. Technol.*, **133/134**, p. 145.
8. Rafaja, D., Valvoda, V., Perry, A.J. & Treglio, J.R., 1997, in *Surf. Coat. Technol.*, **92**, p. 1135-141.
9. Dopita, M., Rafaja, D., Wüstefeld, Ch., Růžička, M., Klemm, V., Heger, D., Schreiber, G. & Šíma, M., *Surf. Coat. Technol.*, submitted.
10. Oliver, W.C. & Pharr, G.M., 1992, in *Mater. Res.*, **7** (4), p. 1564.
11. Rafaja, D., Wüstefeld, Ch., Dopita, M., Růžička, M., Klemm, V., Schreiber, G., Heger, D. & Šíma, M., 2007, in *Surf. Coat. Technol.*, **201**, p. 9476.
12. Rafaja, D., Dopita, M., Růžička, M., Klemm, V., Schreiber, G., Heger, D. & Šíma, M., 2006, in *Surf. Coat. Technol.*, **201**, p. 2835.
13. Attar, F. & Johannenesson, T., 1995, in *Thin Solid Films*, **258**, p.205.
14. Inorganic Crystal Structure Database, ICSD on CD-ROM, Version 1.4.1, 2006, FIZ Karlsruhe, 2006.
15. Dopita, M., Rafaja, D., Wüstefeld, Ch., Klemm, V. & Schreiber, G., 2007, *Z. Kristallogr.aphie*, this issue.
16. Rafaja, D., Poklad, A., Schreiber, G., Klemm, V., Heger, D. & Šíma, M., 2005, in *Z. Metall.kd.*, **95**, p. 736.
17. Rafaja, D., Klemm, V., Wüstefeld, Ch., Motylenko, M., Dopita, M., Schwarz, M., Barsukova, T. & Kroke, E., 2007, *Z. Kristallogr.aphie*, this issue.
18. Rafaja, D., Poklad, A., Klemm, V., Schreiber, G., Heger, D., Šíma, M. & Dopita, M., 2006, in *Thin Solid Films.*, **514**, p. 240.
19. Rafaja, D., Klemm, V., Schreiber, G., Knapp, M. & Kužel, R., 2004, in *J. Appl. Crystallogr.*, **37**, p. 613.
20. Vepřek, S., Reiprich, S. & Shizhi, L., 1995, in *Appl. Phys. Lett.*, **66** (20), p.2640.
21. Frank, F..C, 1950, in *Report on the Symposium on Plastic Deformation of Crystalline Solids*, Carnegie Institute of Technology and Office of Naval Research, U.S. Government Printing Office, Washington, p.150.

Acknowledgements

The authors would like to thank SHM Ltd. for the coatings deposition and Dr. T. Modes from the FEP Dresden for the indentation measurement. The authors appreciate the financial support of the project #RA-1050/9 through the German Scientific Council (DFG). HRTEM Jeol 2010FEF was purchased from the funds of the DFG Priority Programme # 1062.

MATERIALS MICROSTRUCTURE

Characterization of hot wall grown N-9-anthracenylidene-1-anthramine films

R. K. Bedi[1,*], S. S. Brar[1], N. Kaur[2], S. Kumar[2], A. Mahajan[3]

[1] Material Science Research Laboratory, Department of Physics, Guru Nanak Dev University, Amritsar, 143005, India
[2] Department of Chemistry, Guru Nanak Dev University, Amritsar, 143005, India
[3] Postgraduate Department of Physics & Electronics, DAV College, Amritsar, 143 001, India
[*] rkbedi@rediffmail.com

Keywords: N-9-anthracenylidene-1-anthramine, thin films, structure, electrical properties, optical properties

Abstract. The compound N-9-anthracenylidene-1-anthramine has been synthesized and films of thickness 1.1 μm were grown by Hot Wall setup on glass substrate kept at different temperatures in vacuum of 10^{-5} torr. The experimental conditions were optimized to obtain better crystallinity of the samples. Nuclear magnetic resonance (NMR) and infrared (IR) spectra of samples obtained from the films appear to be same as that of pure starting compound. Observations reveal that the current-voltage characteristics of the films show ohmic behaviour within the investigated field and temperature range and the conduction appears to take place by thermally activated hopping mechanism. The film deposited at higher substrate temperature shows comparatively more intense X-ray diffraction peaks suggesting a higher degree of crystallinity. Sharp cuboid crystallites as wide as 0.7 μm are obtained for films deposited at 340 K. The optical absorption studies indicate that the inter band transition energy of films prepared at different substrate temperatures has been found to lie in range 2.33 -2.41eV.

1. Introduction

The development of molecular electronic devices requires the efficient construction of effective molecular architectures, which are obtained form synthetic materials capable of directional and long range electron transport. The important control of both electronic and structural features of the substituents on the individual molecules in forming efficient structural architectures and the use of different techniques in developing efficient materials (thin films, wires etc.) has achieved paramount significance. Organic semiconductors are an emerging class of electronic materials that offer interesting properties for low cost organic devices [1-4]. Organic compound derived from anthracene have attracted a great deal of interest in recent years because of their optical and semiconducting properties [5-10]. The organic compounds derived from anthracene are used in dye sensitized nanocrystalline solar cells, sensors, diodes and organic thin film transistors (OTFTs) [11-15].

A number of potential devices which may arise from anthracene depend upon the fabrication of the anthracene as thin films. Accordingly, there is continuing interest in depositing these compounds in this type of formulation, with research focused on the design of suitable derivatives for providing ordered and reproducible films. The growth and properties of derivatives of some organic films deposited under different experimental conditions have been reported [16, 17] and found that the improvement in grain size and reduced structural defects occurs when deposition is done under thermodynamical equilibrium conditions. Hot Wall technique studied by Lopez Otero [18] permits film deposition near thermodynamical equilibrium.

In this paper we report the synthesis, structural, electrical and optical properties of N-9-anthracenylidene-1-anthramine films prepared by hot wall set up onto the glass substrate kept at different temperatures.

2. Experimental

2.1 General Procedure

^{1}H Nuclear magnetic resonance (NMR) spectra have been recorded on a Bruker AC 200 MHz machine by using tetramethylislane as reference. IR spectra were taken in absorption mode in the 500-4000cm^{-1} range using FTIR 8300 (Shimadzu, Japan) spectrophotometer.

2.2 Synthesis of N-9-anthracenylidene-1-anthramine compound

9-Anthraldehyde (495 mg, 2.4 mmol) was added to clear solution of 1-aminoanthracene (386 mg, 2 mmol) in ethanol (25 ml). The mixture was stirred at 333 K for 6-8 hours. During this period yellow solid was separated. The completion of the reaction was noted by TLC. After completion of the reaction, the solid separated was filtered and was crystallized from ethanol to give pure N-9-anthracenylidene-1-anthramine yellow solid with melting point 508 K. The yield of this process was 80%. FAB mass spectra shows M$^+$ *m/z* 382 (M$^+$+1).^{1}H NMR (CDCl$_3$) data of N-9-anthracenylidene-1-anthramine : δ 7.43-7.65 (m, 8H, ArH), 7.96-8.05 (m, 3H, ArH), 8.11 (d, J = 7.5 Hz, 2H, ArH), 8.49 (s, 1H, ArH), 8.65 (s, 1H, ArH), 8.95 (s, 1H, ArH), 8.98 (s, 2H, ArH), 9.89 (s,1H, CH); ^{13}C NMR (normal / DEPT-135) (CDCl$_3$): δ 111.37 (CH), 123.24 (CH), 123.54 (C), 124.89 (CH), 125.35 (CH), 125.49 (C), 125.52 (CH), 125.71 (CH), 126.04 (CH), 126.46 (CH), 127.46 (CH), 127.95 (C), 128.78 (CH), 129.12 (CH), 130.96 (CH), 131.02 (C), 131.40 (C), 131.68 (C), 132.14 (C), 150.73 (C), 159.73 (CH). The structure of compound is shown in figure 1.

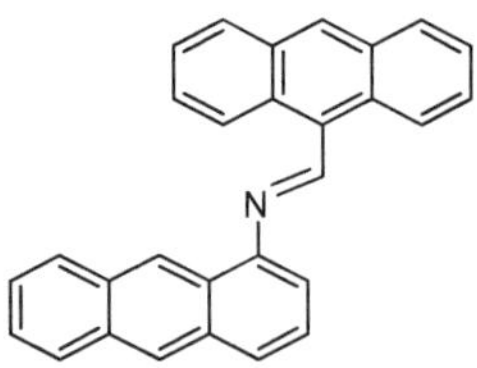

Figure 1. Structure of N-9-anthracenylidene-1-anthramine.

2.3 Preparation of films from N-9-anthracenylidene-1-anthramine compound

The N-9-anthracenylidene-1-anthramine powder was taken in a single-evaporation zone hot wall set up described earlier [19]. Two separate resistive windings of nicrome were used to heat the evaporation port and walls of the column. A radiant heater was used to heat the substrate in the temperature range 310 to 340 K. The temperatures of the different zones were controlled by PID temperature controllers using K-type thermocouple obtained from Omega Eng. (USA). Chemically and ultrasonically cleaned glass slides were used as substrate material. The whole apparatus was fitted onto the base plate of the vacuum chamber of Hind High Vacuum coating unit 12A4H. The chamber was then evacuated to pressure of order less than 10^{-5} Torr after mounting the substrate and making other desired adjustments. Before carrying out the deposition, the vacuum chamber was baked at 473 K for one hour. During the degassing, the substrate and wall heaters were also switched on. After attaining the desired experimental conditions, the source heater was switched on and the substrate holder was positioned on the top of the tube to make it a closed system. The temperature of the substrate is monitored by a copper constantan thermocouple. After depositing films at particular substrate temperature the system is allowed to attain room temperature (298 K) and then samples are removed for further studies.

2.4 Measurements

The thickness of the films was measured using depth profiler (Dektek 3030 - XT). To study the surface topography of films, the Hitachi ion Sputter coater E-1010 is used for gold coating and SEM of the films are taken using a Hitachi (S-3400N) scanning electron microscope with beam voltage 30kV. The XRD patterns of the films are recorded using Philips diffractrometer using Cu Kα (wavelength = 1.54 A°) and step size (°2θ) = 0.060 per sec in the 2θ range 5-70°. To study the optical properties, the transmittance and absorbance spectra of the samples were obtained in the photo energy range 1.12-4.13 eV by using UV-1601PC (Shimadzu, Japan) spectrophotometer. All structural and optical measurements were performed at room temperature (298 K). The electrical conductivity of the films are determined by Kithley electrometer 6517A using two probe technique in the temperature range 290 - 400 K under a pressure of 10^{-4} Torr. Silver contacts are used to connect thin copper wires to the films and verified to be ohmic from the symmetric straight line of I-V characteristics passing through the origin.

3. Results and discussion

3.1 Identification of N-9-anthracenylidene-1-anthramine films

The comparison of FTIR spectra of N-9-anthracenylidene-1-anthramine powder and sample scrapped from glass slides shows that the compound remains unaffected during vacuum deposition and annealing processes because of unchanged vibrational frequencies. Moreover, the comparison of ^{1}H NMR spectral data of the powder and sample obtained from film show that sample remained unchanged under thermal evaporation.

3.2 Structure characterization

The deposited film thickness was in the range of 1.0-1.1 μm. The scanning electron micrograph of N-9-anthracenylidene-1-anthramine film deposited at substrate temperature of 340 K is shown in figure 2. The surface topography in these micrographs delineated the

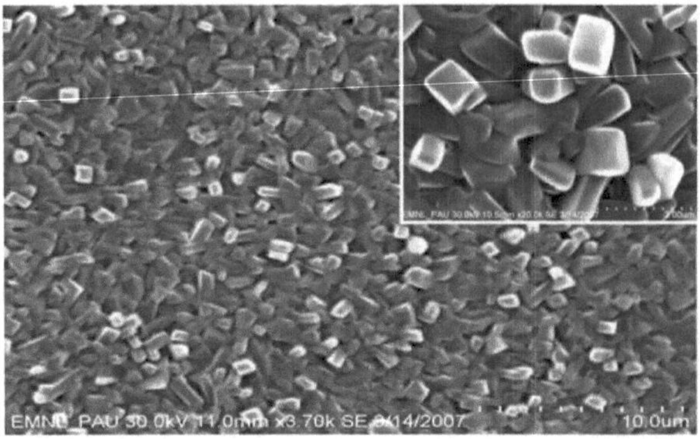

Figure 2. Surface topography of N-9-anthracenylidene-1-anthramine film deposited at substrate temperature 340 K (inset magnified view of few grains)

distribution, shape and size of crystallites. The SEM image depicts a unique microstructure with cuboidal crystallites. Crystallites as wide as 0.71μm are obtained for films deposited at 340 K. These crystallites are pointing along outward direction with respect to substrate plane. An appreciable increase in crystallite from 0.35 to 0.71 μm has been found with increase in substrate temperature from 310K to 340 K because of the fact that at higher substrate temperatures, crystallites agglomerate to form bigger crystallites. These SEM results can be understood in the light of Otero model of hot wall technique [18]. Here, the close tube mode upgrades the quality of films and favours the condition for the attainment of high quality crystals. The growth of the films near to thermodynamical equilibrium condition serves to provide crystals with the required degree of crystalline perfection. Moreover, at elevated substrate temperatures, the atomic surface mobility is large and large crystals with lower lattice faults are formed.

The X-ray diffractogram obtained for N-9-anthracenylidene-1-anthramine films (figure 3)

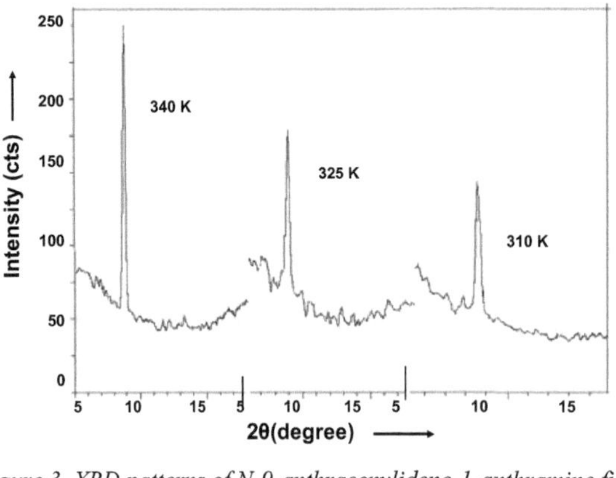

Figure 3. XRD patterns of N-9-anthracenylidene-1-anthramine films deposited at different substrate temperatures.

shows the presence of a sharp and well resolved diffraction peak, which confirms its crystalline behaviour. Typical results obtained from diffractrometer traces are given in table 1.

Table 1. XRD data of N-9-anthracenylidene-1-anthramine films

Substrate temperature(K)	Angle (° 2θ)	d-value(Å)	Peak intensity (Counts)
310	8.84	9.98	94
325	8.75	10.09	98
340	8.80	10.03	193

The d-values corresponding to diffraction peaks are calculated using Bragg's equation

$$2d\sin\theta = n\lambda \qquad (1)$$

for films. The films deposited at higher substrate temperature show comparatively intense diffraction peaks. Moreover, a decrease in full width measured at half maxima (FWHM) of diffraction peaks has been noticed, thus suggesting a higher degree of crystallinity and increase in particle size. Thus, the substrate temperature appears to be important parameters for

3.3 Electrical Properties

Figure 4 shows the Arrhenius plots of temperature dependence of electrical conductivity for the N-9-anthracenylidene-1-anthramine films deposited on to the glass substrate kept at

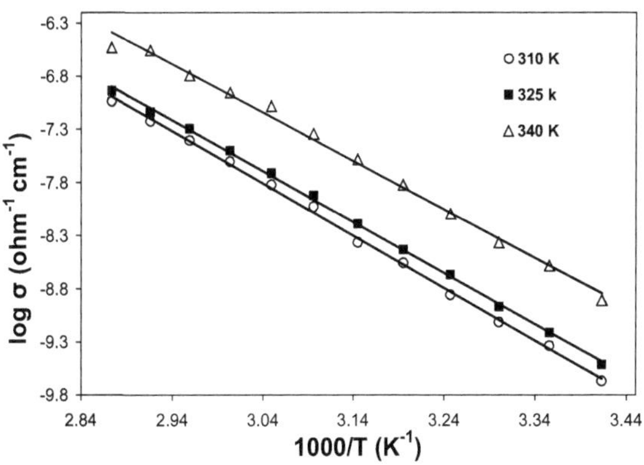

Figure 4. Variation of log of conductivity with inverse absolute temperature for N-9-anthracenylidene-1-anthramine films deposited at different temperatures

different temperatures. Observations reveal that, the conductivity of the films increases with increase in temperature, within the experimental range (290-400 K) investigated. The temperature dependence of log σ shows almost a straight line, indicating the applicability of the well-known Arrhenius relation

$$\sigma = \sigma_0 e^{(-\Delta E / 2kT)} \tag{2}$$

for these films, where ΔE is the activation energy and k is Boltzmann's constant.
It has been observed that the electrical conductivity increases with increase in the substrate temperature. Since the crystallites do not grow sufficiently large at low temperatures, the inter-crystalline regions are wide, offering a high resistance to the movement of charge carriers. The higher substrate temperatures results in large crystallites, which ultimately decrease the barrier size. The charge carriers have to cross narrow inter-crystalline barriers and therefore may cause increase in conductivity. Electrical conductivity of hot wall grown films found to be increase by an order of 10^1 approximately as compared to thermally evaporated films. The activation energy of charge carriers responsible for the conduction has been calculated from the slope of log of conductivity vs. 1000/T plots and is found to lie in 0.39-0.42 eV. It is found that the activation energy decreases with the increase in the substrate temperature, indicating the formation of more ordered films at higher substrate temperatures. These results are in good agreement with SEM and XRD pattern of these films.
Moreover, it has been suggested [20] that films of aromatic organic compounds can be considered as an array of potential wells separated by intermolecular barriers and here it appears that charge motion occurs by series of hops from one molecule to adjacent one. Electrical conductivity can be assumed to take place by thermally activated hopping, which can be expressed as equation (3)

$$\sigma = \sigma_o e^{\left(-\Delta W_{hop} / 2kT\right)} \tag{3}$$

where ΔW_{hop} is the hopping energy needed to surmount the barrier between two molecules and appears to be same as that of conduction activation energy which lies between 0.39-0.42 eV.

3.4 Optical properties

The optical absorption coefficient (α) of N-9-anthracenylidene-1-anthramine films is determined over the energy range 2.14 to 2.7 eV at room temperature 296K. This absorption region has been examined in terms of an interband transition and the dependence of absorption coefficient upon energy [20] of incident photon is given as equation (4)

$$\alpha h\nu = B\left(h\nu - E_g\right)^{1/2} \tag{4}$$

where B is the constant, E_g is the band gap and ν is the frequency of the incident light. The plot of square of the absorption coefficient vs. photon energy of the films prepared at different substrate temperatures is shown in the figure 5.

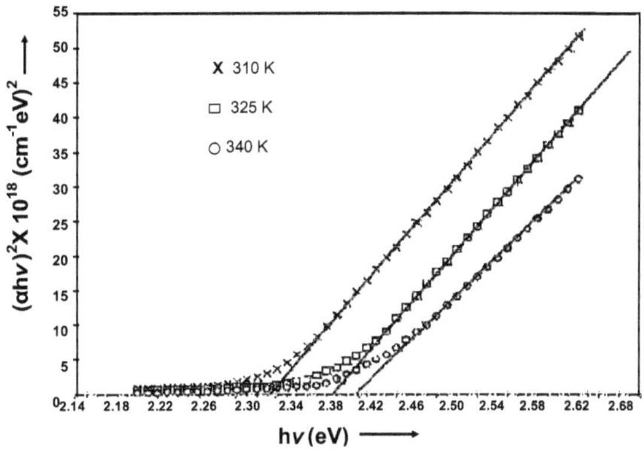

Figure 5. Spectral variation of N-9-anthracenylidene-1-anthramine films deposited at different temperatures films.

The extrapolation of the straight line to $\alpha h\nu = 0$ give the optical transitions which lie in the range 2.33-2.41 eV. The results indicate that band gap energy of films increases with the increase in the substrate temperature which may be due to the fact that the films deposited at higher substrate temperatures have large size crystallites, making it more ordered and thus leading to less absorption.

4. Conclusions

The comparison of IR and NMR spectra of N-9-anthracenylidene-1-anthramine films with compound show that the compound does not decompose during the Hot Wall deposition technique. The films are found to be crystalline in nature having cuboidal crystallites and crystals as wide as 0.71 μm are obtained for films deposited at 340 K. The substrate temperature appears to be important parameters for molecular packing in the solid state structure and influence the properties of the films. Films deposited at elevated temperatures show comparatively higher electrical conductivity and optical band gap. The optical band gap energy of film samples lie in the range 2.33-2.41 eV.

References

1. Mrwa, A., Friedrich, M. & Hofman, A.,1995, *Sens. Actuators B*, **24-25**, 596.
2. Suzuki, T., Kuwabara, Y., Nima, N., Sirota, Y. & Willis, M.R., 1996, *Jpn. J. Appl. Phys.*, **35,** L447.
3. Tominga, T., Hayashi, K. & Toshims, N., 1997, *Appl.Phys. Lett.*, **70**, 762.
4. Bach, V.,Lupo, D., Comte, P., Morse, J.E., Weissortel, F., Salbeck, J. & Gratysel, M., 1998, *Nature,* **395**, 583.
5. Joseph, Shinar, 1998, *Organic Light Emitting Devices* (Taylor & Francis Group).
6. Gupta, S.C., *Optoelectronic Devices and Systems* (New Delhi, India: Prentice Hall).

7. Shi, J. & Tang S.W., 2002, *Appl. Phys. Lett.*, **80**, 3201.
8. Tao, S., Hong, Z., Peng, Z., Lu, W., Wang, P., Wu, S. & Lee. S., 2004, *Chem. Phys. Lett.*, **397**, 1.
9. Lee, M., Chen, H., Liao, C., Tsai, C. & Chen, C.H., 2004, *Appl. Phys. Lett.*, **148**, 205.
10. Hara, K.,Tachibana, Y., Ohga, Y., Shinpo, A., Suga, S., Sayama, K., Sugihara, H., & Arakawa, H., 2003, *Solar Energy Materials and Solar Cells*, **77**, 89.
11. Staneva, D. & Betcheva , R., 2007, *Dyes and Pigments*, **74**, 148.
12. Qing, G.-Y., He, Y.-B., Chen, Z.-H., Wu, X.-J. & Meng, L.-I., 2006, *Tetrahedron*, **17**, 3144.
13. Jiang, X., Zhang, Z., Zheng, X., Wu, Y. & Xu, S., 2001, *Thin Solid Films*, **401**, 251.
14. Inoue, Y., Tokito, S., Ito, K. & Suzuki, T., *J. Appl. Phys.*, **95**, 5795.
15. Inour, Y., Suzuki, T., Ito, K. & Tokito, S., 2003, *IEEE International Conference on Consumer Electronic*, 230.
16. Mahajan , A., Bedi, R.K. & Kumar, S., 2001, *J. Appl. Phys.*, **89**, 7866.
17. Mahajan , A., Bedi, R.K., Pramila & Kumar, S., 2001, *Thin Solid Films*, **398**, 82.
18. Otero. L.A., 1978, *Thin Solid Films*, **49**, 1.
19. Singh, A., Bedi R.K., 2001, *Thin Solid Films*, **398-399**, 442.
20. Nasrallah, T.B., Bardeen, J.C. & Jamali, M., 1995, *Adv. Mater. Opt. Electron.*, **5**, 31.
21. Berdeen, J., Blatt, F.J. & Hall, L.H., 1995, *Photoconductivity conference,* edited by R. Breekeuridge, B. Bessel, T. Hahn (New York: Wiley).

Acknowledgements. Authors are grateful to S. Jagtar Singh RSIC, Punjab University, Chandigrah for doing XRD studies. Thanks are also due to Er. I.P. Singh Gill and Dr. V.K. Preshad Punab Agricaultur University, Ludihana for providing SEM facilities.

Structural and optical properties of $AgInSe_2$ films

R. K. Bedi[1,*], D. Pathak[1], Deepak[2], D. Kaur[2]

[1] Material Science Research Laboratory, Department of Physics, Guru Nanak Dev University, Amritsar, India
[2] Department of Physics and Center of Nanotechnology, Indian Institute of Technology, Roorkee, India
[*] rkbedi@rediffmail.com

Keywords: thin films, $AgInSe_2$, AFM, optical properties

Abstract. Silver Indium selenide ($AgInSe_2$) films have been prepared by thermal evaporation technique onto the glass substrate kept in the range (473-573 K) at pressure of $1.3x10^{-5}$ Torr. The films so prepared were studied for structural, electrical and optical properties. The films appear to be n-type indicating electrons as dominant charge carriers and show increase in electrical conductivity with corresponding increase in substrate temperature .X-ray diffraction pattern reflects $AgInSe_2$ single phase with predominant (112) orientation. Scanning electron micrographs show an increase in grain size with substrate temperature which may be attributed to more ordered structure. The films show an allowed transition near the fundamental absorption edge (Eg_1) in addition to a transition originating from crystal field split levels (Eg_2). The optical band gap of the films lie in the range 1.19-1.98 eV.

1. Introduction

Thin films solar cells made from ternary chalcopyrite compounds, such as $CuInSe_2$ and its analogous alloys like $CuInGaSe_2$ have begun to fulfill the promise of low cost, power generation from low polluting, clean energy sources. Because of direct band gap and high absorption coefficient, these semiconductors are particularly attracted for solar cell applications.$CuInGaSe_2$ solar cell energy conversion efficiency of more than 19.5 % has been reported [1]. Melting point and band gap energy of $AgInSe_2$ is better than $CuInSe_2$ for solar cell applications. $AgInSe_2$ (AIS) is one of the potential candidates of this family of alloys due to its band gap 1.2 eV [2], which makes it ideal absorber material for solar cell. $AgInSe_2$ crystallizes in the chalcopyrite structure which is closely related to zinc blende structure with the c/a ratio approximately equal to 2. However unit cell is tetragonally distorted with a distortion parameter x=2-c/a [3].The upper most valence bands of these materials are influenced by proximity of noble metal d levels in the valence bands [4]. An efficiency of 7.5 % was reported for solar cell which used a $AgInSe_2$ absorber on p-$AgInSe_2$ /n-CdS [5]. How ever, few papers have been reported in literature on the optical properties of $AgInSe_2$ films.

The present research deals with the synthesis of high quality $AgInSe_2$ films on glass substrates by vacuum evaporation and to investigate the effect of substrate temperature on their structural and optical properties.

2. Experimental procedure

$AgInSe_2$ films were grown by an evaporation method .The starting material was stoichiometric mixture of Ag, In and Se powder(99.999 % pure). $AgInSe_2$ ingot was prepared by direct fusion of stoichiometric proportion of the constituent element in vacuum sealed quartz tube. At first furnace temperature was slowly raised (1K /min) to 1100 K with continuous vibrational shaking to ensure homogeneity of the sample. This temperature was maintained for 48 hours and then furnace was turned off and tube was left inside until it reached to room temperature. The films were prepared from the ingot material by conventional thermal evaporation technique using vacuum coating unit (High vacuum coating unit 12A4H) and onto the cleaned glass substrate kept at nearly 300 K ,473 K,523 K and 573 K. Vacuum system, which consist of an oil diffusion pump coupled with rotary pump ,was used for the preparation of films .Molybdenum boats were used for evaporation .The vacuum of the order of 10^{-5} m bar was maintained in the chamber throughout the deposition. Cleaned optically flat glass substrates were used for the preparation of films. The glass substrates (Blue Star PIC 1) were first cleaned with acetone and alcohol and then washed in an ultrasonic bath. The glass substrates were mounted on the substrate holder with heating arrangement and the temperature was measured with the help of K-type thermocouple obtained from Omega Engineering Inc. (USA). The film thickness was controlled during deposition by means of quartz crystal monitor and was kept near 1.5 μm.

The absorption spectra, transmission and reflection spectra of $AgInSe_2$ films deposited at 473 -573 K on glass substrate were measured in the wavelength range 200-1100 nm using UV-VIS spectrophotometer (Perkin, Elmer). SEM were recorded by Jeol JSM - 6100 (Japan) scanning electron microscope .XRD scan of the films were taken using Cu Kα (wavelength =1.5505 A^0) radiations in 2θ range 20^0-80^0 by Bruker diffractrometer.

3. Results and discussion

Figure 1(a) show the XRD pattern of powder bulk sample. This figure illustrate that the

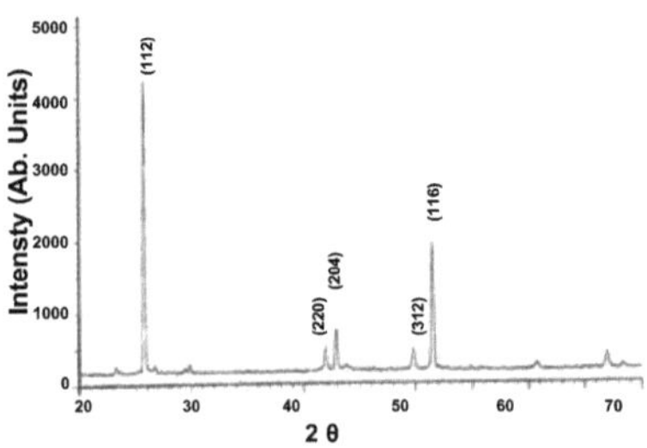

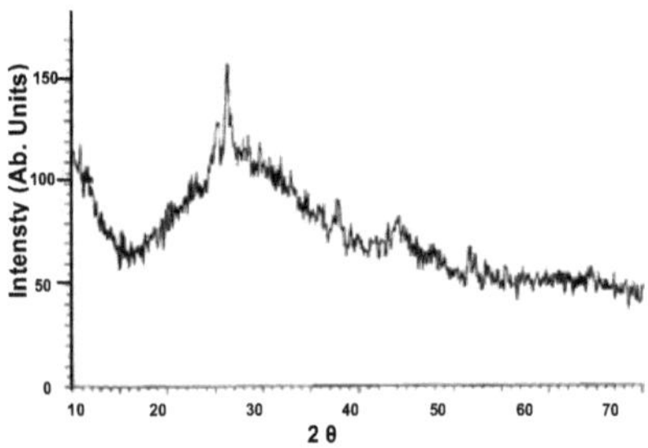

Figure 1. XRD patterns of (a) as prepared powder (b) the film prepared at 300 K.(RT)

$AgInSe_2$ chalcopyrite material is polycrystalline (tetragonal phase) .However films deposited at room temperature 300 K are amorphous as characterized by the absence of diffraction peaks (Figure 1(b)). The XRD of powder sample exhibit peaks due to the (112) as dominant peak and (204), (220), (312) and (116) are other phases of this chalcopyrite. An intense peak corresponding to (112) is also obtained for films deposited at higher substrate temperature (Figure 2). This indicates that film as well as powder is highly oriented in (112) direction.

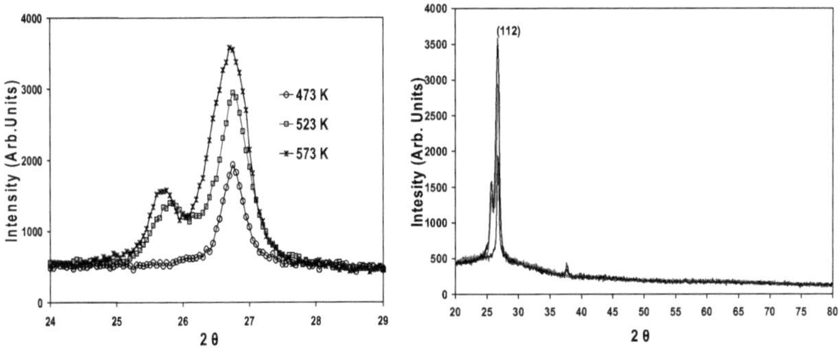

Figure 2. XRD pattern of films prepared at Ts (473 K-573K)

The lattice parameter calculated were (c=11.645 A° and a = 6.10 A°) which were very close to reported values. Also c/a ratio was found to be less that 2. These observations are in good agreement with those observed by A.H. Ammer et. al [3]. Crystallite size has been calculated from FWHM value of the (112) peak by using Scherrer's equation and is observed 14 nm for the films prepared at 573 K. In literature crystallite size of 45 nm has been reported for the films prepared with Ag_2Se and In_2Se_3 as starting material [6]. Observations reveal that intensity of the peaks increases with increase in substrate temperature thus indicating an improvement in crystallinity.

Figure 3 shows the SEM of $AgInSe_2$ films deposited on glass substrate kept at 573K. The

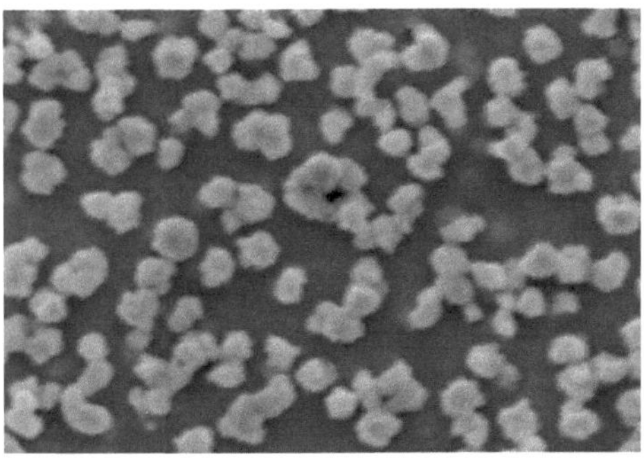

Figure 3. SEM of $AgInSe_2$ films deposited on glass substrate kept at 573 K

surface topography shown in the micrograph delineates the distribution and shape of crystallites. In general the micrograph indicates uniformly distributed microstructure. It has been observed that the films deposited at higher temperature show more ordered and densely packed arrangements. The surface morphology of the films studied from AFM has

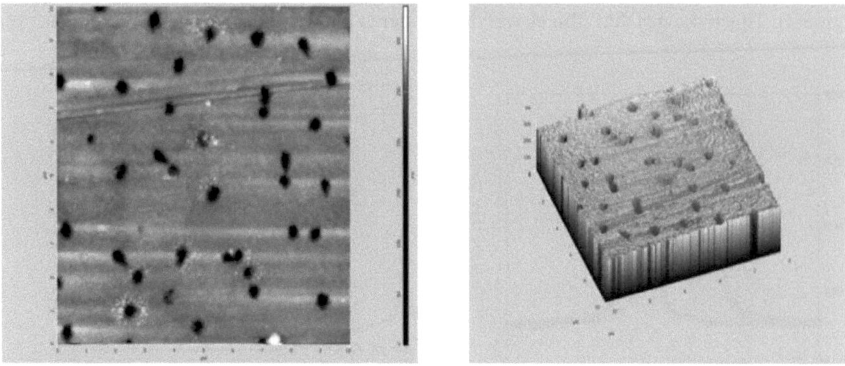

Figure 4. 2D and 3D AFM images of $AgInSe_2$ films deposited at 473 K.

been shown in Figure 4. It depicts nano well type structures with depth of 250 nm.
The optical spectra of films recorded in the wavelength range 200 to 1100 nm show two absorption humps (Figure 5). The first hump occur in the photon energy range 1.15 >hν >1.4 eV and has been examined in terms of allowed direct transition from the top of valance band to the conduction band minimum at the center of brilloun zone using equation of Bardeen [7]

$$(\alpha h \upsilon) = A (h \upsilon - E_g)^{1/2} \qquad (1)$$

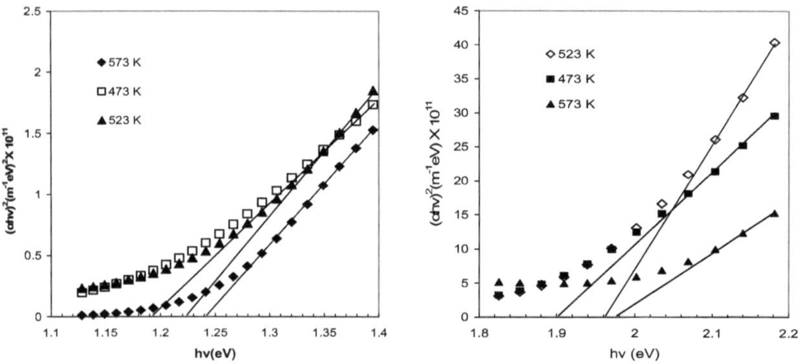

Figure 5. Spectral variation of α for $AgInSe_2$ films in energy range(a) 1.1-1.4 eV and (b) 1.8-2.2eV.

where Eg is optical band gap , A is a constant and υ is the frequency of incident light. The extrapolation of linear portion to αhv = 0 gives the value of optical band gap Eg_1 under direct allowed transition for $AgInSe_2$ films. The second hump has been observed in the energy range 1.8 > hv > 2.2 with energy gap Eg_2 (Figure 5(b)) which is due to the transition from crystal field split valance band that result from interaction between the valance electron and the non-cubic crystal field of chalcopyrite to the conduction band minimum. The linear nature of the plot near the absorption edge is due to the semiconducting behavior of the material [8].This increase in optical band gap may be attributed to the fact that the films deposited at higher substrate temperature exhibit larger grain size and show more ordered structure which give comparatively less contribution to the absorption. It can be deduced that two characteristics band gaps for AIS were found, Eg_1 and Eg_2. These observed results are in good agreement with earlier observed for other chalcopyrite like $AgGaSe_2$ [9] $AgInTe_2$ [10].

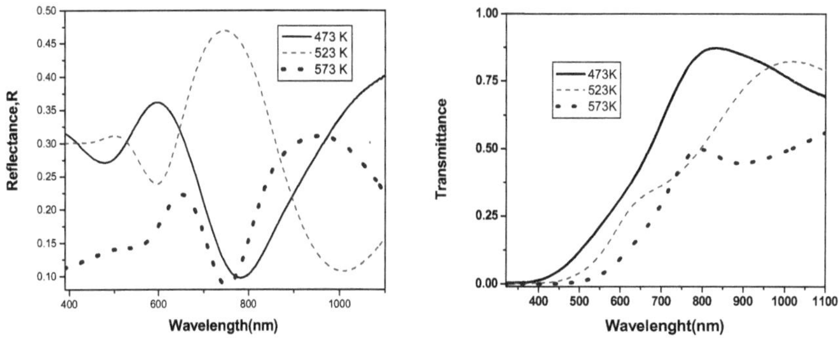

Figure 6. Variation of (a) Reflectance (R) and (b) Transmittance (T) with wavelength (nm)

Figure 6(a) and 6(b) shows the spectral behaviour of the Reflectance R and Transmittance T in the wavelength range 400-1100 nm for $AgInSe_2$ films. The transmission was found to be maximum of 80 % for the films deposited at the substrate temperature of 473 K. With increase in substrate temperature, the transmission was found to decrease from the value 0.8 to 0.53 which may be assigned to the fact that the surface of the films becomes rough and hence results in more scattered light and less transmittance and reflectance [11].
It has been observed that the electrical conductivity increases with increase in substrate temperature with in the experimental temperature range 290K-380K and can be represented by equation

$$\sigma = \sigma_0 \exp\left(-E_g/2kT\right) \qquad (2)$$

where E_a is thermal activation energy .The temperature dependence of log σ shows an almost straight line (Figure 7) indicating the applicability of the well known Arrhenius relation. The increase of electrical conductivity with increase in substrate temperature can be explained using Petritz 's barrier model [12] , Since the grain size do not grow sufficiently large at low temperature the intercrystallite region are wide, offering a high resistance to the movement of charge carriers. The formation of fewer nucleation centres at higher substrate temperature results in large crystallite sizes which ultimately decreases the barrier

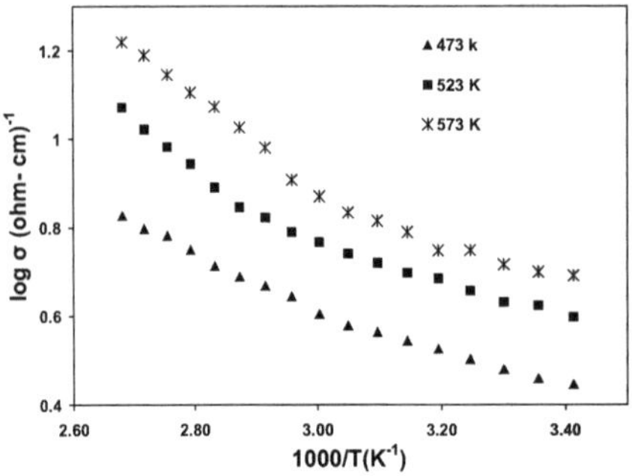

Figure 7. Inverse absolute temperature dependence of log conductivity of films deposited at different temperatures.

size.The activation energy obtained from the slopes of the plot lie in the range 0.045-0.065 eV.There is no evidence of any discontinuity in the slopes of the log σ vs 1000/T plots, thus there is not any reason to assume the presence of more than one conduction mechanism with in this temperature range. It was found that activation energy decreases with increase in substrate temperature, indicating the formation of more ordered films at higher substrate temperature. These results are in good agreement with XRD/SEM data of these samples. The variation of crystallite size, optical band gap and activation energy of $AgInSe_2$ films prepared at various substrate temperatures is recorded in table 1.

Table 1. Experimental data of crystallite size optical band gap and activation energy of $AgInSe_2$ films.

Substrate Temperature	Crystallite size	Activation Energy(eV)	Optical band gap(eV) Eg_1	Optical band gap(eV) Eg_2
473 K	9 nm	0.065	1.19	1.90
523 K	12 nm	0.053	1.22	1.96
573 K	14 nm	0.045	1.24	1.98

For Hall measurements indium contacts were used onto the samples .All samples show n type conduction. Carrier Concentration and mobility observed is around 5×10^{19} m^{-3} and 6.2 $m^2V^{-1}s^{-1}$ at room temperature (300K) respectively for the prepared samples.

4. Conclusions

High quality single phase Silver Indium Selenide films were successfully prepared by thermal evaporation technique. Optical studies show that $AgInSe_2$ show an allowed direct transition near fundamental absorption edge (Eg_1) in addition to a transition originating from crystal field split levels (Eg_2). The substrate temperature appears to affect the crystallinity of films. AFM studies depict nano well type structures with depth of 250 nm. Crystallite size varies from 9-14 nm for the films deposited in the range 473-573 K. Samples show n type conduction. The carrier concentration and mobility are found to be around 5 X 10^{19} m^{-3} and 6.2 $m^2V^{-1}s^{-1}$ respectively for the prepared samples.

References

1. Contreras, M.A., Ramanathan, K., Abushama, J., Hasoon, F., Keane, J., Young D.L., Egaas, B. & Noufi, R., 2005, *Prog. Photovolt. Res. Appl.*, **13**, 209.
2. Tell, B., Shay, T.J. & Kasper, H.M., 1972, *J. Appl. Phys.*, **246**, 43.
3. Ammar, A.H, Farid, A.M. & Seyam, M.A.M., 2002, *Vacuum*, **66**, 27.
4. Shay, J.L., Tell, B., Kasper, H.M. & Schiavone, L.M., 1972, *Phys. Rev. B*, **5**, 5003.
5. Ramesh, P.P., Hussain, O.M., Uthanna, S., Naidu, B.S. & Reddy, P.J., 1998, *Mater. Lett.*, **34**, 217.
6. Matsuo, H., Yoshino, K. & Ikari, T., 2006, *Phys. Stat. Sol.*, **8**, 2644.
7. Berdeen, J., Blatt, F.J., Hall, L.H, Breekwuridge, R., Russel, B. & Hahn, T., (Eds.), 1956, *Photoconductivity Conference* (New York: Wiley).
8. Pankove, J.I., 1971, *Optical Processes in Semiconductors,* (New York: Dover).
9. Soliman H.S., 1994, *J. Phys. D: Appl. Phys.*, **28**, 764.
10. Singh, A. & Bedi, R.K., 2001, *Thin Solid Films,* **398**, 427.
11. Kumar, A., Singh , P., Kulkarni, N. & Kaur D., *Thin Solid Films*, in press
12. Petritz, R.L., 1956, *Phys. Rev.*, **104**, 1508.

Acknowledgements. Authors are grateful to Inter University Accelerator Center (Formerly Nuclear Science Center), an autonomous center of UGC, New Delhi for providing financial assistance for carrying out this project.

In situ synchrotron X-ray diffraction study of formation mechanism of $Rh_{0.33}Re_{0.67}$ nanoalloy powder upon thermal decomposition of complex precursor

E. Filatov[1,*], Yu. Shubin[1], M. Sharafutdinov[2]

[1] Nikolaev Institute of Inorganic Chemistry SB RAS, Lavrentyev Ave. 3, 630090 Novosibirsk, Russia
[2] Institute of Solid State Chemistry and Mechanochemistry SB RAS, Kutateladze str. 18, 630090 Novosibirsk, Russia
[*] decan@che.nsk.su

Keywords: in situ powder diffraction, synchrotron, nanoalloy, complex precursor

Abstract. This paper presents step-by-step and *in situ* thermal decomposition of the $[Rh(NH_3)_5Cl](ReO_4)_2$ complex compound in hydrogen atmosphere. Main decomposition stages of this compound and development of the $Rh_{0.33}Re_{0.67}$ solid solution have been revealed. It has been demonstrated in the *in situ* experiments, that the first stage of reduction of the starting compound affords nanocrystalline metallic rhodium. Further, rhenium is reduced and its atoms are intercalated in the lattice of primary nuclei yielding the Rh_xRe_{1-x} solid solution with the hcp lattice.

1. Introduction

Nowadays the studies in the field of nanoalloy powders gain growing impact on both fundamental and applied sciences because of their unique features contrasting to those of bulk materials [1-6]. A convenient approach to fine bimetallic powders of predetermined composition is the reductive thermal decomposition of double complex compounds containing the target metals in desired ratio within complex ions [7, 8]. This makes it possible to achieve "mixing" of the metals at the atomic level and to obtain solid solutions, rather than individual metals, as decomposition products, at the temperatures much smaller than used for conventional alloying.

Tuning the ligand environment and the charge of the central atoms of the complex cations and anions, one can vary the ratio of metals in the precursor compound, and, therefore, achieve the desired metal composition. The range of double complex compounds, suitable for preparation of nanoalloy powders, is rather wide. Metal reduction can be achieved both with an additional reagent and a reducing ligand. [7].

2. Experimental

Preparation of the precursors

Coordination compound $[Rh(NH_3)_5Cl](ReO_4)_2$ was prepared by precipitation from water solution and characterized by IR spectroscopy and XRD phase analysis. Synthesis, crystal structure and thermal properties of $[M(NH_3)_5Cl](ReO_4)_2$ complexes, M = Co, Cr, Ru, Rh, Ir, were reported in detail earlier [8].

Instrumentation

Powder X-ray diffraction studies of thermolysis products were carried out on a DRON-RM4 diffractometer (CuK_α-radiation, graphite monochromator, ambient temperature). The refinement of lattice parameters was performed by the full profile technique applied to full-range diffraction data using PowderCell 2.4 program [9]. Crystallite sizes of the metal phases were determined by Fourier decomposition of profiles of single diffraction peaks, and with the Scherrer equation (WINFIT 1.2.1 [10]). EDX analysis was carried out by scanning electron microscope JSM 6700F (accelerating voltage 50 keV) equipped with energy dispersive spectrometer EX-23000BU.

The *in situ* experiments were carried out in a high temperature X-ray chamber on an X-ray diffraction station using time resolved synchrotron radiation (VEPP-3 ring storage in the Budker Institute of Nuclear Physics of the Siberian Branch of the Russian Academy of Sciences, 5b beamline) and OD-3 one-coordinate detector. The detector has 3328 channels and operates in the angle range of 30 degree [11]. The synchrotron radiation wavelength was adjusted to $\lambda = 1.517$ Å. Samples were prepared by pressing the starting compound into a cavity of a stainless steel cell.

3. Results

Determination of the solid solutions composition

Rhodium metal belongs to the structural type of copper, i.e., has the face-centered cubic lattice (fcc). Rhenium belongs to the structural type of magnesium, i.e., has the hcp lattice. Therefore, composition of the solids solutions can be determined on the basis of additivity of atomic volumes in solid solutions (Retgers rule): $V/z = x_1 \cdot V_1/z_1 + x_2 \cdot V_2/z_2$, where V – unit cell volume, z – number of atoms in the unit cell, x – fraction of the components [12]. The validity of this approximate approach was confirmed by additional determinations of compositions of single-phase solid solutions with EDX technique, and by independent literature data [13, 14]. Figure 1 and table 1 illustrate the experimental dependence of this parameter on the composition of the solid solution. Literature data are depicted with open circles, filled circles correspond to the data of the present work. It is noteworthy that the rhodium-rhenium system is characterized by a significant negative deviation of the atomic volume from the linear relationship (11%).

Step-by-step thermolysis of precursors in hydrogen atmosphere

Weighted samples of complexes in quartz combustion boats were put in a flow reactor and heated in hydrogen stream to a preset temperature, additional 1 h exposure being given afterwards. After cooling in a helium stream the samples were weighted, and their powder diffraction patterns were recorded in air at ambient temperature. Figure 2 presents the evolution of the phase composition on reduction of the $[Rh(NH_3)_5Cl](ReO_4)_2$. It was found that the complex decomposes in four continuous stages. At *the first stage* (190°C), partial reduction of the cationic and anionic fragments occurs resulting in the decrease of intensities of diffraction lines of the complex compound, and development of a very broad reflection at the angles corresponding to the most pronounced reflections of rhenium (100, 002, 101) and rhodium (111, 200). On further heating (*the second stage* - 200°C), the reflections of the complex disappear entirely, and the peaks of NH_4ReO_4, NH_4Cl and $[Rh(NH_3)_5Cl]Cl_2$ emerge in the pattern. At 220°C (*the third stage*), the pattern manifests only reflections of NH_4ReO_4

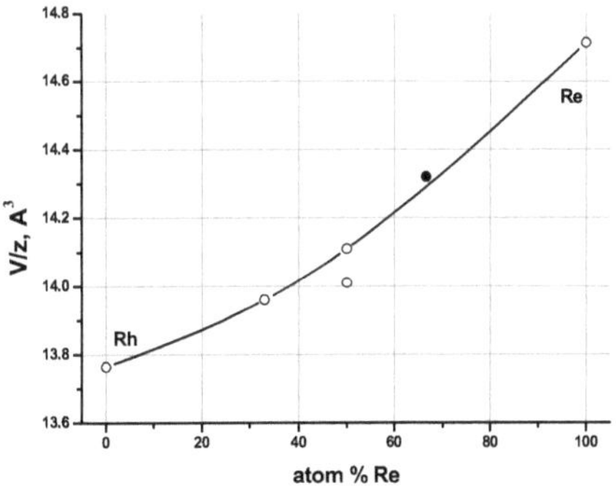

Figure 1. Calibrations curves presenting dependence of atomic volume on the composition of the solid solution. Open circles – literature data, filled circles – this work.

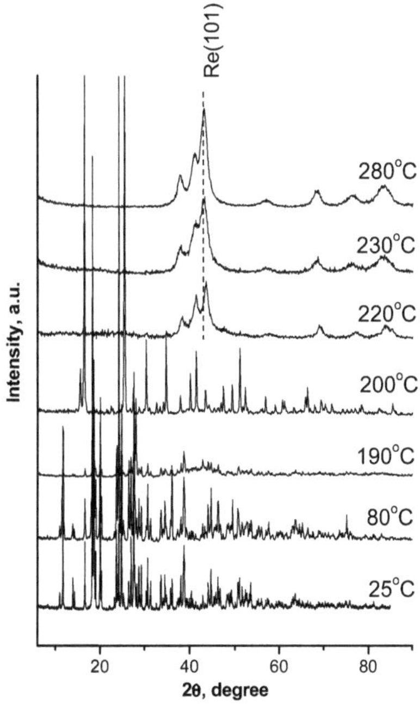

Figure 2. Evolution of the phase composition during the reduction of $[Rh(NH_3)_5Cl](ReO_4)_2$ in hydrogen (taken on quenched samples).

Table 1. X-ray data of the bimetallic phases.

Phase	Conditions of preparation	Lattice parameters, Å		Atomic volume, $Å^3$	Space group
		a	*c*		
Rh #5-685 [15]	-	3.8031	-	13.764(1)	Fm-3m
$Rh_{0.33}Re_{0.67}$ [8]	H_2, 600°C	2.740(5)	4.370(6)	13.96(8)	$P6_3/mmc$
$Rh_{0.50}Re_{0.50}$ [13]	H_2, 600°C	2.733(5)	4.364(6)	14.11(8)	$P6_3/mmc$
$Rh_{0.50}Re_{0.50}$ [14]	Ar, 550°C	2.727(5)	4.352(6)	14.01(8)	$P6_3/mmc$
$Rh_{0.67}Re_{0.33}$ [14]	H_2, 700°C	2.722(5)	4.350(6)	13.96(8)	$P6_3/mmc$
Re #5-702 [15]	-	2.760	4.458	14.713(4)	$P6_3/mmc$

and wide peaks of the solid solution $Rh_{0.70}Re_{0.30}$ having hcp lattice parameters: a = 2.72(1) Å, c = 4.36(2) Å. Therefore, it is evident that the phase of the solid solution developing at the first stage is deficient in rhenium. At *the fourth stage* (220-280°C), NH_4ReO_4 is reduced, and resulting atoms of rhenium are included in small particles of the existing solid solution. This is accompanied by the rearrangement of the crystal lattice and formation of a solid solution of different composition. As the result, the parameters of the solid solution steadily grow to the values a = 2.75(1) Å, c = 4.39(2) Å corresponding to the $Rh_{0.30}Re_{0.70}$ composition. This composition is very close to $Rh_{0.33}Re_{0.67}$ predetermined by the stoichiometry of the precursor complex. Crystallite sizes are 100-150 Å after annealing at 300°C .

***In situ* thermolysis of precursors in hydrogen atmosphere**

The above experiments carried out with quenched samples indicated some discrepancy with the conclusions formulated in refs. [16, 17]. In the cited papers, there is reported that during decomposition of single-molecule precursors in the initial stage the nuclei formed comprise only one metal. Therefore, to get better insight according of the development of the solid solution, *in situ* hydrogen reduction of the complex $[Rh(NH_3)_5Cl](ReO_4)_2$ was undertaken (figure 3). However, the experiments performed failed to provide a sound basis for unambiguous identification of reflections belonging to pure rhodium or rhenium. It was found that very wide reflections appear at *the first stage* (200°C). Further heating causes a decrease in intensities of diffraction lines of the starting compound followed by their complete disappearance. At *the second stage* – 204-206 °C (figure 4), the development of a solid solution $Rh_{0.40}Re_{0.60}$ based on the hcp lattice of rhenium is detected (a = 2.75(3) Å, c = 4.39(6) Å). Further temperature rise to 340°C is accompanied by extension of unit cell dimensions to a = 2.759(2) Å, c = 4.422(4) Å (table 2) related to development of the solid solution $Rh_{0.33}Re_{0.67}$. In parallel, crystallite sizes increase from 45-50 Å at 206°C to 110-125 Å at 340°C. Thus, considering the evolution of the composition and keeping in mind the data of refs [16, 17], it is possible to suppose that at the first stage of the reaction metallic rhodium develops as an amorphous non-diffracting phase. Rhenium is formed at the second stage of the process and it penetrates into nanosized rhodium particles resulting in development of larger particles of the hcp-based solid solution and in the change of the composition. Similarly to the experiment with step-by-step reduction followed by quenching, the final product of decomposition of the $[Rh(NH_3)_5Cl](ReO_4)_2$ precursor is a solid solution having metal ratio corresponding to that in the starting compound (Rh:Re - 1:2). In contrast to the quenching experiment, no intermediate products of thermolysis of the precursor - NH_4ReO_4, NH_4Cl and $[Rh(NH_3)_5Cl]Cl_2$ – were observed

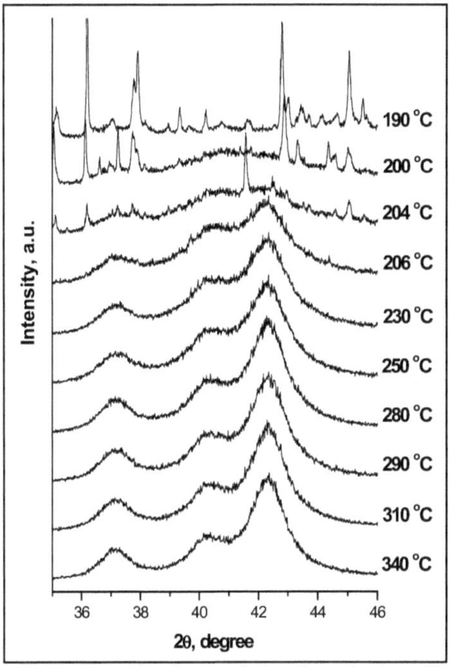

Figure 3. Evolution of the phase composition ***in situ*** *during the reduction of* $[Rh(NH_3)_5Cl](ReO_4)_2$ *in hydrogen.*

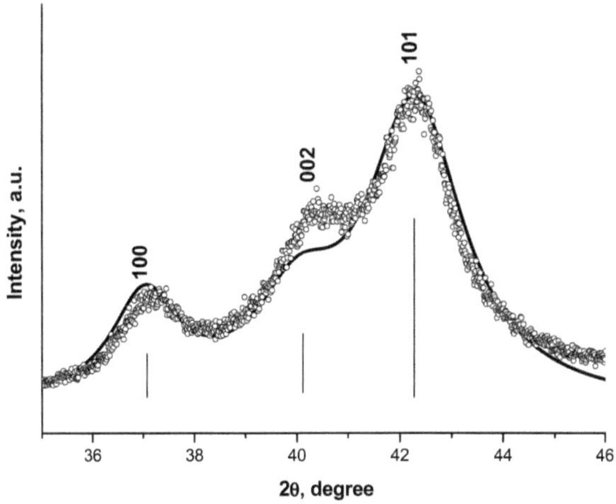

Figure 4. Approximation of peaks profile of the hcp solid solution phase (206 °C). Open circles – experimental data, solid line – calculated profile.

Table 2. Lattice parameters and crystallite size of metal phases of ***in situ*** *$[Rh(NH_3)_5Cl](ReO_4)_2$ decomposition in hydrogen.*

T, °C	Lattice parameters *, Å		V/z	Composition	Crystallite size, Å
	a				
206	2.75(3)	14.37(12)	14.37(12)	$Rh_{0.40}Re_{0.60}$	45-50
230	2.75(2)	14.44(8)	14.44(8)	$Rh_{0.35}Re_{0.65}$	50-60
250	2.755(3)	14.52(2)	14.52(2)	$Rh_{0.34}Re_{0.66}$	55-70
280	2.756(3)	14.54(2)	14.54(2)	$Rh_{0.33}Re_{0.67}$	60-85
290	2.756(3)	14.54(2)	14.54(2)	$Rh_{0.33}Re_{0.67}$	65-85
310	2.756(3)	14.54(2)	14.54(2)	$Rh_{0.33}Re_{0.67}$	80-100
340	2.759(2)	14.58(1)	14.58(1)	$Rh_{0.33}Re_{0.67}$	110-125

* The parameters listed in the table are adjusted to room temperature using known thermal expansion laws of rhodium and rhenium lattices [18, 19].

Relation $Rh_{0.33}Re_{0.67}$ of solid solution composition with Rh-Re phase diagram

Mapping of the composition of the single-phase solid solution on the phase diagram of the system Rh-Re [20] (figure 5) suggests that the prepared sample is in the two-phase field and should be non-equilibrium. To determine if it is correct, the sample prepared was annealed in an evacuated quartz ampoule for 300 h at 950 °C, however, the phase did not separate in two solid solutions. To clarify whether the $Rh_{0.33}Re_{0.67}$ solid solution is the equilibrium phase at the annealing temperature, or the decomposition was hampered by kinetic hindrances, a mixture of ultrafine powders of rhodium and rhenium (crystallite sizes are 40 and 30 Å, respectively) was subjected to sintering under ratio 1:2. The sintering was performed in an evacuated quartz ampoule at 950 °C for 400 h. It yielded a single-phase sample of the $Rh_{0.33}Re_{0.67}$ solid solution with the parameters identical to those of the solution obtained on thermolysis of the $[Rh(NH_3)_5Cl](ReO_4)_2$ complex. This result may suggest some inaccuracy in determination of the boundaries of the two-phase field in the state diagram of the Rh-Re [20] and necessity for their redetermination.

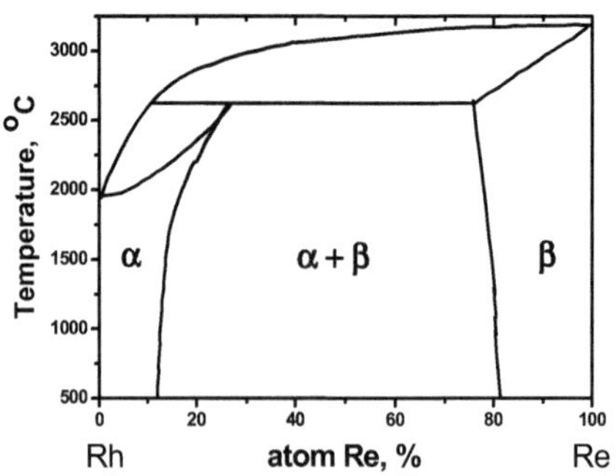

Figure 5. Rh-Re phase diagram according to the date [20].

4. Discussion

The development of solid solutions of such refractory metals as rhodium and rhenium at relatively moderate temperature is explained by high diffusion activity of atoms in nanosized particles formed on precursor decomposition. This property is governed by excess surface energy of nanosized particles, which is substantiated by the data on decreasing melting temperature of small particles occurring in parallel to reduction of their size [21, 22].

References

1. Sode, A., Li, W., Yang, Y., Wong, P.C., Gyenge, E., Mitchell, K.A. & Bizzotto, D., 2006, *J. Phys. Chem. B: Condens. Matter. Mater. Surf. Interfaces Biophys.*, **110**, 8715.
2. Silvestre-Albero, J., Serrano-Ruiz, J.C.& Sepúlveda-Escribano, A., 2005, *Appl. Catal. A: General*, **292**, 244.
3. Ito, S., Suwa, Y., Kondo, S., Kameoka, S., Tomishige, K. & Kunimori, K., 2003, *Catal. Comm.*, **4**, 499.
4. Yang, H., Coutanceau, C. & Léger, J.-M., 2005, *J. of Electroanal. Chem.*, **576**, 305.
5. Lenarda, M., Moretti, E., Storaro, L., Patrono, P., Pinzari, F., Rodríguez-Castellón, E., Jiménez-López, A., Busca, G., Finocchio, E., Montanari T. & Frattini, R., 2006, *Appl. Catal. A: General*, **312**, 220.
6. Ogihara, H., Takenaka, S. & Yamanaka, I., 2006, *J. Catal.*, **238**, 353.
7. Korenev, S.V., Venediktov, A.B., Shubin, Yu.V., Gromilov, S.A. & Ysenko, K.V., 2003, *Russ. J. Struct. Chem.*, **44**, 58.
8. Shubin, Yu.V., Filatov, E.Yu., Baidina, I.A., Yusenko, K.V., Zadesenes, A.V. & Korenev, S.V., 2006, *Russ. J. Struct. Chem.*, **6**, 1114.
9. Kraus, W. & Nolze, G., 2000, *PowderCell 2.4, Program for the representation and manipulation of crystal structures and calculation of the resulting X-ray powder patterns* (Berlin: Federal Institute for Materials Research and Testing).
10. Krumm, S., 1994, *Mater. Sci. Forum*, **228-231**, 183.
11. Baru, S.E., Aulchenko, V.M., Babichev, E.A., Dubrovin, M.S., Groshev, V.R., Khabakhpashev, A.G., Nekhanevich, E.L., Porosev, V.V., Savinov, G.A., Shekhtman, L.I., Smykov, L., Talyshev, A.A. Titov, V.M. Ukraintsev, Y.G. Velikzhanin, Y.S. & Zanevsky, Y.V., 1997, *Nucl. Instr. Meth. Phys. Res. A*, **392**, 12.
12. Goldschmidt, V.M., 1934, *Kristallchemie* (Jena: Verlag von Gustav Fischer).
13. Gromilov, S.A., Korenev, S.V., Baydina, I.A., Korolkov, I.V. & Yusenko, K.V., 2002, *Russ. J. Struct. Chem.*, **43**, 527.
14. Yusenko, K.V., Gromilov, S.A., Korolkov, I.V., Baydina, I.A., Romanenko, G.V. & Korenev, S.V., 2004, *Russ. J. Inorg. Chem.*, **49**, 568.
15. PCPDFWin, Vers. 1.30, 1997, JCPDS ICDD, Swarthmore, PA, USA.
16. Shubin, Yu.V., Korenev, S.V. & Sharafutdinov, M.R., 2006, *Russ. Chem. Bull.*, **7**, 1109.

17. Plyusnin, P.E., Baidina, I.A., Shubin, Y.V. & Korenev, S.V., 2007, *Russ. J. Inorg. Chem.*, **52**, 371.
18. Wasilewski, R.J., 1961, *Trans. Metall. Soc. AIME*, **221**, 1081.
19. Ross, R.G. & Hume-Rothery, W., 1963, *J. Less-Common Metals.*, **258**, 258-270.
20. Rh-Re, 1990, in *Binary Alloy Phase Diagrams, Second Edition*, edited by T.B. Massalski (Ohio: ASM International, Materials Park), p. 3199.
21. Zhang, Z., 2004, Ph.D. Dissertation, University of Notre Dame, Indiana (US), p.67.
22. Moisala, A, Nasibulin, A.G. & Kauppinen, E., 2003, *J. Phys.: Condens. Matter*, **15**, S3011.

Acknowledgements. This work was financially supported by the program of Presidium of Russian Academy of Science “Development of preparation techniques of novel chemical substances and design of new materials” (project № 8-17) and by the Russian Foundation for Basic Research, grant no. 07-03-01038-a and 00-15-97448.

Crystallite size and micro- and macro-strain changes during layer exchange upon annealing amorphous/crystalline aluminium

D. He[1], J. Y. Wang[1,*], E. J. Mittemeijer[1]

[1] Max Planck Institute for Metals Research, Heisenbergstr. 3, D-70569 Stuttgart, Germany
[*] j.y.wang@mf.mpg.de

Keywords: crystallite size, stress, strain, layer exchange

Abstract: Crystallite size and micro- and macro-strain changes during layer exchange upon annealing amorphous silicon/crystalline aluminium (a-Si/c-Al) bilayers at 165°C and at 250°C for various times were evaluated by X-ray diffraction analysis (analysis of both diffraction-line shift and diffraction-line broadening). As revealed by Auger electron spectroscopic depth profiling, c-Al→c-Si layer exchange occurs upon annealing at 250°C (more or less completely) and at 165°C (partially). Upon layer exchange, the Al {111} fibre texture in the bilayer becomes stronger, the Al crystallites grow in both lateral and perpendicular directions, and both the macrostress and the microstrain in the Al sublayer are released. The amorphous Si sublayer crystallizes into an aggregate of nanocrystals with {111} planes parallel to the surface, exhibiting very large microstrains of 0.7-2.6%, depending on annealing temperature and time.

1. Introduction

Aluminium-induced layer exchange (ALILE) for the formation of continuous polycrystalline silicon (c-Si) film on glass substrate has been regarded as a promising method to produce low-cost solar cells [1]. The ALILE process involves that, upon annealing an amorphous (a)-Si/crystalline (c)-Al bilayer, the original Al sublayer is replaced by a continuous c-Si sublayer (c-Al→c-Si) [2].

Our previous work on Al/a-Si and a-Si/Al bilayers has shown that a layer-exchange (in particular c-Al→c-Si) process can occur upon annealing in both Al/a-Si bilayers and a-Si/Al bilayers, i.e. irrespective of crystalline Al or amorphous Si being the top layer, and that the layer exchange can be driven by the energy change in bulk Al phase (release of macrostress and microstrain, increase of grain size) [3-5].

The morphology of the resulting c-Si film depends strongly on the annealing conditions, as well as the deposition technique and the thickness of the initial Si and Al sublayers [6-8]. It was recently shown that a larger grain size of crystallized Si can be achieved by decreasing the annealing temperature [9]. However, the general microstructural development, expressed in terms of crystallite size and micro- and macro-strains associated with the c-Al→c-Si layer exchange, has not been investigated. The present work is intended to provide such data.

2. Experimental procedures

The Si/Al bilayers were deposited onto single crystal Si(510) substrates (500 μm thickness; 100 mm diameter) covered with a 50 nm thin SiO_2 thermally-oxidized film, which acts as a diffusion barrier. The Si(150 nm)/Al(50 nm) bilayer was obtained by depositing first 50 nm Al and then 150 nm Si with a direct current (*dc*) magnetron sputtering apparatus. The Si/Al bilayers were enclosed in glass capsules containing argon gas and then were isothermally annealed at 165°C and 250°C for a certain time and thereafter cooled down to room temperature. After having performed X-ray diffraction (XRD) analysis on this specimen, the same specimen was encapsulated again and annealed for an additional period of time at the same temperature. This procedure was repeated up to a desired total, cumulative annealing time.

A Philips MRD diffractometer was employed to perform phase analysis and texture, residual stress, crystallite size and microstrain determinations. The diffractometer was equipped with an Eulerian cradle, a copper tube operating at 1.8 kW (45kV; 40mA), an X-ray lens and a graphite monochromator in the diffracted beam to select Cu K_α radiation. Aluminium powder, with a purity of 99.999 wt% and particle diameter less than 50 μm, was annealed at 165°C for 3 hours and then used for the correction of instrumental line-broadening in the X-ray diffraction analysis.

To analyze the phases present in the as-prepared and annealed bilayers, diffraction patterns were recorded by performing continuous θ-2θ scanning from 2θ = 20° to 100° (where 2θ is the diffraction angle) with a scan speed of 2θ equal to 0.04°/min and a collection time of 20 minutes for each step. To study the texture of the as-prepared and annealed Al sublayers, a ϕ step scan from 0° to 360° (ϕ=angle of rotation around the surface normal) with a step size of 2° and a counting time of 10s per step, for the {111} reflection of the Al layers, was performed at specimen tilt angles ψ from 0° to 89° in steps of 1° (ψ is the tilt angle of the diffraction vector with respect to the specimen surface). For the determination of macrostress, crystallite size and microstrain, θ-2θ step scans were performed for the Al {111} reflection of the as-deposited bilayers (2θ range of 36 to 41° with a step size of 0.04°), and for the Al {111} reflection and Si {111} reflection (2θ range of 26 to 31° with a step size of 0.04°) of the annealed bilayers, at various ψ angles. The peak parameters, as peak-maximum position, full width at half maximum, integral peak area, and integral breadth, were determined by fitting a Pearson VII function to the peaks in the measured diffraction patterns using the Philips Profit 1.0c software.

The compositional characterization of the as-prepared and annealed Si/Al bilayers was performed using a JEOL 7830 scanning Auger microscope at a base pressure below 8.0×10^{-8} Pa. For details of XRD and AES as performed in this study, see Refs. 4 and 5.

3. Results and evaluation

3.1 Compositional redistribution

The AES depth profiles of the Si/Al bilayers annealed at 165°C and 250°C for various times are shown in figures 1a and b, respectively. The as-deposited Si/Al bilayer exhibited a sharp interface between the top, amorphous a-Si sublayer and the bottom, crystalline Al sublayer.

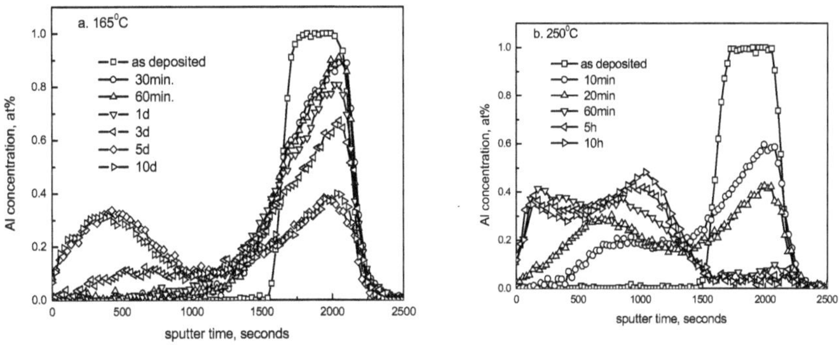

Figure 1. Al concentration-depth profiles of the a-Si(150nm)/Al(50nm) bilayers as-deposited and annealed at 165°C (a) and 250°C (b) for various times.

Upon annealing the Si/Al bilayer at 165°C, the Al concentration-depth profile after annealing for 30 minutes showed that a significant amount of Al atoms had transferred into the Si layer. Further annealing for 1 hour or 1 day did not cause pronounced, further change in the Al concentration-depth profile, i.e. a retardation of mass transfer was observed. After annealing for 3 days, a plateau region in the Al concentration-depth profile had appeared at the location of the original a-Si sublayer. The mass-transfer process proceeded pronouncedly upon increasing the annealing time up to 5 days; a maximum in the Al concentration-depth profile appeared and at this stage more than half of all Al atoms in the original Al sublayer had moved to locations initially occupied by the original Si sublayer. Further annealing up to 10 days didn't cause further, appreciable mass transfer.

Upon annealing at 250°C, after only 10 minutes about half of the Al atoms had already moved to locations initially occupied by the original Si sublayer. After annealing for one hour, more than 92% of the Al atoms had moved to the top sublayer; practically, layer exchange in the Si/Al bilayer had taken place. Further annealing, from 1 hour to 25 hours didn't cause further significant elemental redistribution. A stable compositional distribution was established: an almost complete c-Si sublayer (see section 3.2 for XRD evidence for crystalline nature of Si sublayer) and a mixed Al/Si sublayer had formed at the original locations of the Al sublayer and of the a-Si sublayer, respectively.

It's remarked that, as follows from the O signal recorded as a function of depth, not shown in the AES compositional depth profiles, no compositional changes had occurred upon annealing in the oxide (SiO_2) layer on top of the Si(510) substrate.

3.2 Phase and texture evolutions

XRD patterns of an as-prepared Si/Al bilayer and of a Si/Al bilayer annealed at 250°C for 25 hours are shown in figure 2a. The Al sublayer exhibits a {111} fibre texture (rotational symmetry of the texture with respect to the surface normal was confirmed by intensity measurements performed at various rotation angles ϕ); the results of the Al {111} ψ scans of the as-prepared and annealed Si/Al bilayers indicated that the {111} fibre texture of the Al sublayer has become slightly stronger upon annealing at 250°C for 25 hours (inset of figure 2a).

The Si sublayer in the as-deposited Si/Al bilayer is amorphous, as evidenced by the broad scattering peak at about $2\theta = 27°$ (figure 2a). After annealing, the a-Si layer crystallizes into c-Si. The crystalline Si shows a {111} fibre texture (rotational symmetry of the texture was confirmed by intensity measurements at various rotation angles ϕ).

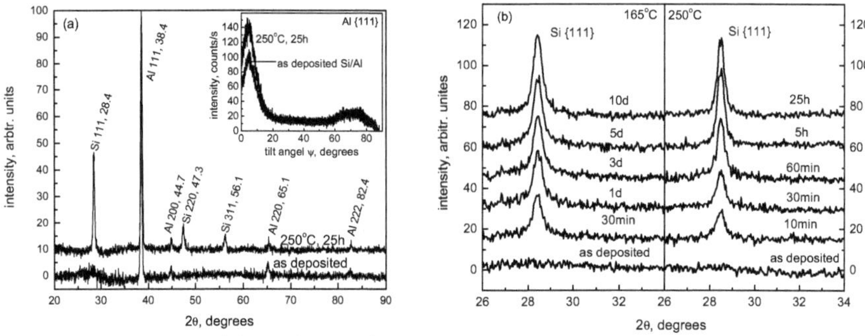

Figure 2. XRD patterns of the as-deposited and annealed a-Si(150nm)/Al(50nm) bilayers: (a) diffraction patterns of the bilayer annealed at 250°C for up to 25 hours, with ψ-scans of Al {111} peaks; (b) Si {111} peaks in the diffraction patterns of annealed Si/Al bilayers, at both 165°C and 250°C. After subtraction of the background as recorded from the "bare" Si(510) substrate. Vertical offsets have been applied to the individual patterns to avoid overlap.

The appearance and evolution of the Si {111} reflection are shown in figure 2b for the Si/Al bilayers annealed at 165°C for up to 10 days and at 250°C for up to 25 hours. After annealing at 165°C for 30min or at 250°C for 10min, Si {111} peaks had emerged, i.e. a-Si had started to crystallize. Evidently, the crystallization proceeded faster at the higher temperature. The intensity of the Si {111} peak increased with increasing annealing time at 165°C and 250°C. With reference to the compositional redistribution as shown in figure 1, it is suggested that the amount of crystallized Si parallels the amount of Si transferred to the location of the original Al sublayer (for discussion, see Ref. [9]).

3.3 Residual macrostress analysis

The X-ray diffraction analysis showed that the as-deposited Al sublayer exhibits a {111} fiber texture (see section 3.2) and possesses a rotationally symmetric biaxial, planar state of stress. For this case, the Al-lattice strain observed at a specimen-tilt angle ψ, ε_ψ, is related to the stress parallel to the surface, $\sigma_{//}$, according to [10,11]:

$$\varepsilon_{\psi}^{\{111\}} = \frac{d_{\{111\}} - d_{\{111\}}^0}{d_{\{111\}}^0} = (2s_{12} + \frac{1}{2}s_{44}\sin^2\Psi + \frac{2}{3}s_0)\sigma_{//} \quad (1)$$

where $s_0 = s_{11} - s_{12} - s_{44}/2$, with s_{11}, s_{12} and s_{44} as the single crystal elastic compliances of the layer material (for Al, s_{11}, s_{12} and s_{44} can be taken as 16.0 TPa^{-1}, -5.8 TPa^{-1} and 35.3 TPa^{-1}, respectively [12]); $d_{\{111\}}$ is the {111} Al-lattice spacing measured for tilt angle ψ, and $d_{\{111\}}^0$ is the strain-free {111} Al-lattice spacing which is equal to 2.338 Å (ICDD-JCPDS card no. 04-0787). Hence, a plot of $d_{\{111\}}$ vs. $\sin^2\psi$ for the Al phase yields a straight line, and $\sigma_{//}$ can be calculated from its slope. Such results are shown in figure 3.

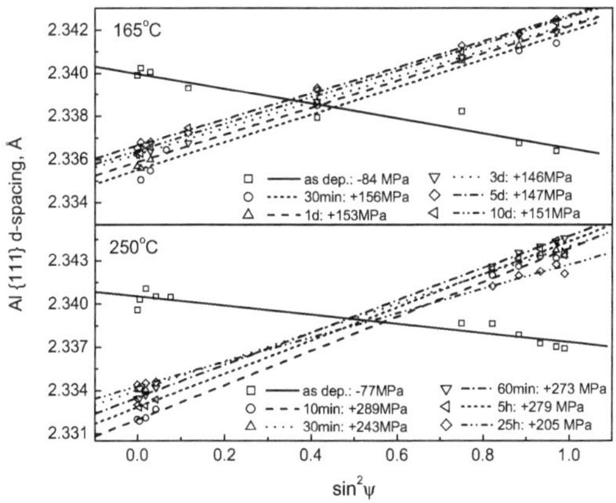

Figure 3. Residual stress of the Al phase of the as-prepared and annealed a-Si(150nm)/Al(50nm) bilayers, at 165°C and at 250°C, as determined in plots of the {111} lattice spacing, $d_{\{111\}}$, as a function of $\sin^2\psi$.

It follows that the as-deposited Al sublayers of the Si/Al bilayers had a compressive stress of -84 MPa parallel to the surface, as measured at room temperature. At the start of the subsequent annealing at elevated temperature, the stress in the bilayer is composed of the stress as observed at room temperature, and the thermal stress caused by the difference in thermal expansion of the substrate and the Al sublayer. Assuming that the thermal misfit is fully accommodated elastically by the Al sublayer, because of the large size of the substrate (10^4 times thicker than the Al film), the thermal stress of the Al sublayer at the beginning of annealing, σ_{th}, can be estimated [5], i.e. -332 MPa and -545 MPa for annealing at 165°C and at 250°C, respectively, if full elastic accommodation of the thermal misfit would occur. Then, the total stress of the Al sublayer at the beginning of the annealing was about -416 MPa at 165°C, and -629 MPa at 250°C for the Si/Al bilayers.

Upon annealing at 165°C, the stress in the Al phase of the annealed Si/Al bilayer attains a value of +156 MPa (tensile stress) after short time (30 minutes) annealing and maintains this stress level (as measured at room temperature) upon prolonged annealing. Upon annealing at

250°C, the stress in the Al phase becomes +205 MPa for the Si/Al bilayer annealed for 25 hours (i.e. after "layer exchange").

Cooling after the annealing would lead to build up of a tensile thermal stress (analogous to the above reasoning). Because the cooling (as compared to the heating up) is very slow (2°C /min) (thermal) stress relaxation during cooling can be appreciable [10]. Hence, the observation of tensile macrostress *at room temperature* after annealing can then be interpreted as that during the annealing the compressive macrostress in the Al sublayer had completely relaxed.

The {111} fibre textured Si phase, as developed upon crystallization, exhibits a rotationally symmetric biaxial, planar stress state, as well. The stress parallel to the surface can then be calculated from $d_{\{111\}}$ vs. $sin^2\psi$ plots following a treatment similar to that for the Al phase. Plots of $d_{\{111\}}$ *vs.* $\sin^2\psi$ for the crystalline Si phase are shown in figure 4. The value of $\sigma_{//}$, calculated from the slope of the fitted straight lines in these plots, indicates the presence of a tensile stress at room temperature.

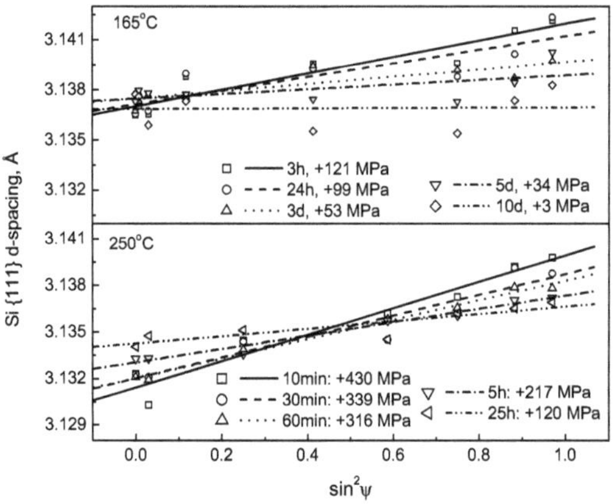

Figure 4. Residual stress of the Si phase of the annealed a-Si(150nm)/Al(50nm) bilayers, at 165°C and at 250°C, as determined in plots of the {111} lattice spacing, $d_{\{111\}}$, as a function of $sin^2\psi$.

The value of the tensile stress for the c-Si phase as measured at room temperature decreased with increasing annealing time: +121 MPa after annealing for 3 hours at 165°C, and +3 MPa after annealing for 10 days at 165°C; +430 MPa after annealing for 10 min at 250°C and +120 MPa after annealing for 25h at 250°C. These stress values can be taken equal to those present after the crystallization *at the annealing temperature* as no thermal misfit develops between the crystalline Si and the (crystalline) Si wafer. The initial development of a tensile stress in the c-Si phase could be ascribed to the volume decrease experienced by the Si phase upon its crystallization; continued annealing then leads to relaxation of this tensile stress (cf. above discussion on stress in the Al phase).

3.4 Crystallite size and microstrain determination;

Experimental diffraction-line profiles are the convolution of the wavelength spectrum with various functions arising from instrumental factors and specimen defects (crystallite size and microstrain) [13, 14]. Here it will be assumed that (i) the measured profile *h* (recorded from the bilayers), the instrumental aberration profile *g* (recorded from a pure annealed Al powder, see section 2) and the only structurally broadened profile *f* are symmetrical, and (ii) can all be described by a Voigt function which is the convolution of Lorentzian and Gaussian components. On this basis the integral breadth of the Lorentzian and Gaussian components of the structural profile *f* can be obtained straightforwardly from the measured breadths parameters of the *h* and *g* profiles. Then also the integral breadth of the *f* profile, β_f, can be simply calculated from the integral breadths of its Lorentzian and Gaussian components [14].

The integral breadths of the components in the *f* profile due to only finite crystallite size and only lattice microstrain, β_{size} and β_{strain}, can be given as (with β expressed in radians on the 2θ scale) [13, 14]:

$$\beta_{size} = \frac{K\lambda}{D\cos\theta}, \qquad (2)$$

$$\beta_{strain} = 4 < \varepsilon^2 >^{1/2} tg\theta, \qquad (3)$$

where λ is the wavelength of the X-rays used, θ is the Bragg angle, D is the "average" crystallite size, K is a constant taken as 1.0 (i.e., 0.89~1.39, see reference [15]), and $< \varepsilon^2 >^{1/2}$ is a measure for the microstrain in the films [16]. Now, according to the so-called single-line Voigt method [13, 14], the integral width of the Lorentzian and Gaussian components of the only structurally broadened profile, *f*, are identified with β_{size} and β_{strain}, respectively, and thus values for crystallite size and lattice strain are obtained by application of Eqs. (2) and (3).

By selecting a number of the specimen tilt angles, ψ, from 0° to close to 90°, this structural information is obtained as a function of orientation within the specimen (ψ=0° corresponds to the direction perpendicular to the surface and ψ=90° corresponds to the direction parallel to the surface). On this basis, the microstructural anisotropy can be investigated; a method applied by Welzel and Mittemeijer [17].

The (anisotropic) data determined for the crystallite size and the microstrain of the Al phase in the as-deposited and annealed (at 165°C for 10 days and 250°C for 25 hours) Si/Al bilayers are presented in Table 1. The value observed in the perpendicular direction for the as-deposited Al sublayer, about 50 nm, is approximately equal to the thickness of the Al sublayer. Upon annealing the average Al grain sizes in the Si/Al bilayer increased in both directions, perpendicular and parallel to the surface: During annealing at 165°C, the Al crystallites had grown from 50 nm to 72 nm in the direction perpendicular to the surface and had grown laterally from 70 nm to 95 nm; correspondingly, during annealing at 250°C, they had grown from 50 nm to 83 nm and from 70 nm to 120 nm, in the perpendicular and lateral directions, respectively. Evidently, as compared to annealing at 165°C, annealing at 250°C leads to larger crystallite sizes of the Al phase in both perpendicular and lateral directions. The crystallite size is generally larger in the lateral than in perpendicular directions. In thin films grain growth in lateral directions is obviously less constrained than in the perpendicular direction due to the presence of (intermediate; cf. section 4.1) interfaces and the surface.

Table 1. Crystallite/grain size and microstrain of the Al phase in the a-Si(150nm)/Al(50nm) bilayers as-deposited and annealed at 165°C for 10 days and at 250°C for 25 hours.

Al phase	crystallite/grain		microstrain	
	perpendicular (Ψ=0°)	lateral (Ψ=90°)	perpendicular (Ψ=0°)	Lateral (Ψ=90°)
as-deposited	~50 nm	~70 nm	~0.5%	~0.4%
165°C (10d)	~72 nm	~95 nm	~0.4%	~0.4%
250°C (25h)	~83 nm	~120 nm	~0.3%	~0.3%

For the as-deposited Si/Al bilayers, the microstrain of the Al sublayer in the direction perpendicular to the layer surface was somewhat larger than that in lateral directions (see Table 1). After annealing (at both 165°C and 250°C), the microstrains had relaxed (in particular in the perpendicular direction) and a practically isotropic state of microstrain in the Al phase was established.

After crystallization, the (anisotropic nature of the) grain size and microstrain for the Si phase could also be determined from the observed diffraction-line broadening, using the same procedure as applied for the Al phase. The thus obtained values for the grain size and the microstrain of the c-Si phase in the Si/Al bilayers annealed at 165°C for 1 day and 10 days and at 250°C for 10 minutes and 25 hours are presented in Table 2.

Table 2. Crystallite/grain size and microstrain of the Si phase in the a-Si(150nm)/Al(50nm) bilayers annealed at 165°C for 1 day and 10 days and at 250°C for 10 minutes and 25 hours.

Si phase	crystallite/grain		microstrain	
	perpendicular (Ψ=0°)	lateral (Ψ=90°)	perpendicular (Ψ=0°)	lateral (Ψ=90°)
165°C(1d)	~18 nm	~7 nm	~1.0%	~2.6%
165°C(10d)	~25 nm	~10 nm	~1.0%	~2.0%
250°C(10min)	~20 nm	~9 nm	~1.3%	~1.9%
250°C(25h)	~57 nm	~31 nm	~0.7%	~0.9%

In general, the crystallites of the c-Si were relatively small, as compared to the Al crystallites. For the bilayers annealed at either 165°C or 250°C, the Si grain size in the perpendicular direction was about twice as large as that in the lateral directions. During the annealing at 165°C, the c-Si phase had a size of about 10 nm in lateral directions independent of annealing time, whereas the size in the perpendicular direction increased from about 18 nm (one day annealing at 165°C) to about 25 nm (10 days annealing at 165°C). After 10 minutes annealing at 250°C, the *c*-Si grain size varied from about 9 nm in lateral directions to about 20 nm in the perpendicular direction. After annealing for up to 25 hours at 250°C (i.e. after complete "layer exchange"), the Si grains had grown in both lateral and perpendicular directions to 31 nm and 57 nm, respectively. Again, as observed for the Al phase (see above), annealing at 250°C leads to larger crystallite sizes in both lateral and perpendicular directions. It is noticed that the crystallite size of the nano-crystalline Si is strongly anisotropic: opposite to the result obtained for the Al phase, the crystallite size in the lateral directions is much smaller than the one in the perpendicular direction. This reflects the constrained nature of lateral growth of crystalline nuclei of Si developing at Al grain boundaries in the original Al sublayer (see section 4).

Evidently, the microstrain in the c-Si as presented in Table 2 was very high, 0.7-2.6% (note: the microstrain in the Al sublayers varied in the range 0.3-0.5%; cf. Table 1). Independent of the annealing temperature, the highest microstrain in c-Si occurred invariably in lateral directions. The microstrain in the c-Si of the bilayers annealed at 250°C, in particular, decreased upon continued annealing and an approximately isotropic state of microstrain was obtained: about 0.7% in the perpendicular direction and 0.9% in lateral directions after annealing 5 hours (and beyond; cf. Table 2). The microstrain of the Si phase is strongly anisotropic in the initial stage of annealing, which can also be interpreted as a consequence of laterally constrained crystallization/growth of c-Si at grain boundaries of the original Al sublayer.

4. General discussion

4.1 Redistribution of Al and Si

On a thermodynamic basis, it has been shown by the present group of authors [18] that the c-Al→c-Si layer exchange proceeds as follows: (i) upon annealing Al GBs in the Al sublayer are wetted by an amorphous Si layer; (ii) when the thickness of this a-Si layer reaches a critical value of ~4 ML of Si atoms, crystallization of Si takes place; (iii) upon continued inward diffusion of Si atoms along the newly formed c-Al/c-Si interfaces, c-Si grains grow laterally in the Al sublayer and eventually the growing c-Si grains have replaced the Al sublayer.

As a consequence of the inward diffusion of Si into the Al sublayer through Al GBs and, thereafter, c-Al/c-Si interfaces, free space is left in the original a-Si sublayer. The intrinsic (initial) and the thermal compressive stress (see section 3.3), cause compressive stress built up in the original Al sublayer upon annealing. Therefore, Al atoms from the original Al sublayer tend to occupy the free space in the original a-Si sublayer to relax the stress. Consequently, Al grains (clusters) may form adjacent to the original a-Si/c-Al interface in front of the wetted Al GBs. These new Al grains obstruct the further transport of Si towards and into the original Al sublayer: a retardation of the material redistribution in the bilayer can occur, as observed (figure 1; section 3.1). In a next stage, the GBs between these new Al grains formed in the original a-Si sublayer may be wetted by Si, originating from the remaining part of the a-Si layer, and then be replaced by small Si crystallites. As a consequence, Al of the new grains is “driven” into the remaining part of the original a-Si sublayer and Si can diffuse into the original Al sublayer along c-Al/c-Si interfaces formed before in the original Al sublayer. By continuation of the above described processes Al is “driven” from its original location: by repeated formation of new Al grains and their subsequent replacement by c-Si grains, Al moves step-by-step towards the surface of the original a-Si sublayer. In the end a mixed, c-Al/c-Si layer has developed at the location of the original a-Si sublayer, whereas a continuous c-Si layer has formed at the location of the original Al layer. For a more detailed discussion of this mechanism, see reference [9].

4.2 Driving force for c-Al→c-Si layer exchange

A sputter deposited thin film system is generally remote from an equilibrium state. Upon annealing, the energy stored in the thin film system can be reduced via various mechanisms, such as mass transport, grain growth (recrystallization) for crystalline material, and crystallization for amorphous material, etc. From a thermodynamic point of view, the Gibbs energy change in the system upon annealing defines a driving force, which determines not only the overall direction of the kinetic process, i.e. change from a non-equilibrium state to an equilibrium state, but also the process rate. For the ALILE process, the total Gibbs energy change in the system upon annealing consists of at least four major contributions [3, 4]: 1) the Gibbs energy change due to crystallization of the a-Si (ΔG_1); 2) the energy change due to the relaxation of macrostress and microstrain and the grain growth of the Al phase (ΔG_2; a quantitative assessment of ΔG_2 can be made on the basis of the macrostress, microstrain and crystallite-size data provided by this paper); 3) the surface energy change (ΔG_3); 4) the sublayer interface-energy change (ΔG_4). For a successful, quantitative approach to interface thermodynamics, see Ref. 19.

The largest gain in energy upon annealing is the result of the crystallization of a-Si, but this energy gain in principle is independent of the occurrence of layer exchange. Energy changes that are sensitive to the layer exchange are the contributions of the relaxation energy of macrostress and microstrain and grain growth for the Al phase upon its (fragmentation and) relocation, and the surface and interface energy changes. Then, because the energy change in bulk energy of the Al phase is negative and the surface- and interface-energy changes are positive, the total Gibbs free energy change, i.e. $-(\Delta G_2+\Delta G_3+\Delta G_4)$, can be defined as a net driving force for the layer exchange. It has been shown that for an anneal of 1 hour a critical temperature of 240°C exists for the occurrence of the layer exchange [20]. This prediction is consistent with the experimental observations as shown in figure 1: the Al atoms in the original Al sublayer had moved only partially into the original Si sublayer upon annealing at 165°C for up to 10 days; but more than 90% of the Al atoms had moved into the original Si sublayer upon annealing at 250°C for 60 minutes. Below 240°C, for example by annealing an a-Si/c-Al bilayer at 165°C for 1 hour (or even for ten days), the net driving force for the layer exchange is (even) negative and layer exchange is not expected to occur. The negative driving force for layer exchange for low-temperature (below 240°C) annealing is mainly due to the smaller energy change due to (the more limited) grain/crystallite growth of the Al phase.

5. Conclusions

- Upon annealing the a-Si/Al bilayer, redistribution of Si and Al takes place in association with the crystallization of amorphous Si. The c-Al→c-Si layer exchange occurs upon annealing at 250°C for more than 1 hour (completely) and at 165°C for up to 10 days (partially).
- Upon annealing the a-Si/Al bilayer, the compressive macrostress parallel to the surface (comprising the intrinsic and thermal stresses) of the Al phase relaxes completely, which is ascribed to the outward diffusion of Al to the location of the original a-Si sublayer.
- The crystallite size of the Al phase increases upon annealing and is generally larger in the lateral directions than in the perpendicular direction due to less constrained grain growth of the Al crystallites in lateral directions. At the same time, the microstrain of the Al phase relaxes.
- Upon annealing the a-Si/Al bilayer, initially a tensile macrostress develops in the nano-crystalline Si phase due to the volume reduction of the Si phase upon its crystallization. Continued annealing causes a relaxation of this stress.
- The crystallite size of the nano-crystalline Si is strongly anisotropic: opposite to the result obtained for the Al phase, the crystallite size in the lateral directions is much smaller than the one in the perpendicular direction. This reflects the constrained nature of lateral growth of crystalline nuclei of Si developing at Al grain boundaries in the original Al sublayer.
- The microstrain of the Si phase is also strongly anisotropic: the larger microstrain values occur in lateral directions, which is (also) interpreted as a consequence of constrained crystallization/growth of c-Si at grain boundaries in the original Al sublayer.

References

1. Gall, S., Schneider, J., Klein, J., Hubener, K., Muske, M., Rau, B., Conrad, E., Sieber, I., Petter, K., Lips, K., Stoger-Pollach, M., Schattschneider, P. & Fuhs, W., 2006, *Thin Solid Films*, **511**, 7.
2. Nast, O., Puzzer, T., Koschier, L.M., Sproul, A.B. & Wenham, S.R., 1998, *Appl. Phys. Lett.*, **73**, 3214.
3. Zhao, Y.H., Wang, J.Y. & Mittemeijer, E.J., 2004, *Appl. Phys. A*, **79**, 681.
4. He, D., Wang, J.Y. & Mittemeijer, E.J., 2005, *Appl. Phys. A*, **80**, 501.
5. He, D., Wang, J.Y. & Mittemeijer, E.J., 2005, *J. Appl. Phys.*, **97**, 093524.
6. Nast, O. & Wenham, S.R., 2000, *J. Appl. Phys.*, **88**, 124.
7. Jenq, K., Chang, S.S., Lian, Y., Pan, G.Z. & Rahmat-Samii, Y., 2004, *Mat. Res. Soc. Symp. Proc.*, **808**, A4.20.1.
8. Khalifa, F.A., Naseem, H.A., Shultz, J.L. & Brown, W.D., 1999, *Thin Solid Films*, **355-356**, 343.
9. Wang, J.Y., Wang, Z.M. & Mittemeijer, E.J., 2007, *J. Appl. Phys.*, **102**, 113523.

10. Vermeulen, A.C., Delhez, R., de Keijser, Th.H. & Mittemeijer, E.J., 1995, *J. Appl. Phys.*, **77**, 5026.
11. Welzel, U., Ligot, J., Lamparter, P., Vermeulen, A.C. & Mittemeijer, E.J., 2005, *J. Appl. Cryst.*, **38**, 1.
12. Brandes, E.A. & Brook, G.B., 1992, *Smithells Metals Reference Book*, 7th ed., (Oxford: Butterworth-Heinemann Ltd) p. 15-5.
13. Delhez, R., de Keijser, Th.H. & Mittemeijer, E.J., 1982, *Fresenius J. Anal. Chem.*, **312**, 1.
14. de Keijser, Th.H., Langford, J.I., Mittemeijer, E.J. & Vogels, A.B.P., 1982, *J. Appl. Cryst.*, **15**, 308.
15. Suryanarayana, C. & Grant Norton, M., 1998, *X-Ray Diffraction* (Plenum Press), p. 213.
16. Delhez, R., de Keijser, Th.H., Mittemeijer, E.J. & Langford, J.L., 1998, *Aust. J. Phys.*, **41**, 213.
17. Welzel, U. & Mittemeijer, E.J., 2005, *Powder Diffraction*, **20**, 376.
18. Wang, J.Y., He, D., Zhao, Y.H. & Mittemeijer, E.J., 2006, *Appl. Phys. Lett.*, **88**, 061910.
19. Sommer, F., Singh, R.N. & Mittemeijer, E.J., *J. Alloys Compds.*, in press.
20. He, D., Wang, J.Y. & Mittemeijer, E.J., 2005, *Scr. Mater.*, **54**, 559.

Acknowledgements. We thank Mr. F. Thiele from the Max Planck Institute for Metals Research for sputter deposition of the layers.

Structural state of $Eu_2(MoO_4)_3$ single crystal after different thermobaric treatments

E. A. Kudrenko[1,*], I. M. Shmyt'ko[1], V. V. Sinitsyn[1], E. G. Ponyatovsky[1], B. N. Red'kin[1]

[1] Institute of Solid State Physics RAS, Chernogolovka, Moscow district 142432, Russia
[*] lenak@issp.ac.ru.

Keywords: rare-earth compounds, high-pressure amorphization, X-ray diffraction, dilatometry study

Abstract. The X-ray and dilatometric studies of $Eu_2(MoO_4)_3$ single crystal after different thermobaric treatments were carried out. The compression of the initial β-phase of $Eu_2(MoO_4)_3$ single crystal to P=9 GPa at room temperature results in new high-pressure phase (*hpp*) formation which is retained at decompression to P=1 atm. The *hpp* phase undergoes phase transition back into β phase around T=860-880 K at heating. The *hpp*–β phase transition is accompanied by the unusual structure state formation when coherent reflections are essentially dropped and all x-ray diffraction intensity of the sample is transformed into the background-like spectrum.

1. Introduction

Solid state amorphization (SSA) under pressure is actively studied last decades [1-3]. It was found more hundred substances which undergo SSA and some of them remain in the amorphous state at the releasing pressure to 1 atm. These investigations have been usually performed on the polycrystalline samples at the quasihydrostatic conditions whereas only a few works were devoted to SSA studies both on polycrystalline and single crystal samples. Particularly unexpected results were found at investigations of $AlPO_4$ (berlinite) single crystal which became amorphous at pressures ~ 15 GPa but transformed back into a single crystal state with the same orientation (memory-glass effect) at decompression to 1 atm [4,5].
Recently the quite similar results were found for $Eu_2(MoO_4)_3$ [6]. The thermodynamically stable phase of the crystals at ambient conditions is the monoclinic α–phase, which usually is not formed by traditional methods and samples are in metastable tetragonal β-phase [7].
The high pressure investigations of polycrystalline β-$Eu_2(MoO_4)_3$ using both "*in situ*" and thermobaric treatments methods revealed amorphization in the compound at $P \approx 7$-9 GPa [8-11]. The amorphous state was retained after releasing pressure to 1 atm and transformed back into crystalline β-phase only at sample heating to ~723 K.

However β-$Eu_2(MoO_4)_3$ single crystal demonstrated unlike behavior. Instead of amorphization the new tetragonal high pressure phase (*hpp*) was observed at pressure treatment ~9 GPa and subsequent decompression to 1 atm [6]. Moreover the samples remained in the single crystal state in spite on essential difference in the specific volumes (~18%) between β and *hpp* phases.
To clarify the thermal stability of the new high pressure phase and structure relationship of the phase with the initial β-phase we performed the x-ray and dilatometric studies of $Eu_2(MoO_4)_3$ single crystals subjected different thermobaric treatments.

2. Results and discussion

The sluggish kinetics of the β - *hpp* phase transition in $Eu_2(MoO_4)_3$ at P=9 GPa allowed us to prepare two-phase (β+*hpp*) or single phase (*hpp*) samples by varying the duration of the high-pressure treatment. These samples were the objects of further x-ray studies at ambient pressure.

A sample of a *c*-cut single crystal was exposed to P=9 GPa at room temperature for 7 days. The diffraction pattern of this sample measured immediately after its decompression to 1 atm is presented in figure 1 (curve b) together with the pattern of the initial β-phase (curve a). One can see that pattern (b) consists of strong reflections of the *hpp* phase and weak reflections of the β-phase. The lines of the *hpp* phase are broadened compared to those of the β-phase. The relative intensities of the β-phase reflections significantly increased during the sample storage at ambient conditions (curves c and d), but these were only of the (00ℓ) type. The presence of reflections only from the planes parallel to the cut surface indicates that the sample consisted of two single-crystalline domains of the β and *hpp* phases. As the β - *hpp* phase transition was accompanied by a large volume effect of about 18%, the structural matching of the domains was an interesting problem. To examine the matching, we constructed 2D patterns for the cross sections in

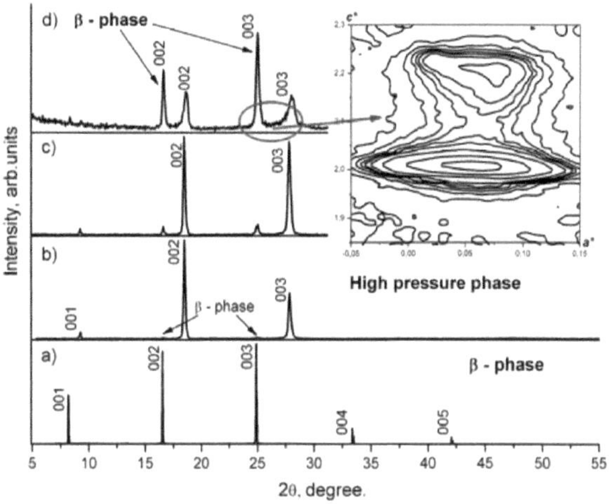

Figure 1. X-ray diffraction patterns of a sample of a c-cut $Eu_2(MoO_4)_3$ single crystal in the initial state of the β-phase (a) and after being exposed to P=9 GPa at room temperature for 7 day and decompressed to 1 atm: (b)- the two phase (β+hpp) state immediately after the decompression; (c)- 2 months later and (d)- 9 months later. The inset shows a 2D section in the reciprocal space in the region of the $(002)_{\beta\,phase}$ and $(002)_{hpp\,phase}$ reflections.

the reciprocal space of the (*00ℓ*) base reflections. As seen from the insert in figure 1, the related reflections of the *β*- and *hpp* phase are connected by a junction region outlined by a few isolines of the diffraction intensity. The presence of the junction region indicates that the transition between the phases proceeds via the formation of transition zones characterized by a continuous and smooth variation of the crystal lattice parameters from $d(002)_{\beta\ \mathrm{phase}}$ to $d(002)_{hpp\ \mathrm{phase}}$. These interphase boundaries are responsible for the pronounced scattering intensity between the related reflections clearly seen in pattern (fig. 1 d).

The pressure treatment of the $Eu_2(MoO_4)_3$ single crystals with an *a*-orientation under the same conditions of P=9 GPa and room temperature resulted in the complete transition of the samples into the *hpp* phase. The samples were investigated by x-ray diffraction and dilatometry in the regime of heating. Figure 2 demonstrates the temperature dependence of the relative thermal expansion ($\Delta l/l$) measured in the temperature range from 353 K to 873 K. The $\Delta l/l$ parameter displays a rather complex behavior at temperatures 820-860 K. It steeply increases by ~6.17% at 820–826 K, drops by ~1.2% at 826 K and then gradually rise at 830-870 K. The volume change accompanying the phase transition is ~ 18-18.6% assuming a similar $\Delta l/l$ effect in the *c* direction. This result well agrees with the earlier x-ray data [6]. The $\Delta l/l$ dependence measured on further cooling showed no anomalies and the x-ray examination at room temperature demonstrated that the sample had completely transformed to the *β*-phase.

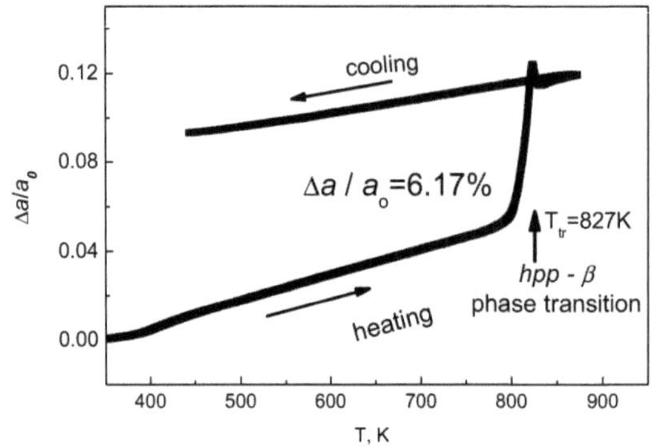

Figure 2. The dilatometric expansion along the a-direction of the hpp phase of $Eu_2(MoO_4)_3$.

Detailed structure investigations of the *hpp–β* phase transition were performed *in situ* in the course of a step-wise sample heating in the high-temperature chamber HTK 2000 (ANTON PAAR). The temperature was increased in steps of about 15 K and the sample was exposed to every fixed temperature for 15 min before the x-ray spectrum was recorded. We were mostly interested in the ratio of the integral intensity $I_{(200)}$ of the (200) reflection to the integral intensity I_b of the background. This ratio gives the degree of the *hpp–β* transition that is important for the analysis of the structure mechanism of the transition.

Some of the x-ray patterns measured during the continuous sample heating in the high-temperature chamber are shown in figure 3. The initial spectrum (figure 3a) corresponds to the *hpp* phase with $I_{(200)}/I_b \approx 2.83$. As the sample temperature is increased to 823 K, the fraction of the crystalline part (producing Bragg's reflections) of the sample decreases by a factor of 47 as compared to the initial spectrum of the *hpp* phase. This state (figure 3b) is characterized by a decrease in the intensity of the (200) reflection of the high-pressure phase virtually to the background value and by a significant increase in the background intensity. In this case, $I_{(200)}/I_b$ is equal only to 0.06, which actually corresponds to the transformation of the reflection intensity into the background. It should be noted that the observed (200) reflections are related to the surface layer with a thickness of ~0.3 μm and the diffraction from practically the whole sample is concealed in the background. The unusual sample state at T=823K might be described as a state without long-range order but it cannot be attributed to the amorphous state because the single crystal of the β–phase with the initial orientation is formed at further heating (similar to the memory-glass effect in $AlPO_4$). Thus the diffraction pattern of the sample in this state is characterized by a nearly uniform intensity distribution throughout the whole range of diffraction angles from 5 to 70^0. Further increase in the sample temperature to 898 K (figure 3c) leads to its transformation to a single crystal of the β–phase with the initial crystallographic orientation.

An analysis of the *hpp* spectrum (figure 3a) allowed us to conclude that the sample possesses a nano-fragmented structure with a characteristic crystallites size of ~30 nm that impairs the scattering coherence and broadens the reflections.

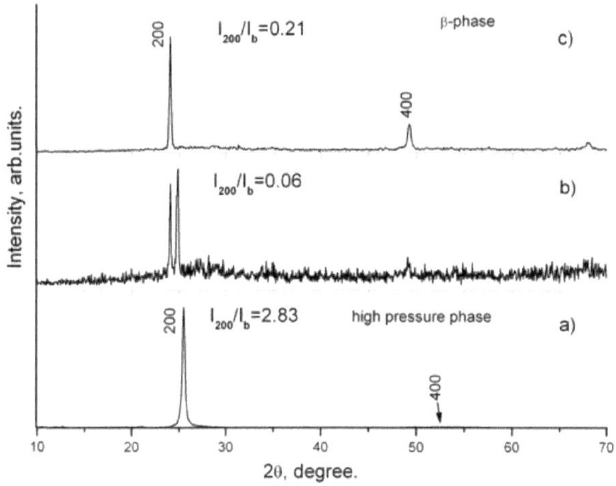

Figure 3. The evolution of the diffraction spectrum of the hpp phase with increasing temperature: (a) 300 K; (b) 823K; (c) 898K.

The volume fraction of grains of the β–phase increases on heating that leads to the gradual transformation of the diffraction pattern to the continuous, background-like spectrum (figure 3b). Taking into account the occurrence of interphase boundaries with a continuous variation of the lattice spacings (see insert in figure 1), this intermediate state of $Eu_2(MoO_4)_3$ can be described as mixture of nano-grains of the *hpp* and β–phases with a random distribution of the interplanar distances inside them. The nano-grains of the *hpp* and β–phases continuously transform to each other within the transitional regions, in which the mean lattice parameters of the β-phase gradually change into the mean lattice parameters of the *hpp* phase. Moreover, the interatomic planes in these nano-grains and intermediate regions are orientated along the same direction (*100*) that is confirmed by the reverse transformation of the sample into a β–phase single crystal with the initial (*100*) orientation (figure 3c). A simulation of the diffraction pattern for such a structure state has shown it to look like a smooth continuous background indeed [12].

3. Concluding remarks

The performed x-ray and dilatometric studies of $Eu_2(MoO_4)_3$ single crystals indicate that the reverse *hpp*-β phase transition displays complex and many steps character. Moreover the initial sample orientation remains the same during whole sequence of transformation:

β-phase $\Rightarrow$ treatment at P=9 GPa, room temperature, one week $\Rightarrow$ hpp phase $\Rightarrow$ heating to 820-840 K at P=1 atm $\Rightarrow$ β-phase

The reversibility of the transition into β–phase with the initial orientation results from the unusual structure state formation when coherent reflections are essentially dropped and all x-ray intensity is transformed into the background-like spectrum. The behavior is explained in the framework of nano-structured state characterizing by the definite orientation of the crystallographic stacks with a random changes of a lengths between them. It is assumed that namely the state revealed as an amorphous spectrum at the pressure treatment of the polycrystalline $Eu_2(MoO_4)_3$ sample. By analogy with the memory-glass effect observed in $AlPO_4$ single crystal, the reversible transformations of $Eu_2(MoO_4)_3$ single crystals under described thermobaric treatments can be called as a memory-crystal effect.

References

1. Ponyatovsky, E.G. & Barkalov, O.I., 1992, *Mater. Sci. Rep.*, **8**, 147.
2. Sharma, S.M. & Sikka, S.K., 1996, *Mater. Sci.,* **40**, 1.
3. Arora, A.K., 2002, in *High Pressure Phenomena*, edited by R.J. Hemley *et al.*, (Amsterdam: IOS Press), p. 545.
4. Kruger, M.B. & Jeanoloz, R., 1990, *Science,* **249**, 647.
5. Gillet, P., Badro, J., Varrel, B. & McMillan, P.F., 1995, *Phys. Rev. B*, **51**, 11262.
6. Shmyt'ko, I.M., Kudrenko, E.A., Sinitsyn, V.V., Red'kin, B.S. & Ponyatovsky, E.G., 2005, *JETP Lett.*, **82** (7), 409.
7. Brixner, L.H., Barkley, J.R. & Jeitschko, W., 1979, *Handbook on the Physics and Chemistry of Rare Earths*, edited by K.A. Gscheidner Jr. & L.R. Eyring, **Ch.30**, p. 610.
8. Pal'nichenko, A.V. Ponyatovsky, E.G. Red'kin, B.S. & Sinitsyn, V.V., 1998, *JETP Letters*, **68**, 657.
9. Dmitriev, V., Sinitsyn, V., Dilanian, R., Machon, D., Kuznetsov, A., Ponyatovsky, E., Lucazeau, G. & Weber, H-P., 2002, *J. Phys. Chem. Solids*, **64**, 307.
10. Shmurak, S.Zh., Kiselev, A.P., Sinitsyn, V.V., Shmitko, I.M., Aronin, A.S., Red'kin, B.S. & Ponyatovsky, E.G., 2006, *Solid State Phys. (rus),* **48**(1), 48.
11. Kiselev, A.P., Shmurak, S.Zh., Red'kin, B.S., Sinitsyn, V.V., Shmitko, I.M., Kudrenko, E.A. & Ponyatovsky, E.G., 2006, *Solid State Phys. (rus)*, **48**(8), 1458.
12. Smitko, I.M., Kudrenko, E.A. & Solovev, A.V., to be published.

Acknowledgements. The work was supported by RFBR № 04-02-017143, 05-02-08302, 06-02-17298 and the RAS Program "Physics and mechanics of high compression substances". One of the authors (EK) thanks Organizing Committee for her financial support.

Z. Kristallogr. Suppl. **27** (2008) 211-218 / **DOI** 10.1524/zksu.2008.0027

Anomalous structure of nanocrystallites of rare-earth compounds produced by sol-gel methods

I. M. Shmytko[1,*], E. A. Kudrenko[1], G. K. Strukova[1], V. V. Kedrov[1], N. V. Klassen[1]

[1] Institute of Solid State Physics of Russian Academy of Sciences, Chernogolovka, Moscow distr. 142432, Russia
[*] shim@issp.ac.ru

Keywords: nanoscaled rare-earth compounds, X-ray diffraction investigations

Abstract. Unusual two-phase state is discovered in nanocrystallites of simple rare-earth oxides Lu_2O_3, Gd_2O_3, Y_2O_3, La_2O_3 and Eu_2O_3, garnet $Y_3Ga_5O_{12}$ and $Na_5Lu_9F_{32}$ at first stages of their formation. X-ray diffraction investigations have shown that both these phases are isomorphous and are realized in the same crystallite. It is concluded that over early nanocrystallization stages nanograins consist of surface- and core-phases and that the surface-phase has a larger lattice parameter, compared to the core-phase parameter.
Unlike the afore mentioned cubic structures nanograins of orthorhombic LuF_3 are plate like at first stages of their formation and have the surface-phase with continuously changing interplanar distance from the outer developed face into the inner part.

1. Introduction

Today physical properties of nanodispersed systems are of keen scientific interest. This is because these systems have unusual properties and structure which cannot be described in the context of currently existing laws. Also, they make possible producing new materials and instruments with unique physical characteristics. An analysis of early publications shows that nanoscience was particularly successful with metals and simple chemical compounds [1, 2]. Now complex chemical systems have become very attractive. Herewith special attention is payed to the complex rare-earth compounds because of their practical importance. Rare-earth garnets ($R_3Me_5O_{12}$), perovskites ($RMeO_3$), borates (RBO_3), molibdates ($R_2(MoO_4)_3$), where R is rare-earth element, and some fluorides (LuF_3 and $Na_5Lu_9F_{32}$) are good scintillators. Moreover, in some cases the ceramics incorporating nanogranules of rare-earth compounds have more effective scintillation characteristics than scintillators made from single crystals [3, 4]. This circumstance has defined our interest to investigation of thin structure features of rare-earth compounds at first stages of the nanocrystallite formation (D~10-40 nm) and change of the structure with their size increasing (D> 100 nm).

2. Samples and investigation procedures

Rare-earth oxides Lu_2O_3, Gd_2O_3, Y_2O_3, La_2O_3 and Eu_2O_3, garnet $Y_3Ga_5O_{12}$, borate $LuBO_3$, molibdate $Eu_2(MoO_4)_3$ and fluorides LuF_3 and $Na_5Lu_9F_{32}$ have been chosen for the investigations. It is known that bulk crystals of simple rare-earth oxides and garnets at room temperature have the cubic cells. For example at room temperature Eu_2O_3 has the cubic structure with space group symmetry (S.G.) *Ia(-)3* and the lattice parameter a=10.85 Å. At 1100^{0}C this structure transforms into monoclinic one with S.G. *C/2m* and the lattice parameters a=14.082Å, b=3.604Å, c=8.778Å, β=100^0, [5].
$Y_3Ga_5O_{12}$ exists in cubic modification with S.G. *Ia3d* only and has the lattice parameter a=12.27 Å, [6].
Lutetium borate $LuBO_3$ undergoes a set of polymorphic transformations, [7]. At room temperature lutetium borate has the equilibrium calcite phase (S.G. *R(-3)c*, a=4.91Å, c=16.21Å), which is stable up to 1310^{0}C. At more elevated temperatures calcite phase transforms into vaterite form with S.G. $P6_3/mmc$ and the lattice parameters a=3.73Å, c=8.71Å.
In $Eu_2(MoO_4)_3$ crystals two stable crystalline modifications are realized: the high temperature β-phase $(P42_1m)$ and the low temperature α-phase *(C2/c)* [8]. At heating $Eu_2(MoO_4)_3$ to 881^{0}C the $\alpha \Rightarrow \beta$ transformation takes place which is accompanied by a significant volume effect ~ 18 %, [8, 9]. So, the great difference in the cell volumes of α- and β-phases retards the kinetics of the inverse $\beta \Rightarrow \alpha$ transition and conserves the β-phase down to 180^{0}C at which europium molibdate undergoes the first order phase transition into metastable β'–phase with S.G. *Pba2*, [8, 10].
Lutetium fluoride has an orthorhombic cell with space symmetry *Pnma* and lattice parameters a=6.15 Å; b=6.76 Å; c=4.47 Å. In air this structure is decomposed into oxyfluorides at 650-700^{0}C.
Sodium-lutetium fluoride $Na_5Lu_9F_{32}$ has the cubic structure with S.G. *Fm3m* up to melting and the lattice parameter a=5.463 Å at room temperature.
The method of synthesis of nanodispersed simple oxides was based on the solution of microscopic powders of simple oxides in nitric acid HNO_3 and consequent fast decomposition of the solvent at high temperature, [11, 12]. As a result, the crystallites of ~10 nm in size were produced. The subsequent high-temperature annealing of the obtained resulted in a growth of dimensions of the synthesized crystallites from nano- to micro-scale.
The synthesis procedure of nanodispersed rare-earth complex oxides was almost the same. The solutions of rare-earth simple oxides or nitrates and other required compounds (Al_2O_3, Ga_2O_3, B_2O_3, MoO_2 and so on) in nitric acid HNO_3 or in a melt of NH_4NO_3 have been heated at elevated temperature up to decomposition of the solvent. As a result, an intermediate state (so called precursor state) was obtained. Electron microscopy and X-ray investigations showed that the precursor state was an amorphous state. Successive annealing of the precursors at increased temperatures results in a growth of the nanocrystallites of the complex oxides.
The methods of synthesis of nanodispersed fluorides LuF_3 and $Na_5Lu_9F_{32}$ were slightly different. The crystalline form of lutetium fluoride was obtained by exchange reaction between water-soluble salts of lutetium (nitrate or chloride of lutetium) and NH_4F. The reaction was carried out in water at room temperature and initial concentration of the lutetium salt within 0.05-2.0 M. Ammonium fluoride twice exceeded the stoichiometric amount required for the

reaction: $Lu(NO_3)_3 + 6(NH_4F) = (NH4)_3\ LuF_6 + 3(NH_4NO_3)$. The resultant sediment of the complex salt of lutetium fluoride was filtered and cleaned in water and pure ethanol and after that dried on a filter. The produced dry powder was kept for 1 hour at 100°C and then annealed at 350-400°C for 3-5 hours. In the process of annealing the complex salt of lutetium fluoride decomposed with the formation of the arid crystalline powder of LuF_3.The size of the crystallites being about 30-40 nm.

The crystalline phase of $Na_5Lu_9F_{32}$ was obtained in the following way. The mixture of the obtained nanodispersed powder of LuF_3 with NH_4F and NaF in stoichiometric ratio of $Na_5Lu_9F_{32}$ was annealed at 900 °C, the annealing time ranging from a few tens of seconds to a few minutes. The longer the annealing time the bigger the size of the crystallites was.

The structural states of the synthesized compounds were investigated by X-ray diffraction and electron-microscopy methods. An X-ray diffractometer D500 (Siemens) with monochromatic Cu-K_α radiation and an electron microscope JEM-100CX II were used. Precise chemical content of the compounds synthesized was done by means of an X-ray spectroscopy analysis on a SUPRA 50VP spectrometer.

3. Results and discussion

The example of structure transformations of simple rare-earth oxides at first stages of nanograin formation emerged after decomposition of the solvent is shown in figure 1. The diffraction spectrum (a) belongs to the powder obtained immediately after decomposition of the solvent at T=300^0 C from a solution of Eu_2O_3 in nitric acid HNO_3. The spectrum corresponds to the crystalline phase of europium nitrate $EuONO_3$. The diffraction spectrum (b) belongs to the phase state obtained after annealing the synthesized europium nitrate for 20 minutes at T=400^0 C. The specific feature of the spectrum is the splitting of the reflections. The fitting of the spectrum onto two sets of the reflections shows that the left-side and right-side components belong to two cubic phases which differ insignificantly in the lattice parameters. Here the right-side set of the components belongs to a cubic lattice known for bulk

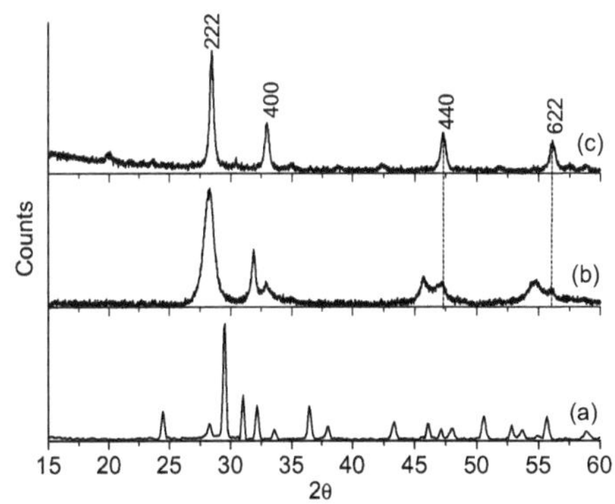

Figure 1. Diffraction spectra of Eu_2O_3 obtained by solvent decomposition method: (a)- immediately after solvent decomposition at 300^{0}C; (b) – after annealing the a-state at 400^{0}C for 20 minutes; (c) – after annealing the a-state at 800^{0}C for 20 minutes.

crystals of Eu_2O_3 and the left-side components of the reflections belong to a cubic phase with larger lattice parameter. It is worthwhile to emphasize that the intensities of the left components are stronger than the intensities of the right components. The fitting parameters for the right set of the components are shown in table 1. Also, the sizes of crystallites calculated from half widths of the reflections by using Selyakov's approximation are shown in the last column. The coincidence of the dimensions of the crystallites calculated for different reflections suggests that the widths of the reflections really display the dimensions of the crystallites.

Table 1. The results of the fitting of the right side components of the diffraction spectrum of the two-phase state of Eu_2O_3,, see spectrum (b) in figure 1.

(hkl)	2θ, degree	d, Å	Δ(2θ),degree	D(size), nm
222	28.205	3.16	0.89	10
400	32.877	2.722	1.32	7
440	46.629	1.95	2.28	10
622	55.212	1.662	2.50	8

The subsequent annealing of the obtained state at 800^0C for 20 minutes, see spectrum (c) in figure 1, results in the disappearance of the left-side components and narrowing of the remained right-side components. The crystallite size concurrently increased to ~ 70 nm. The lattice parameter of this state corresponds to the microcrystalline state of Eu_2O_3.
The two-phase states were also observed in Gd_2O_3, La_2O_3 и Lu_2O_3. The structure states for which the both components joined into one very wide reflection were registered herewith. In this case it was impossible to evaluate the size of the emerged crystallites. The successive annealing resulted in the splitting of the reflection into two components and transformation of the state into the one-phase state in the end.
Additional features of the first stages of nanocrystallization were obtained in the process of $Na_5Lu_9F_{32}$ synthesis. Figure 2 displays the evolution of the profile of the diffraction reflection (311) of the synthesized phase vs the annealing time at 900 °C of the mixture of LuF_3 with NH_4F and NaF in stoichiometric ratio of $Na_5Lu_9F_{32}$ phase. The same evolution is observed for all reflections of the synthesized phase. It is very clearly seen that after annealing the mixture for 30 and 45 seconds the reflections consist of two components. Here the intensity of the left-side component obtained after a shorter annealing is stronger than the right-side component but as the annealing time is increased the relation of the intensities is opposite. Additional annealing at the same temperature results in a further decrease of the intensity of the left-side components which finally are converted into asymmetrical left shoulder of the remaining right-side reflections. The final structure state corresponds to one-phase state with lattice parameters known for bulk crystals of $Na_5Lu_9F_{32}$. It is worthwhile to emphasize that in the process of fine structure reconstruction over the early stages of crystallite formation the positions and half-widths of the both reflection components change to the half-width and position of the final-state reflection. Here the left-side component has always a larger parameter and the right-side component has a smaller parameter than the final struc-

ture state. At successive stages the both-side components approach the final-state component however the intensity of the left-side component is drastically decreased and the right-side-component intensity conversely is increased sharply and converted into the final state component. The rest of the left-side component is transformed into the more intensive left-side shoulder of the final reflection.

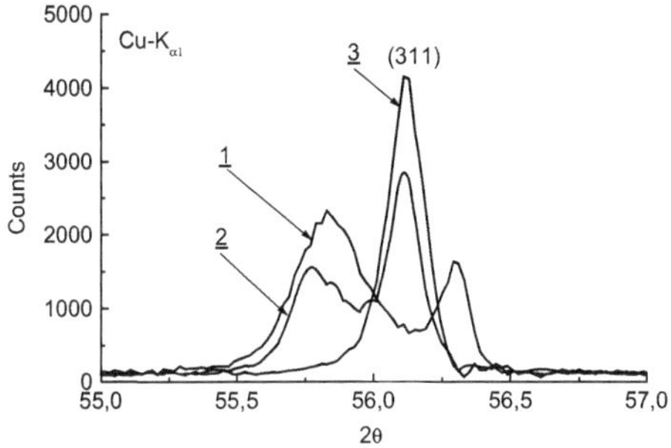

Figure 2. Evolution of the separate reflection of $Na_5Lu_9F_{32}$ in the process of annealing at 900 °C: 1-30 seconds; 2-45 seconds; 3-3minutes.

A question arises: "Do these two phases which are represented in diffraction spectra by the two components of the reflections belong to different nanograins or do they exist in the same grain?" The answer to this question was derived from the measurement of the chemical content of the two- and single-phase states of $Y_3Ga_5O_{12}$ which were produced via a long annealing of the precursor state at 1250 °C, see figure 3. The fitting of the spectrum (a) shows that the integral intensity of their left-side components constitutes ~ 30 % of the integral intensity of the whole spectrum. The lattice parameter difference of the components constitutes ~ 0.8 %. The half- widths of the both components for different reflections can be presented as

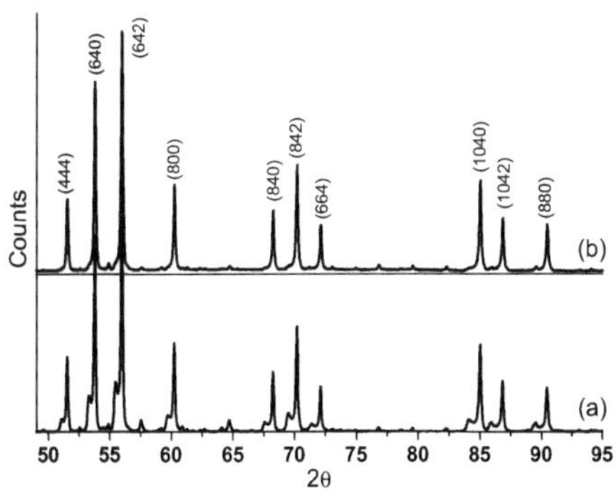

Figure 3. High angular part of the diffraction spectra of $Y_3Ga_5O_{12}$ in two-phase state (a)obtained at annealing of the precursor at 1250 °C for 10 hours and single-phase state (b)obtained after additional annealing of the phase (a) at the same temperature for 15 hours in air.

reflections from two systems of sizes: *D*=25 nm and *D*=60 nm for left-side and right-side components, correspondingly. This means that if the both phases belong to different nanograins the chemical contents of single- and two-phase states have to be different. However the precise (with accuracy of 1%) content measurements by means of SUPRA 50VP spectrometer show that the percent contents of Y, Al and O atoms are the same for single- and two-phase states and fit the $Y_3Ga_5O_{12}$ content, see table 2.

Table 2. Atomic content of the two- (a) and single-phase (b) states of $Y_3Ga_5O_{12}$.

Element	Weigh t%	Weight % Sigma	Atomic %	Weight %	Weight % Sigma	Atomi c%
O	23.67	1.49	59.75	23.93	1.48	60.07
Ga	44.51	1.72	25.79	44.78	1.74	25.80
Y	32.34	1.64	14.46	31.28	1.63	14.13
Totals			100.00			100.00
Sample	(a)	(a)	(a)	(b)	(b)	(b)

This means that these two phases exist in the same nanograins. In turn this means that an individual nanocrystallite consists of surface- and core-phases at the first stages of formation. The parameter of the surface-phase is always larger than the parameter of the core-phase. Following this model the transformation of the two-phase states into the single-phase state presented in figures 1, 2, 3 means that the sizes of nanograins are increased with temperature and annealing time increasing. Herewith a relative part of surface atoms and, consequently, the amount of the surface-phase drops down with crystallite growth.

The upper results on the two-phase state of nanocrystallites of rare-earth compounds at the first formation stages were derived for cubic structures. A little different but analogous result was derived for orthorhombic structure, see figure 4. In this figure the diffraction spectra of LuF_3 are shown for different annealing temperatures of the precursor. It is very well seen that for the low annealing temperature (T=350°C) the (002) reflection is wide and symmetrical and its gravity centre has a lower diffraction angle than the spectrum of the final state, see curve (3). This means that at early stages of nanocrystallite formation the lattice parameter "*c*" of LuF_3 is larger than that for micro-crystalline state. At higher annealing temperature (T=650°C) the spectrum transforms into the "triangular" curve, see spectrum (2). Such form of the reflection characterizes the continuous change of interplanar distances. An additional annealing of the state (2) at higher temperatures (T> 650°C) results in a sharp increase of the reflection intensity which becomes symmetrical again, see spectrum (3). We believe that all these transformations occur in the same nanocrystallites analogously to the garnet $Y_3Ga_5O_{12}$. Importantly, the described transformations were observed for (00l) reflection only. The forms of the (h00) and (k00) reflections were symmetrical at all annealing stages. The gravity centre positions and the half-widths of these reflections were changed with increasing of annealing temperature only. Here the sizes of the nanocrystallites produced at the annealing temperature T=350°C equal 750 Å and 380 Å along [100] and [001] directions, correspondingly. This means that the growing crystallites are plate-like in form. Such conclusion is supported by optical observation of the form of the micro-crystallites finally obtained after a long annealing at 750 °C.

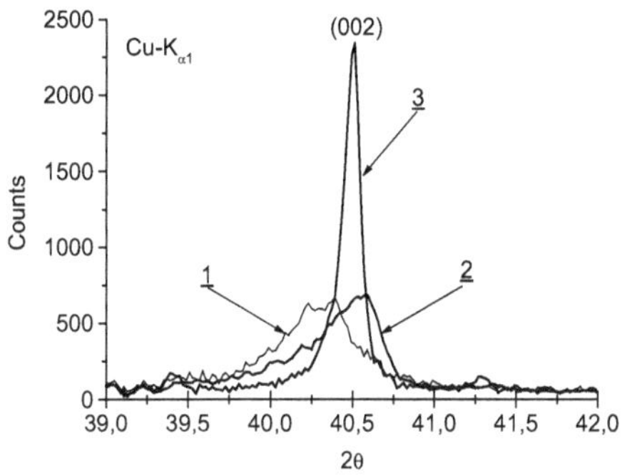

Figure 4. Evolution of (002) reflection of LuF_3 in the process of annealing at 350 °C (1), 650 °C (2) and 750 °C (3) for 3 hours.

Taking it into account the "triangular" spectrum (2) and more intensive left-side shoulder of the spectrum (3) the structure of the plate-like nanograins of LuF_3 can be described by a surface-phase with a continuously changing lattice parameter "*c*" and a core-phase in the inner part of the grain. At early stages of crystallite formation the structure state can be represented as the surface- phase only. This state gives the "triangular" form of the (002) reflection. The growth of the grains results in decreasing of the surface-phase relatively to the core-phase and in the transformation of the "triangular" reflection into more intensive left shoulder of the reflection for the microcrystalline state.

A question now arises as to the physical origin of the two-phase states emerged at the early stages of crystallite formation. Our opinion is that the answer to this question lies in the sizes of nanocrystallites themselves. It is clear that, owing to the additional surface energy, the surface atoms should possess a considerably higher free energy density as compared to the volume atoms. Therefore, in a number of cases, the mean energy of the atoms constituting surface layers may turn out to be equal to the mean energy of the atoms at high temperatures. The higher the temperature the larger the lattice parameters are. This explains why the reflections of the surface-phase are placed at lower diffraction angles. However it is true for nanosized crystallites for which the relative part of surface atoms is big enough and hence the intensity of the surface-phase reflections has to be comparable with reflection intensity of volume (core) phase. It is also clear that a growth of the crystallites results in a decrease of the relative part of surface atoms. In turn the decrease results in lowering of the intensity of the left-side components of the initial two-phase state which finally are converted into the more intensive left-shoulder of the reflections for the microcrystalline state.

The above mentioned mechanism of the origin of the two-phase states emerged at early stages of crystallite formation is supported by emergence the analogous two-phase state in powders of $LuBO_3$ and $Eu_2(MoO_4)_3$ subjected to a long milling. In this case the surface-phase is defined by increased surface energy caused by surface defects arising in the milling.

4. Concluding remarks

In conclusion it is worthwhile to emphasize that the discovered two-phase state of nanocrystallites does not depend on the method of rare-earth compound synthesis. We suppose that for dielectrics to which rare-earth compounds belong the two-phase state is a certain fundamental property of the nanosized crystallites emerged at the first stages of crystallization.

References

1. Gusev, A.I. & Rempel, A.A., 2001, *Nanocrystalline Materials*, (Moscow: Fizmatlit); 2004, Cambridge Int. Sci., Cambridge). Gusev A.I., 2005, *Nanomaterials, Nanostructures, Nanotechnologies* (Moscow: Fizmatlit). [in Russian].
2. Serveev, G.M., 2003, *Nanochemistry* (Moscow: Moscow State University); 2006, Elsevier, Amsterdam).
3. Shmurak, S.Z., Strukova, G.K., Smyt'ko, I.M., et al., 2005, *Nucl. Instrum. Methods Phys. Res., Sect. A*, **537**, 149.
4. Klassen, N.V., Snmurak, S.Z., Shmyt'ko, I.M., et al., 2005, *Nucl. Instrum. Methods Phys. Res., Sect.* A ,**537**, 144.
5. Arsen'ev, A.P., Kovbova, L.M., Bagdasarov, Kh.S., et. al, 1983, in *Rare Earth Compounds. Systems with oxides of I-III Group Elements,* edited by I.V. Tananaev (Moscow: Science), p. 280. [in Russian].
6. Portnoi, K.I. & Tomofeev, N.I., 1986, *Oxygen-Containing Compounds of Rare Earth Elements* (Moscow: Metallurgia). [in Russian].
7. Levin, E.M., Roth, R.S. & Martin, J.B., 1961, *Am. Mineral.*, **46**, 1030.
8. Brixner, L.H., Barkley, J.R. & Jeitschko, W., 1979, in *Handbook on the Physics and Chemistry of Rare Earths,* edited by K.A. Gschneidner, Jr. Eyring & L. Eyring (Amsterdam: North-Holland), **3**, p. 609.
9. Brixner, L.H., 1972, *Mat. Res. Bull.*, **7**, 879.
10. Fouskova, A., 1969, *J. Phys. Soc. Japan*, **27**, 1699.
11. Strukova, G.K., Red'kin, B.S., Kurlov, V.N., et al., 1992, *Ferroelectrics*, **130**, 333.
12. Strukova, G.K, Smirnova, I.S., Bazhenov, A.V. & Dilanyan, R.A., 1996, *Physica C* (Amsterdam) **267**, 67.

Acknowledgements. This work was supported by the Russian Foundation for Basic Research (project nos. 06-02-17298) and Programs of Russian Academy of Sciences "Support of Innovations" and "Structures and Properties of Condensed Matter".

Refinement of extinction-affected X-ray reflection profile of textures

I. Tomov[1,*], S. Vassilev[2]

[1] Jordan Malinowski Central Laboratory for Photographic Processes, Bulgarian Academy of Sciences, Acad. G. Bonchev Str., Block 109, Sofia 1113, Bulgaria
[2] Institute of Electrochemistry and Energy Systems, Bulgarian Academy of Sciences, Acad. G. Bonchev Str., Block 10, Sofia 1113, Bulgaria
[*] iv_tomov@yahoo.co.uk

Keywords: XRD, extinction, texture, integral breadth, FWHM

Abstract. Using first-order approximation of secondary extinction (SE) correction, a method is described for conversion of extinction-affected reflection profile into kinematical (extinction-free) one. In this connection, changes of the two reflection profile parameters integral breadth and FWHM are investigated by means of change of the level of interaction between X-rays and crystal medium. To this end, a controlled reduction of the incident-beam intensity is carried out by means of the transmission factor of a foil crossed by the incident beam. The investigation is confined to 111 and 222 reflections corresponding to the ideal <111> direction of the main texture component of electrodeposited silver coating appearing infinitely thick to the X-rays. It is found that the extinction-induced reflection broadening is controlled by the reflectivity Q, texture factor P and incident-beam intensity I_0.

1. Introduction

Whereas in single-crystal diffractometry the effect of extinction is subject to extensive research in characterizing the domain size and shape of real crystals [1-4], there is only little interest in this topic in conventional X-ray diffractometry despite that extinction is an intrinsic property of textures [5,6]. Neglecting the secondary extinction (SE) effect on the reflection profile distortion of textures is not justified, because this compromises the correct assessment of the microstructure parameters. As a first step to adequate characterization of the microstructure, one needs a method for eliminating the extinction-induced profile distortion, since this distortion causes a respective reflection enlargement. That is why, the aim of the present study is to convert the measured (extinction-affected) profile into a kinematical (extinction-free) one and thus to correctly assess the physical effect due to size-strain broadening controlled by both the integral breadth and the FWHM. To this end, one needs to introduce the SE correction in the intensity measured at each point inside the reflection range.

2. Basic definitions of extinction theory and texture analysis

The formalism considered here is valid for the symmetrical Bragg geometry with a plane-parallel plate sample appearing infinitely thick to the X-rays. According to theory [1-3,7,8], extinction decreases the measured intensity I_m of a reflection with the extinction factor y defined by

$$I_m = yI_{kin}, \tag{1}$$

where I_{kin} is the intensity that a Bragg reflection would have if kinematical theory would apply exactly to the system being examined. The kinematical intensity I_{kin} is expressed as

$$I_{kin} = PI_0QS/2\mu . \tag{2}$$

Here P is the texture factor (pole density) defined by the volume fraction of crystallites whose <*hkl*> poles contribute within the reflection range [9], I_0 is the intensity of the incident beam, S is the cross section of the beam, Q is the reflectivity per unit crystal volume, and μ is the ordinary linear absorption coefficient. As defined, the factor y is always less than unity and has a value averaged over the reflection range. Chandrasekhar [10] gave the following expression for y:

$$y = \mu/\mu_\varepsilon , \tag{3}$$

where μ_ε is an effective absorption coefficient. In the symmetrical Bragg geometry with a plane parallel plate specimen one should use as a first order approximation for the SE correction ε a definition given below [1]:

$$\mu_\varepsilon = \mu + gQ\left(p_2 / p_1^2\right) = \mu + \varepsilon . \tag{4}$$

Here g is the SE coefficient which depends on the crystallographic orientation and microstructure in the probing direction of the textured sample. The symbol p_n denotes the polarization factor for the incident X-ray beam [1]:

$$p_n = \left[1 + \cos^{2n}(2\theta_B)\right]/2 , \tag{5}$$

where n=1,2,..., and θ_B is the Bragg angle of reflection. Equation (4) leads to the formula of Darwin [7] for the SE correction ε. Later [1,10], the polarization p_2/p_1^2 of the incident X-ray beam was incorporated in this formula which can be represented as

$$\varepsilon = gQ\left(p_2/p_1^2\right). \tag{6}$$

Replacing Q with the help of (2) transforms (6) into

$$\varepsilon = kI_{kin}\left(p_2/p_1^2\right), \tag{7}$$

where k, given by

$$k = 2g\mu/PI_0S , \tag{8}$$

is is an *empirical extinction coefficient* having dimension of reciprocal volume [11]. The difference in the behaviour of the coefficients g and k is discussed elsewhere [12].

3. Conversion of the extinction-affected reflection profile into a kinematic one

According to (1), the measured intensity $I_m(2\theta)$ at angle 2θ within the reflection range is defined as:

$$I_m(2\theta) = y(2\theta)I_{kin}(2\theta). \tag{9}$$

The variation of the intensity $I_m(2\theta)$ results in variation of the extinction parameters $y(2\theta)$, $\mu_\varepsilon(2\theta)$ and $\varepsilon(2\theta)$ (see (3), (4) and (7), respectively:

$$y(2\theta) = \mu/\mu_\varepsilon(2\theta), \tag{9a}$$

$$\mu_\varepsilon(2\theta) = \mu + \varepsilon(2\theta), \tag{9b}$$

$$\varepsilon(2\theta) = k(2\theta)I_{kin}(2\theta)(p_2/p_1^2). \tag{9c}$$

Substituting $I_{kin}(2\theta)$ from (9) into (9c) and accounting for (9a) and (9b) reveals that the SE correction $\varepsilon(2\theta)$ depends on the measured intensity $I_m(2\theta)$ as follows:

$$\varepsilon(2\theta) = \mu k(2\theta)I_m(2\theta)(p_2/p_1^2)/[\mu - k(2\theta)I_m(2\theta)(p_2/p_1^2)]. \tag{10}$$

Equation (10) says that the SE correction attains its highest value $\varepsilon(2\theta_B)$ at the reflection maximum and, hence, the intensity $I_m(2\theta_B)$ suffers highest reduction among all of the measured intensities $I_m(2\theta)$. Therefore, due to this reduction the integral breadth B_m and FWHM$_m$ of the extinction-affected profile are larger than the true integral breadth B and FWHM corresponding to the kinematical profile. To refine B_m and FWHM$_m$, a technique is developed for converting the extinction-affected profile into the kinematical one. In this respect, the single reflection method [12] provides an appropriate procedure for such a refinement. Suppose the couple of intensities $I_m(2\theta)$ and $I_{m,t*}(2\theta$, at angle 2θ inside the reflection range is measured at incident-beam intensities I_0 and $I_{0,t*} = I_0\exp(-t*\mu*)$, respectively (Figure 1).

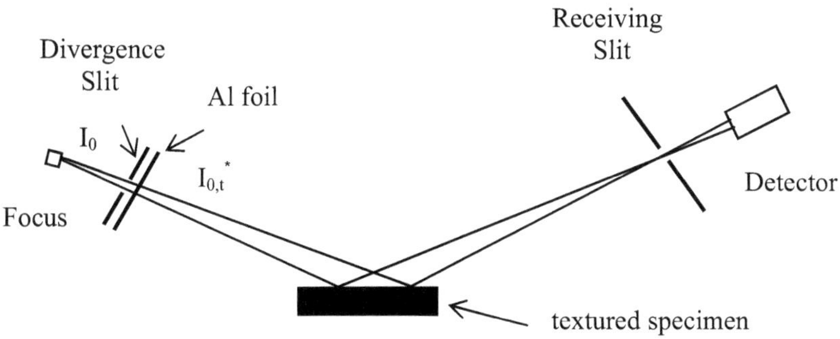

Figure 1. Simplified sketch of Bragg-Brentano diffractometer adapted for controlled reduction of the incident-beam intensity from I_0 to $I_{0,t*}$ by means of a thin foil of known thickness.

Here $t*$ and $\mu*$ are, respectively, the thickness and the linear absorption coefficient of a foil crossed by the incident beam. With (9a), (9b), and (9c) introduced in succession into (9), the intensity $I_m(2\theta)$ of the first measurement is expressed by

$$I_m(2\theta) = \{\mu/[\mu + k(2\theta)I_{kin}(2\theta)(p_2/p_1^2)]\}I_{kin}(2\theta). \tag{11a}$$

To distinguish between the parameters of the first and the second measurement, the parameters of the second one are additionally designated with subscript $t*$, i.e.

$$I_{m,t*}(2\theta) = \{\mu/[\mu + k(2\theta)I_{kin,t*}(2\theta)(p_2/p_1^2)]\}I_{kin,t*}(2\theta). \tag{11b}$$

Since the pole-density distribution $P(2\theta)$ inside the reflection range is the same, the following relation holds at angle 2θ:

$$I_0/I_{0,t*} = I_{kin}(2\theta)/I_{kin,t*}(2\theta) = \exp(\mu * t*). \tag{12}$$

From (11), for $I_{kin}(2\theta)$ and $I_{kin,t*}(2\theta)$ we have the expressions

$$I_{kin}(2\theta)=\{\mu/[\mu-k(2\theta)I_m(2\theta)(p_2/p_1^2)]\}I_m(2\theta), \quad (13a)$$

$$I_{kin,t*}(2\theta)=\{\mu/[\mu-kI_{m,t*}(2\theta)(p_2/p_1^2)]\}I_{m,t*}(2\theta). \quad (13b)$$

Dividing these expressions and taking into account (12) defines $k(2\theta)$ by means of measurement data corresponding to angle 2θ within the reflection range, i.e.

$$k(2\theta)=\mu[\exp(\mu*t*)-(I_m(2\theta)/I_{m,t*}(2\theta))]/I_m(2\theta)[(p_2/p_1^2)\exp(\mu*t*)-1]. \quad (14)$$

4. Experimental

Electrodeposited silver coating (denoted Ag2RE) with a thickness of about 40 μm was used as a model sample. It represents moderate texture whose texture factor in the ideal <111> direction has the value of P=23.5 determined by a method described elsewhere [11]. The XRD measurements of the 111 and 222 reflections were scanned with 0.01° step in 2θ direction by Brucker θ–θ diffractometer using CuK_α-radiation separated by nickel filter. The focusing slit system included Soller slits, divergent and antiscatter slits (0.1°), and a receiving slit (0.1 mm). A solid-state detector (SOLX D8 Advance) was used for detection of the X-radiation. The data-collecting time at each measured point of the XRD pattern was 32 s. As a result of this strategy, the statistical accuracy in the measurement of a particular intensity was higher at stronger extinction suffered by this intensity. Thus, trading time for accuracy, the statistical error in the measurement of a point inside the reflection range may vary from about 1 to 0.1% (or even less).

To measure the same interval in the scattering space corresponding to each peak of a reflection pair, the scanned area was defined in $\sin\theta$ scale, i.e.

$$\sin\theta_2' - \sin\theta_1' = \sin\theta_2'' - \sin\theta_1''. \quad (15)$$

Here the magnitude of the angle interval from θ_1 to θ_2 is selected so that the intensity returns to background level before the two ends of the interval. The superscripts "prime" and "double prime" denote the first- and the second-order peaks, respectively. To reduce the intensity from I_0 to $I_{0,t*}$, Al foil of thickness $t*$=54.5 μm was crossed by the incident beam. This foil was mounted tightly behind the divergent slit.

5. Results and discussion

The procedure for conversion of the extinction-affected intensities into the kinematical ones is presented in Figures 2-4. After striping the $K\alpha_2$ component from the $K\alpha$ doublet, the $K\alpha_1$ component has to be symmetrical. The texture sharpness (P=23.5) in the 111 direction of the sample predetermines a high counting rate for each measured point of the reflection profile. Since the selected data-collecting time of 32 s ensures a very high counting statistics, the measurement data within the reflection range are remarkably unaffected by statistical noise (see Figure 2). This figure shows an example for striping the 111 reflection profile $I_m(2\theta)$ from the $K\alpha_2$ component. Analogous procedure is applied to the $I_{m,t*}(2\theta)$ profile measured with absorbing foil of thickness $t*$. Each couple of intensities corresponding to the same angle 2θ inside the profiles is used for computing the empirical extinction coefficient $k(2\theta)$ (see (14)). Once determined, the coefficient $k(2\theta)$ is further used for calculating the kinematical intensities $I_{kin}(2\theta)$ and $I_{kin,t*}(2\theta)$ that correspond to the measured intensities $I_m(2\theta)$ and $I_{m,t*}(2\theta)$, respectively (see Figures 3 and 4). Since the reflection profile $I_{m,t*}(2\theta)$ is less influenced by extinction than $I_m(2\theta)$, the respective extinction corrections for the kinematical profile $I_{kin,t^*}(2\theta)$ (Figure 4) are smaller than those obtained for the $I_{kin}(2\theta)$ profile (Figure 3).

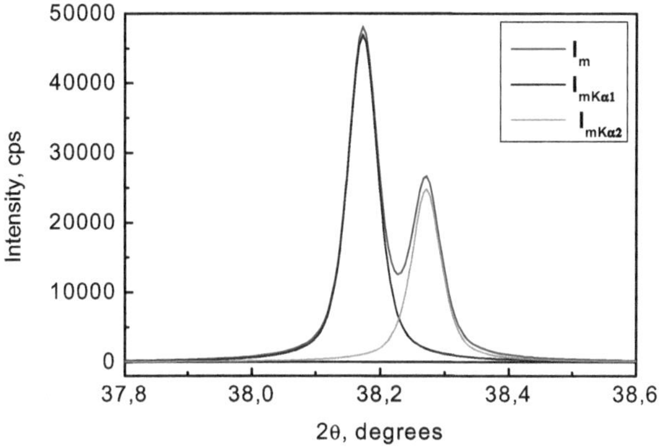

Figure 2. Striping the measured Kα doublet ($I_m(2\theta)$ profile) of the 111 reflection from the $K\alpha_2$ component. After striping, the $K\alpha_1$ component has to be symmetrical.

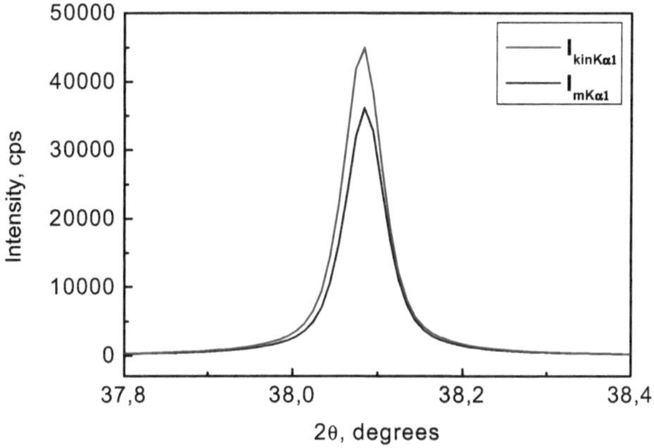

Figure 3. Converting the measured $I_m(2\theta)$ profile of the $K\alpha_1$ component of the 111 reflection into the kinematical $I_{kin}(2\theta)$ profile.

Table 1 compares the integral breadth and FWHM of the measured (denoted by subscript m) and the kinematical (without subscript) intensities of the $K\alpha_1$ component. Integral breadth and FWHM values, listed in the table, are obtained by numerical integration, the Simpson method, and a fitting procedure [13], respectively. In all the cases, integral breadths B_m determined by the numerical integration are systematically smaller than those determined by the fitting procedure. The deviations between the so-obtained values vary from about 0.5% to 1.3% for the 111 and 222 reflections, respectively.

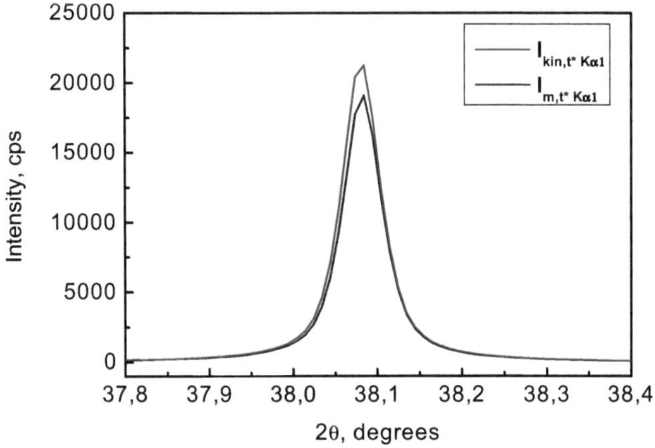

Figure 4. Converting the measured $I_{m,t*}(2\theta)$ profile of the $K\alpha_1$ component of the 111 reflection into the kinematical $I_{kin,t*}(2\theta)$ profile.

Obviously, the values of the parameters B_m and $FWHM_m$ of the profile I_m that is strongly affected by extinction are larger than those corresponding to the $I_{m,t*}$ profile which is weakly affected by extinction. This effect amounts to about 2%. The decrease of the parameters B and FWHM, corresponding to the I_{kin} and $I_{kin,t*}$ profiles, amounts respectively to about 4% against B_m and $FWHM_m$. Due to the weak extinction effect suffered by the 222 reflection, a difference between the parameters of the extinction-affected profile and the kinematical one is practically not observed. The results considered so far show that under a given texture factor P the extinction-induced enlargement of the profile parameters is controlled by the reflectivity Q and the incident-beam intensity I_0.

Table 1. Integral breadth B and FWHM of the measured (denoted by subscript m) and the kinematical (without subscript) profiles of the $K\alpha_1$ component of the 111 and 222 reflections.

	I_m		$I_{m,t*}$		I_{kin}		$I_{kin,t*}$	
hkl	$FWHM_m$	B_m	$FWHM_{m,t*}$	$B_{m,t*}$	FWHM	B	FWHM	B
111	0.0545	0.0727	0.0533	0.0710	0.0523	0.0698	0.0521	0.0694
222	0.1003	0.1401	0.1003	0.1399	0.1006	0.1403	0.1003	0.1400

6. Concluding remarks

Extinction-induced profile distortion due to continuous change of the SE correction along the profile of the strong Bragg reflections is investigated. These reflections correspond to the direction of the main texture component. It is shown that in case of a moderate texture of a silver coating, the extinction-induced enlargement of the integral breadth and FWHM of the 111 reflection amounts to a few percent, while the same parameters of the 222 reflection do not change their values. Moreover, the results reveal that the enlargement of B_m and FWHM_m is controlled simultaneously by the reflectivity Q, the texture factor P and the incident-beam intensity I_0.

References

1. Zachariasen, W.H., 1963, *Acta Cryst.*, **16**, 1139.
2. Backer, P.J. & Coppens, P., 1974, *Acta Cryst.*, **A30**, 129.
3. Sabine, T.M., 1992, *International Tables for Crystallography* (Dordrecht: Kluwer), Vol. C.
4. Rossmanith, E., 2002. *Acta Cryst.*, **A58**, 12.
5. Masimov, M., *J. Appl. Cryst.*, **40,** 990.
6. Tomov, I., 2001, *Size-Strain III Conference, Analysis of microstructure and residual stress by diffraction methods,* Trento, Italy, Dec. 2-5, p. 125.
7. Darwin, C.G., 1922, *Phil. Mag.*, **43**, 800.
8. James, R.W., 1965, *The Optical Principles of the Diffraction of X-rays* (London: G. Bell & Sons, Ltd.).
9. Bunge, H.J., 1997, *Text. Microstruct.*, **29**, 1.
10. Chandrasekhar, S., 1960, *Adv. Phys.*, **9**, 363.
11. Yamakov, V. & Tomov, I., 1999, *J. Appl. Cryst.*, **32**, 300.
12. Tomov, I., 2007, *Z. Kristallogr. Suppl.*, 131.
13. Krumm, S., 1996, *Mater. Sci. Forum*, **228-231**, 183.

Acknowledgments. This work was partially supported by the National Science Fund at the Ministry of Education and Science of Bulgaria (Contract No. X-1507/2005).

STRESSES AND STRAINS

Diffraction analysis of elastic strains in micro- and nanostructures

O. Thomas[1,2,*]

[1] Aix-Marseille Université, IM2NP
[2] CNRS, IM2NP (UMR 6242), Faculté des Sciences et Techniques, Campus de St Jérome, F-13397 Marseille Cedex, France
[*] olivier.thomas@univ-cezanne.fr

Keywords: stress, strain, thin film, coherent diffraction, microdiffraction, nanoobjects

Abstract. Very high stresses arise in thin films and in nanosized structures (lines, dots) because of the constraint of the substrate to which they are attached. The mechanical behavior of these small structures can deviate significantly from scaling laws developed for bulk materials. Moreover, the origins and magnitudes of these stresses are of great interest in technology as many fabrication and reliability problems are stress related. X-ray diffraction has at least three distinct advantages with respect to strain measurements: 1) it is very sensitive to atomic displacements; 2) it is non-destructive; 3) it can yield the six elements of the strain tensor. In objects with small dimensions stress gradients can arise from the proximity of surfaces or interfaces. The knowledge of strain distributions in such objects is necessary in order to understand their properties. The need to characterize displacement fields in nanostructures together with the advent of third generation synchrotron radiation sources has generated new and powerful methods (coherent diffraction, micro-diffraction, ...). This paper reviews some of the recent and promising results in the field of strain measurements in small dimensions via X-ray diffraction.

Full paper published in Z. Kristallogr. **223** (2008) 569-574 / **DOI** 10.1524/zkri.2008.1215 (open access).

Determination of residual stress at weld interruptions by neutron diffraction

M. Turski[1], J. A. Francis[1], P. J. Withers[1,*]

[1] School of Materials, University of Manchester, Grosvenor Street, Manchester, M1 7HS, United Kingdom
[*] Philip.Withers@manchester.ac.uk

Keywords: neutron diffraction, structural integrity, intergranular stress, macrostress length scales, welding, microstresses

Abstract. All too often those making welding stress measurements fail to consider what length scales and types of residual stress are important from a materials or component performance viewpoint. Unwittingly this can lead to inappropriate measurements and interpretations. Traditionally neutron diffraction has been used to infer macro (type I)-residual stresses, but type II and type III residual stress information is also contained within the diffraction peak shifts and widths. In this paper we examine the residual stresses generated at different scale levels within an austenitic stainless steel welded plate containing weld stop-start features such as those typically encountered in industry. Significant increases in the macro-stresses are found at these locations. In addition, calibration samples that were free of macro-stress but had undergone known amounts of plastic strain were also measured, allowing diffraction peak broadening measurements to be discussed in terms of plastic strain (type III stress). Finally, an analysis of the individual diffraction peaks allowed the level of type II intergranular stresses to be determined and thereby the extent of plastic strain inferred and related to the welding process. This stresses can be important as weld metal itself is anisotropic in nature and weld measurements based on single diffraction peak measurements can be significantly affected by type II intergranular stresses, leading to stress levels deviating significantly from the underlying macrostress.

1. Introduction

Welding introduces stress at a range of scales. Structural integrity assessments of pressure vessel components require information about the peaks in macro residual stresses commonly encountered in the vicinity of welds. Furthermore, while computational models simulating the welding process are available and may assist in determining the magnitude and the location of large residual stresses within a component, they necessarily involve many oversimplifications. This is because the complexity of the analyses and the disparity between the length scales of the torch and the component preclude detailed models as they would be computationally too demanding. One simplifying assumption is that of a 'steady state' welding condition such as might be found at the mid-length of a welding run. While such assumptions may be valid in many cases this neglects transient stresses such as those that are associated with weld stop-start effects. Current evidence suggests residual stresses above those in the steady state region exist in the vicinity of weld interruptions [1].

This paper describes measurements carried out on a 20 mm thick 304 austenitic stainless steel welded plate, containing two different types of weld stop-start effects. One interruption involves a gradual ramp down before stopping and the other involves an abrupt interruption. As such interruptions are representative of stop-start effects commonly encountered within multi-pass welds, it is important that the effects of these interruptions in terms of residual stress are well understood.

Measurements were carried out on the pulse overlap diffraction instrument (POLDI) at PSI, Switzerland, allowing a number of diffraction peaks to be measured at each measurement location. This paper examines the results in terms of the evolution of residual stress at each of the three main scale levels (macro, type II and type III) and discusses their implications.

2. Specimen manufacture

2.1 Welded plate

This paper is concerned with measurements within a 5 pass weld containing 2 different types of weld stop-start effects. Groove weld plate specimens were produced from type 304 stainless steel machined down to 200 x 150 x 20 mm blanks. A 10 mm deep groove was then machined down the centre of each specimen. After machining, the specimens were solution annealed (in a vacuum furnace) at 1050 °C, then gas fan quenched to relieve machining residual stresses from the blank. Each treated blank was subsequently welded by laying down weld metal within the machined groove. Welding was carried out with a type 308L filler wire, 0.9 mm in diameter, using the tungsten inert gas (TIG) welding process with argon as the shielding gas. The tungsten electrode was the cathode and a pulsed current was used with the peak current ranging from 180-200 A and with the background current in the range between 130-150 A. The pulse frequency was 1 Hz with the peak current being applied for 50% of the duration of each cycle. The welding speed was 75 mm/min, with the nominal heat input being approximately 1.3 kJ/min. The plates were unrestrained during welding and the interpass temperature did not exceed 70 °C for any of the weld passes.

The groove was initially filled with 4 weld passes each in the form of stringer weld beads. The first pass was placed along the groove centreline, filling the base of the groove, with subsequent passes being laid down at either side of the centreline. The 5th (final) weld pass was interrupted at 2 locations. The first interruption comprised a ramping down in welding current over 10 mm of travel starting at 65 mm from the beginning of the plate. The pass was then restarted 55 mm from the beginning of the plate, thus providing 10 mm of overlap. The second interruption occurred abruptly at 135 mm from the beginning of the plate. In this case the pass was restarted at the same location as the abrupt stop, so that there was no overlap.

2.2 Plastic strain samples

A number of tensile and compression test specimens were manufactured to produce samples with known levels of tensile or compressive plastic strain. These specimens were produced to allow correlations to be made between plastic strain and intergranular (type II) stress as well as peak broadening (type III stress).

Tensile and compression test samples were both extracted from parent metal from the same batch of type 304 austenitic stainless steel plate as was used to fabricate the welded plate specimen. Flat tensile test samples were cut parallel to the longitudinal direction within the plate, having a 37mm gauge length, with the loading axis being orientated parallel to the longitudinal direction of the plate (figure 1). A total of 10 tensile test samples were cut; each sample being strained to different levels of plastic strain. Samples were strained under a constant displacement rate of 1mm/min to 0.5, 1, 2, 4, 7, 10, 15 or 20 % total strain.

After plastic straining each test piece was sectioned to produce a set of four 6 x 5 x 4 mm cuboids. Each set of cuboids were glued together, while maintaining orientation specific information, to produce a 12 x 8 x 5 mm plastic strain sample for each level of plastic work.

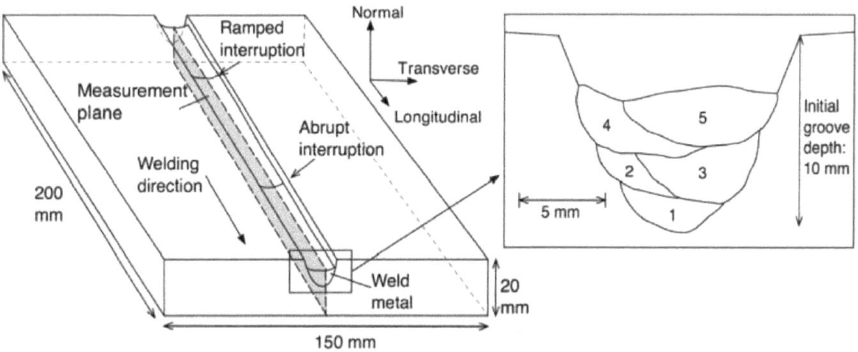

Figure 1. Geometry of welded grooved plate with the 3 components of residual strain measurements labelled, showing the locations of the individual weld beads taken from a similar TIG test weld with identical groove and plate geometry. The weld beads have been numbered in order of deposition sequence.

3. Neutron diffraction measurements

Neutron diffraction measurements were carried out at the POLDI instrument (Pulse OverLap DIffractometer) [2, 3] at the SINQ facility located at the Paul Scherrer Institut, Switzerland. POLDI is a type of time-of-flight (TOF) diffractometer, allowing a complete diffraction pattern to be acquired at a fixed diffraction angle, 2θ. Conventional TOF diffractometers use separate pulses to determine the velocity and hence wavelength of a diffracted neutron. However, this requires the time between pulses to be long enough so that the fastest neutrons cannot catch up with the slowest ones from the previous pulse. Consequently this 'waiting time' between pulses limits the effective flux and hence increases the measurement time required at the instrument. In contrast, POLDI overlaps individual pulses allowing an increase in effective flux but retaining the intrinsic advantages of a TOF instrument. The overlapping TOF diffraction patterns are deconvolved by recording both the time of flight and the scattering angle of the diffracted neutrons [2, 3]. Neutron counts are collected by a time-focused He^3 position sensitive detector over a 30° range in scattering angle.

3.1 Measurements within welded plate

Measurements were carried out using a scattering angle $2\theta = 90°$. The measurement gauge volume was defined by horizontal and vertical incident slit sizes of 5 x 2.5 mm for measurements within the welded plate and 4 x 2.5 mm for measurements in stress free 'd_0' reference samples. In all cases a 3.8 mm collimator was used, mounted in front of the detector, resulting in 47.5 mm^3 and 38 mm^3 gauge volumes within the welded plate and d_0 samples, respectively. Measurements of residual strain were made along 3 orthogonal directions (see figure 1) assuming that these corresponded to the principal stress directions. Residual stress was determined using a generalised version of Hooke's law.

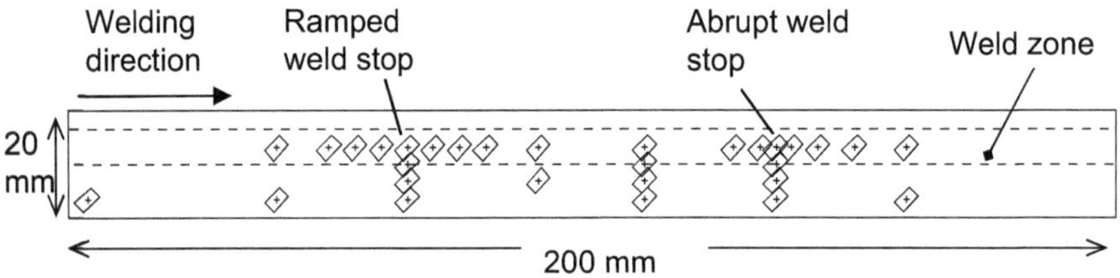

Figure 2. Position of neutron diffraction measurements within the measurement plane of the welded plate sample. The size and shape of the nominal gauge volumes along with the measurement plane have all been drawn to scale.

All measurements within the welded plate were made along the plane of the weld centreline, which passes through the middle of the welded groove and runs parallel to the welding direction (see figure 1). Figure 2 shows the positions of the neutron diffraction gauge volumes within the welded plate measurement plane. The origin of the measurement positions is defined by the intersection of the bottom surface (y = 0) and the weld start end of the plate (x = 0). Measurements were made along a line within the weld metal (y = 13 mm) in order to identify variations in residual stress due to the different weld stop-start effects as well as along through-thickness lines at x = 65, 110, and 135 mm. The x = 65 and 135 mm positions correspond to the ramped weld interruption and the abrupt weld interruption respectively while it is expected that x = 110 mm corresponds to a 'steady-state' welding condition between the two weld interruptions. The through thickness line at the middle of the plate (x = 100 mm) was not selected due to the presence of a drilled thermocouple hole at that location. Additional measurements were also added to allow a map of residual stress to be generated over the measurement plane.

3.2 Measurement of strain-free reference sample 'd_0'

Strain-free references, known as 'd_0' samples, are required for the determination of residual strain according to the following equation:

$$\varepsilon_{el}^{hkl} = \frac{d^{hkl} - d_0^{hkl}}{d_0^{hkl}}, \tag{1}$$

where ε_{el}^{hkl} is the residual (elastic) strain for a given set of lattice planes with indices *hkl,* and where d^{hkl} and d_0^{hkl} are respectively the average measured *d*-spacing within the measurement gauge volume, and the reference *d*-spacing for the same set of lattice planes. Consequently the accuracy of the measured residual strain will depend on both the accuracy of the measured *d*-value and the measured d_0 value. In many cases the microstructure within a measured component can be assumed to be homogeneous and a single value of d_0 is simply required to represent any position within the component in the stress free condition. However, in the case of welded components the d_0 measurement must also account for variations in local microstructure as well as being assumed to be stress free. This is because the weld metal often has a different chemical composition to the parent material, and any changes in microstructures either in or near to the weld metal may also result in a different *d*-spacing to the parent material. For this reason it is important to carry out representative d_0 measurements for each region within the component [4]. In this work a duplicate plate manufactured to the same geometry and from the same batch of material, but with 8 weld passes, was sectioned for the extraction of suitable d_0 samples. These were 4 x 4 x 5 mm cuboids cut from the central through-thickness plane of the plate; such dimensions were judged to be sufficient for relaxation of macro stress within the material. The cuboids were cut such that the 5 mm dimension was parallel to the normal direction of the plate, as this direction was expected to carry the least stress. The remaining faces of the cubes were colour coded after cutting to allow orientation specific information to be retained for each cuboid. The approach used for this investigation was to extract a number of cuboids from different locations along the weld bead to allow an 'average' weld metal d_0 to be measured. A total of 4 cuboids were extracted from the final passes of the weld metal at 8 mm intervals along the welding direction. These 4 cuboids were subsequently joined together to produce 8 x 8 x 5 mm cuboid, taking care to maintain orientation specific information within this cuboid 'composite'. A similar approach was adopted for the parent material. Composite cuboids of this size were produced to allow the same collimator to be used for both *d* and d_0 measurements, thus eliminating the need for corrections between different experimental arrangements.

Direction specific (i.e. longitudinal, transverse and normal) d_0 measurements were made in both parent and weld metal. In the parent metal the variation in the direction specific d_0 was within the uncertainty of the peak fitting. However, significant variations in d_0 were measured between each direction, in some cases differences up to ~250×10^{-6} were measured, while the peak fitting uncertainty was below 50×10^{-6}. Consequently direction specific d_0 values were used for subsequent determination of residual strains. Direction specific variations in d_0 are likely to have been caused by the presence of intergranular strains in the d_0 specimens. Different levels of intergranular strains will tend to accumulate in different directions within polycrystalline material in response to the plastic deformation generated within the original specimen. This is discussed in greater detail in section 4.2.

4. Data analysis

As discussed in section 3, both the diffraction angle and the TOF of neutrons are measured at POLDI, in order that overlapped neutron pulses may be deconvolved. This deconvolution is carried out using a correlation method which has been described in detail elsewhere [3]. The background levels for diffraction patterns at POLDI are typically much higher than for a conventional TOF diffractometer, however the gain in intensity of the diffracted peaks due to the higher effective flux at POLDI compensates for this higher background, resulting in acceptable peak fitting uncertainties. Peak fitting of the correlation diffraction pattern was carried out using peak fitting routines supplied at POLDI. The fitting routines essentially identify individual peaks based on a nominal lattice parameter value and fit each peak independently (effectively single peak fitting) using a Gaussian peak function. The peak fitting uncertainty for the 4 strongest peaks (111, 200, 220, 311) was typically below 100×10^{-6} for measurements made within the welded plate and below 50×10^{-6} for d_0 measurements.

4.1 Measurement of Type I residual stresses

The determination of residual stress within the welded plate was based upon shifts in the 311 diffraction peak. The 311 peak was chosen because the stress strain response of this lattice plane has been shown to be the most representative of austenitic stainless steel bulk properties [5]. Plane specific Young's modulus and Poisson's ratio values (see table 1) were used to convert residual strain measurements based on the 311 peak to residual stress. These values are based on the Kröner model [6].

4.2 Measurement of Type II residual stresses

In order to determine intergranular stress within both the plastically strained samples and the welded plate, single peak fits from the 111 and 200 planes were analysed. These lattice planes were chosen as they represent the two extremes of behaviour in terms of strain anisotropy, with the 111 peak being the stiffest and the 200 peak being the most compliant. For the welded plate, measurements of residual stress were determined for the 111, 200 and 311 planes to compare the difference in response for each reflection. Again, lattice plane specific elastic constants were used to convert residual strains to stress (see table 1).

Table 1. Plane specific elastic constants used for 304 austenitic stainless steel based on the Kröner model [6].

	(200)	**(311)**	**(111)**
Young's modulus	149.1	183.5	247.9
Poisson's ratio	0.34	0.31	0.24

5. Results

5.1 Macro (type I) stress measurement

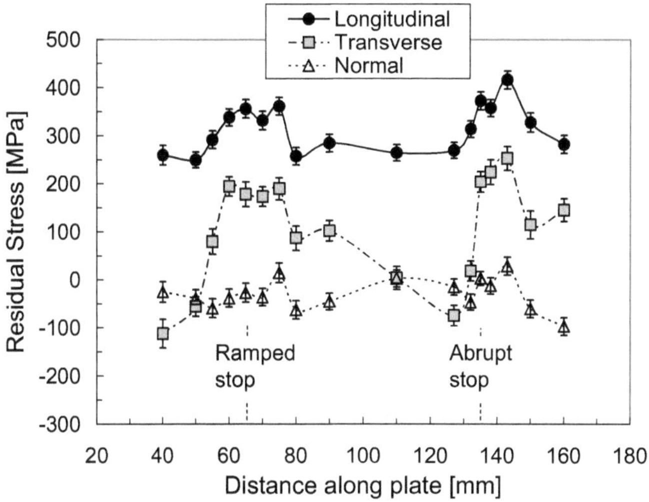

Figure 3. Residual stress measured within weld metal at y=13 mm (see figure 1)

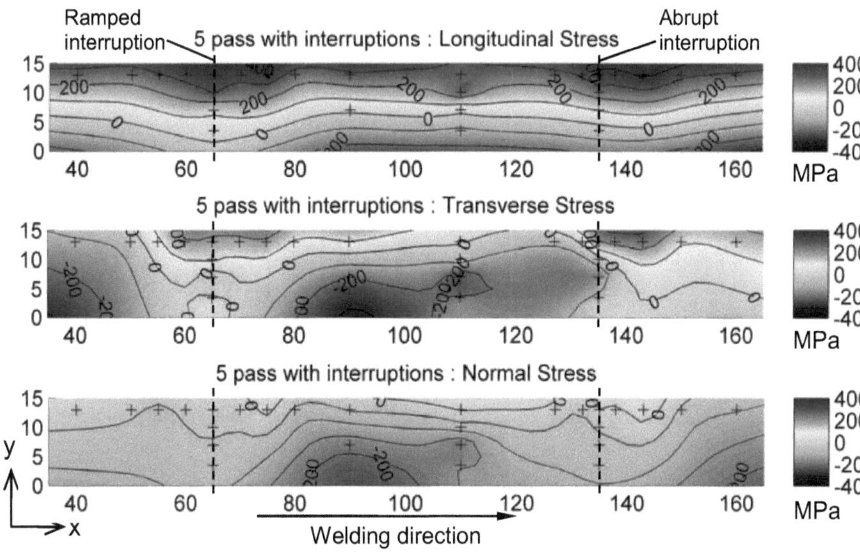

Figure 4. Residual stress maps for the measurement plane (see figure 1) based on the measurement points (crosses) shown in figure 2. The nominal locations of two weld stop-starts have also been added.

Figure 3 shows the variation in residual stress along the welding direction within the welded zone at y = 13 mm. All the components of residual stress jump at both the ramped and abrupt weld interruptions. If the region between x = 90 and 110 mm can be taken to represent steady state welding conditions then, for the longitudinal component, the weld interruptions can be seen to increase the stress by ~ 100 MPa from ~ 260 MPa at the steady state condition to ~360 MPa at the ramped weld interruption. For the abrupt weld interruption the stress is approximately 150 MPa higher than for the steady state condition. Figure 4 shows maps of the three components of residual stress in the transverse, through-thickness plane. The 'hot spots' of stress generated by the weld stop-start features are clearly evident.

5.2 Determination of intergranular (type II) stress

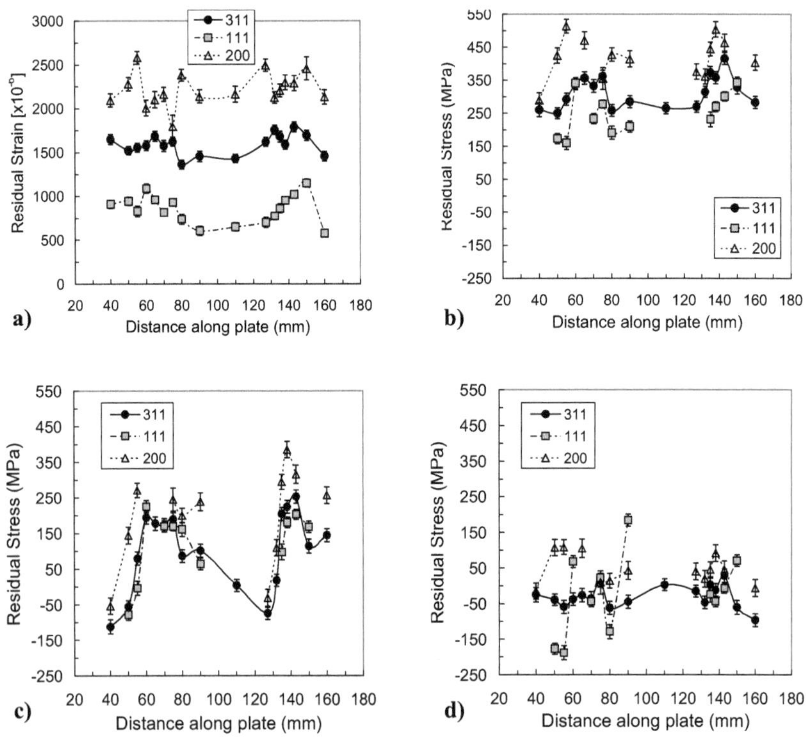

Figure 5. Residual a) longitudinal strain, b) longitudinal stress c) transverse stress and d) normal stress measured along line y=13 mm as determined by 311, 111 and 200 diffraction peaks.

The contour plots in figure 4 generally appear to be plausible, especially for the longitudinal and transverse components of stress. The transverse stress should balance over the measurement plane and it appears to do so. There are some anomalous lobes of compression at the bottom surface of the plate for the normal stress. These suggest that the interpolation procedure may be making inappropriate extrapolations in these regions, which is not surprising, given the low magnitudes of the normal stresses and the sparsity of measurements in these regions.

Figure 5a shows the different levels of residual strain as measured by the 111, 200 and the 311 diffraction peaks, with the 200 peak showing the highest strains and the 111 peak the lowest. This is broadly as expected as a consequence of elastic anisotropy, since 200 is the most compliant crystal plane and 111 the most stiff. Were the differences due to elastic anisotropy alone, they would be expected to disappear on converting strains to stress using the appropriate elastic constants (Table 1). It is clear from Fig 5b-d that while the differences between the 3 residual stress profiles are less than for the residual strain profiles, significant differences remain. In all cases the 311 residual stresses fall between the 200 and 111 residual stress profiles. In most cases the residual stress measured with the 200 is higher than the 311, while the 111 stress profile is lower. The largest differences in stress are found in the longitudinal component, with differences between 100-150 MPa being present in the vicinity of the weld interruptions. Unfortunately not all measurements are available between the weld interruptions due to texture in the weld metal, resulting in no 111 or 200 reflections being detected in either the transverse or normal directions. Close agreement between the stress levels obtained with each crystal plane (within measurement uncertainty) can be seen for the transverse component of stress, with the exception of the 200 reflections at the abrupt weld interruption. Close agreement is also found for the normal component of stress, with the exception of stresses close to the ramped weld interruption.

Figure 6 shows the residual strains that were measured with the uniformly plastically strained samples. The residual strains are calculated by comparing the lattice parameter in each of the strained samples with a sample left in the unstrained condition. The plot shows a trend for increasing residual elastic strain with increasing tensile plastic strain. These measurements represent a measure of the intergranular (type II) stresses since no macro stresses would be expected within these samples. Intergranular stresses would be expected to increase with plastic strain due to the preferential yielding of suitably orientated grains relative to their neighbours, leading to strain redistribution. What is interesting is the magnitude of the intergranular strains for each lattice plane, with larger strains being measured in the 200 plane and the lowest being measured in the 111 plane with the 311 plane being in between. The observed relationship corresponds to that found in the residual stress profiles within figures 5b-d, strongly suggesting that for the welded plate the differences between the estimated residual stresses, as obtained from each crystal plane, are attributable to plastically generated intergranular stresses. This explanation is also consistent with the fact that in welded joints the plastic strain component is generally largest in the longitudinal direction.

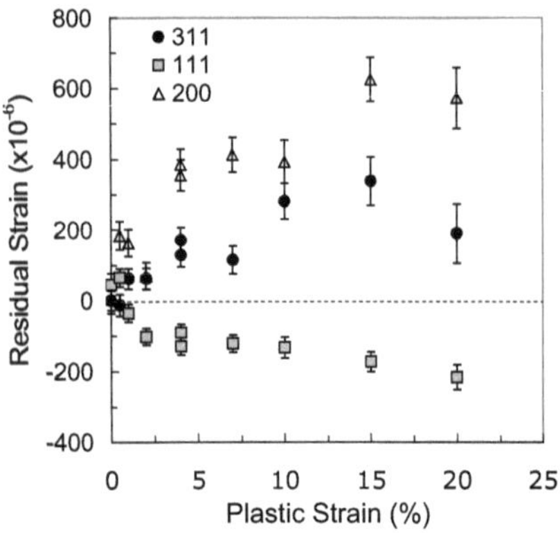

Figure 6. Variation in measured intergranular residual strain with different levels of tensile plastic work

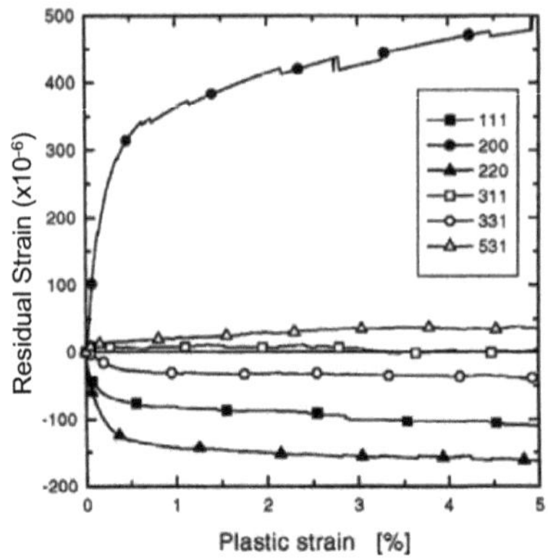

Figure 7. Intergranular residual strains predicted by an elasto-plastic self consistent model for stainless steel, parallel to the loading direction (taken from [5]).

Figure 7 shows results taken from Clausen et al [5] for an elasto-plastic self consistent (EPSC) model of stainless steel. The work is based on Hutchinson's self-consistent model [7] and represents a sample with an initial orientation distribution of 5000 grains having random texture. The residual strains are those predicted, parallel to the loading direction, in the unloaded state, after plastic deformation under uniaxial testing conditions. The measured results in figure 6 show many similarities with the EPSC predictions in figure 7. There is close agreement between the modelling predictions and measurements, for both the magnitude and sign of the residual strains for the 200 and 111 lattice planes. However, for the 311 lattice plane, the EPSC model predicts close to zero residual strain after 5% plastic strain, while measurements indicate 170×10^{-6} residual strain at 4% plastic strain, rising to 340×10^{-6} residual strain at 15% plastic strain.

5.3 Type III strains

Increased peak broadening is generally associated with increasing dislocation density, hence plastic work. However, peak broadening may also arise from size broadening due to the decreasing size of the diffracting volumes or from other sources of inhomogenous elastic strain within the diffracting grains. Figure 8 shows a plot of peak broadening expressed as the full width half maximum (FWHM) for the 311 reflection as a function of plastic work, as measured within the uniformly plastically deformed samples. This plot shows a linear trend with FWHM increasing with plastic strain. Figure 9 shows the variation in FWHM for the 311 reflection through the thickness of the welded plate at the abrupt weld interruption. By referring to figure 8 as a calibration plot the level of plastic work in figure 9 can be inferred. Of course this assumes that the primary cause of broadening is due to dislocations: in view of the thermal excursion of the weld metal and HAZ this may be an oversimplification. Nevertheless Figure 9 exhibits a maximum in FWHM (~4°) at the nominal interface between the weld metal and parent metal, falling to 3.5° within the final weld pass. These values of FWHM would suggest up to ~10% equivalent plastic work has taken place within the bottom weld passes and ~5% close to the final weld pass.

Higher levels of plastic work could be expected in the initial weld passes when compared to the final weld passes, due to the accumulation of plastic work during thermal cycles within the bottom weld beads as subsequent weld beads are laid down. Indeed, measurements made within a multi-pass manual metal arc welded groove weld specimen indicated ~4% more strain within the first weld pass compared to the weld capping pass [8]. These findings again suggest the development of significant intergranular strains due to plastic flow within and near the weld metal, leading to the observed differences in the residual stress profiles shown in figure 5(b-d).

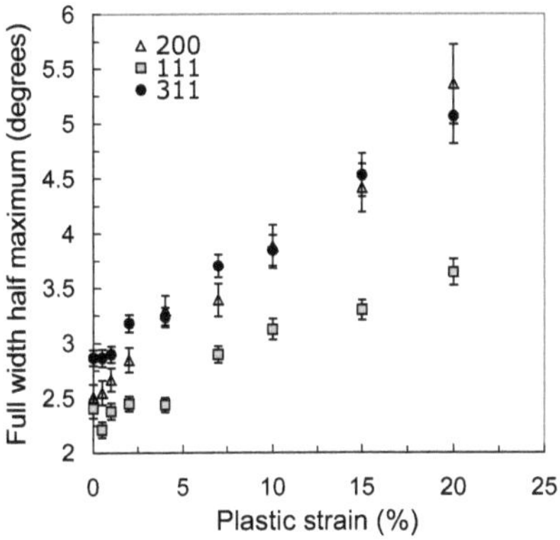

Figure 8. Plot of diffraction peak FWHM as a function of plastic strain.

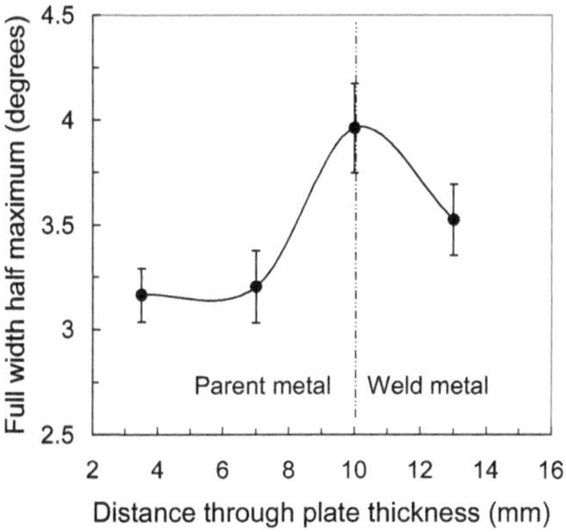

Figure 9. Through thickness variation in FWHM at the abrupt weld interruption (y=135 mm).

Accurate estimates of macrostresses at weld interruptions can only be obtained if d_0 specimens are prepared from a companion weld in such a way that the variations in plastic strain history throughout the weld are accounted for. By definition, at interruptions, welding is not taking place under steady state conditions. As such, it is not sufficient to obtain estimates for d_0 on a single transverse-normal plane as a function of distance from the weld bead. Instead, it is also necessary to characterise the variation in d_0 as a function of distance along the weld, with the greatest density of d_0 measurements in the vicinity of the interruptions. In this way the peak to peak d_0 variations will include the type II plastic misfit stresses so that the derived stresses are representative of the macrostress alone.

6. Conclusions

- Stop-start features such as those that commonly arise in multipass welds upon changing electrodes or spools of filler metal give rise to localised peaks in residual stress. Such peaks in stress, > 100 MPa higher than the steady state longitudinal residual stress, were detected by neutron diffraction measurements based on the reflections obtained from 3 different crystallographic planes. As such, localised peaks in residual stress at stop-start features need to be accounted for in structural integrity assessments on engineering components. Significant variations in the transverse stress were also observed.

- Neutron diffraction measurements on coupons that had been plastically pre-strained to specified levels (but at the time of measurement were free of any macroscopic stress) confirmed that a plastic strain history will lead to a shift in the effective value of d_0 for a given crystal plane due to significant intergranular stresses. Thus, estimates for residual stresses under transient welding conditions (such as is the case at stop-start features) will need to account for variability in intergranular stresses as a function of distance along the weld if a measure of the macrostress is of primary interest.
- The estimates for transverse stresses that were obtained from each of the (111), (311) and (200) reflections were in reasonable agreement. However, there were systematic differences between the estimates for longitudinal stresses as obtained with each of the reflections. This is possibly related to the fact that the longitudinal component of stress is generally associated with the highest levels of plastic strain. However, austenitic stainless steel weld metals are also known to exhibit strong texture. Under such circumstances, it may be necessary to obtain orientation-specific estimates for d_0 that also account for the local plastic strain history in order to achieve agreement between estimates for stress as obtained from different crystal planes. Further work is required to resolve these discrepancies.
- The full width half maximum (FWHM) parameter, which is associated with type III residual strains, has allowed estimations of plastic work to be made within the welded plate indicative of plastic strain levels around 5-10% being largest in the first weld passes and lowest in the final pass.

References

1. Bouchard, P.J., Santisteban, J.R., Edwards, L., Turski, M., James, J. & Withers, P.J., 2005, *Proc. Vessels & Piping Conf.* (ASME Press), **6**, 853.
2. Stuhr, U., 2005, Nucl. Instr. Meth. Phys. Res. A **545**, 319.
3. Stuhr, U., Grosse, M. & Wagner, W., 2006, *Mater. Sci. Eng. A*, **437**, 134.
4. Krawitz, A.D. & Winholtz, R.A., 1994, *Mater. Sci. Eng. A*, **185**, 123.
5. Clausen, B., Lorentzen, T. & Leffers, T., 1998, *Acta Mater.*, **46**, 3087.
6. Kröner, E., 1958, *Z.Physik*, **5**, 504.
7. Hutchinson, J.W., *Proc. R. Soc. London A*, 1970, **319**, 247.
8. Turski, M., Smith, M., Bouchard, P.J., Edwards, L. & Withers, P.J., 2007, *Proc. Vessels & Piping Conf.* (ASME Press), PVP2007-**26613.**

Acknowledgements. The authors would like to acknowledge Paul Scherrer Institute for beam time granted at POLDI and Dr U. Stuhr and Dr G. Schuck for help and support during diffraction measurements. MT and JAF are also grateful for support from Rolls-Royce Marine.

Z. Kristallogr. Suppl. **27** (2008) 245-252 / **DOI** 10.1524/zksu.2008.0031

Residual stress and elastic anisotropy in the Ti-Al-(Si-)N and Cr-Al-(Si-)N nanocomposites deposited by cathodic arc evaporation

M. Dopita[1,2,*], Ch. Wüstefeld[1], V. Klemm[1], G. Schreiber[1], D. Heger[1], M. Růžička[3], D. Rafaja[1]

[1] Institute of Materials Science, TU Bergakademie Freiberg, D-09599 Freiberg, Germany
[2] Department of Condensed Matter Physics, Faculty of Mathematics and Physics, Charles University, Ke Karlovu 5, CZ-121 16 Prague, Czech Republic
[3] SHM Ltd., Průmyslová 3, CZ-787 01 Šumperk, Czech Republic
[*] dopita@gmail.com

Keywords: thin coatings, nanocomposites, residual stress, elastic anisotropy, texture

Abstract. The elastic anisotropy of the Cr-Al-(Si-)N and Ti-Al-(Si-)N thin film nanocomposites with different aluminium and silicon contents deposited by the cathodic arc evaporation was investigated using complementary X-ray diffraction methods. Glancing angle X-ray diffraction (GAXRD) was employed to get the first information about the anisotropy of the elastic constants. The measurements done on an X-ray diffractometer equipped with an Eulerian cradle were used for calculation of the lattice strain and residual stress using the Crystallite Group Method and for determination of the preferred orientation of crystallites. In most samples, the degree of the elastic anisotropy decreased with increasing aluminium and silicon contents. Still, depending on the transition metal (Cr and/or Ti), different dependences of the elastic anisotropy on the aluminium and silicon contents were observed that could be related to the phase stability regions of the cubic phase in the Cr-Al-(Si-)N and Ti-Al-(Si-)N thin film nanocomposites.

1. Introduction

Protective coatings deposited by various PVD techniques are used for decades to improve the mechanical and tribological properties of cutting tools. During the last years, the importance of nanocomposite coatings has increased rapidly, because of their significantly improved hardness. Nowadays, the addition of a certain amount of aluminium and silicon into the chromium and titanium nitrides is a well-established method for the production of nanocrystalline coatings in which the cubic face-centred (fcc) phase dominates. As shown in [1], the solubility of aluminium in the fcc chromium and titanium nitrides is 77 at % for Cr and 65 at % for Ti. For higher aluminium concentrations, the wurtzitic AlN forms in the

coatings as a second crystalline phase that supports the development of the nanocomposites. The phase decomposition of the nano-scale is further supported by the segregation of superfluous silicon from the host structure of the chromium and titanium nitrides.
Besides the small crystallite size, one of the most important parameters which influences the hardness of the coatings significantly is the magnitude of the residual stress. For a correct determination of the residual stress in polycrystalline samples, the choice of a correct crystallite interaction model and the knowledge of the elastic constants are necessary. Unfortunately, rather deficient information about the elastic properties of nanocomposites is available in the literature. The main problems are the unknown dependence of X-ray elastic constants (XECs) on composition, unknown anisotropy of XECs and the unknown dependence of the anisotropy of XECs on composition, i.e. on the aluminium and silicon contents in the fcc Cr-Al-(Si-)N and Ti-Al-(Si-)N compounds. A strong anisotropy of the XECs in the systems under study was already indicated by our previous results [2]. Therefore, the main goal of this contribution is to describe the anisotropy of XECs in the Ti-Al-(Si-)N and Cr-Al-(Si-)N nanocomposites as a function of their chemical composition.

2. Experimental details

2.1 Sample preparation

Four series of coatings, Cr-Al-N, Cr-Al-Si-N, Ti-Al-N and Ti-Al-Si-N, were deposited using cathodic arc evaporation (CAE) in nitrogen atmosphere from two laterally rotating arc cathodes (π-80 from PLATIT) [3]. One cathode was made from the transition metal (Cr or Ti), the second one from aluminium for the sample series without silicon, i.e. Cr-Al-N and Ti-Al-N, and from aluminium containing approximately 11 at. % of silicon for the sample series containing silicon, i.e. Cr-Al-Si-N and Ti-Al-Si-N, respectively. To eliminate the influence of the parameters of the deposition process on the microstructure of the samples, the working pressure, ion current, bias voltage, base pressure and deposition temperature were kept identical for all samples. The working pressure was 1.3 Pa. The ion current on the transition metal cathode (Cr or Ti) was 80 A, and the ion current on the Al or Al-Si cathode was 120 A. The bias voltage was -75 V. The base pressure was 5×10^{-3} Pa and the deposition temperature was approximately 450°C. Polished plates of cemented carbide having random orientation of crystallites were used as substrates. In contrast to the commercially produced coatings, the samples were not rotated during the deposition process. Thus, the preferred orientation of crystallites was not superimposed by the sample rotation. Moreover, a set of coatings with different chemical compositions can be obtained in one deposition run, as the chemical composition of the individual samples depends on the position of the substrate with respect to both cathodes. For each sample series, seven samples differing in composition were prepared. However, only four Ti-Al-(Si-)N and five Cr-Al-(Si-)N coatings containing the cubic nitrides as the dominant phase were selected for a detailed microstructure investigation. As we showed recently [2, 10], the samples with the dominant fcc phase possess the highest hardness and are most important for industrial applications.

2.2 Investigation methods

The overall chemical composition of the coatings was determined using electron probe microanalysis with wavelength-dispersive spectroscopy (EPMA/WDX) at 40 points randomly distributed across each sample surface. The nitrogen concentration was 50 at. % within the experimental accuracy (for Ti-Al-(Si-)N, this was also checked by the glow-

discharge optical emission spectroscopy) and the maximum oxygen concentration was below the detection limit of the EPMA/WDX (lower than 0.1 at %) in all coatings under study. The overall chemical composition of the samples under consideration is given in table 1.
XRD experiments were performed using several diffraction geometries. Glancing-angle X-ray diffraction (GAXRD) measurements with a constant angle of incidence of the primary beam were done using a D8 diffractometer (Bruker, AXS) equipped with a parabolic Göbel mirror in the primary beam and a Soller collimator with acceptance angle of 0.12° and with a flat LiF monochromator in the diffracted beam. Data from GAXRD measurements were used mainly for the phase analysis, for the determination of the crystallite size and for the determination of the strain-free lattice parameter. Moreover, GAXRD data are suitable for fast visualisation of two effects. First, because the lattice parameters are measured on different crystallographic planes and at different inclinations of the sample from the symmetrical position, one can immediately get information about the crystal anisotropy from one GAXRD scan. Second, varying the angle of incidence of the primary beam has in consequence different penetration depths of X-rays, and thus different information depth of diffracted data. GAXRD measurements with several different angles of incidence can provide information about the presence of residual stress gradients as well as the strain-free lattice parameter depth profile can be calculated.

Table 1. Numbers of the samples and the overall chemical compositions of the investigated coatings.

Sample	$Cr_{1-x}Al_xN$	$Cr_{1-x-y}Al_xSi_yN$	$Ti_{1-x}Al_xN$	$Ti_{1-x-y}Al_xSi_yN$
1	$Cr_{0.92}Al_{0.08}N$	$Cr_{0.91}Al_{0.08}Si_{0.01}N$	$Ti_{0.96}Al_{0.04}N$	$Ti_{0.96}Al_{0.04}Si_{0.00}N$
2	$Cr_{0.83}Al_{0.17}N$	$Cr_{0.84}Al_{0.15}Si_{0.01}N$	$Ti_{0.85}Al_{0.15}N$	$Ti_{0.85}Al_{0.14}Si_{0.01}N$
3	$Cr_{0.75}Al_{0.25}N$	$Cr_{0.69}Al_{0.28}Si_{0.03}N$	$Ti_{0.72}Al_{0.28}N$	$Ti_{0.71}Al_{0.26}Si_{0.03}N$
4	$Cr_{0.54}Al_{0.46}N$	$Cr_{0.52}Al_{0.43}Si_{0.05}N$	$Ti_{0.52}Al_{0.48}N$	$Ti_{0.52}Al_{0.43}Si_{0.05}N$
5	$Cr_{0.46}Al_{0.54}N$	$Cr_{0.40}Al_{0.52}Si_{0.08}N$		

The texture measurements were done on a PTS 3000 diffractometer (Seifert) equipped with an Eulerian cradle and a point focus monocapillary optics. For each sample, the pole figures (111), (200) and (220) of the face centred cubic phase were recorded. The strain measurements were performed on a D8 diffractometer equipped with an Eulerian cradle and polycapillary optics located in the primary beam. Selected reflections were measured at three different azimuthal angles ϕ. For each ϕ setting, θ-2θ scans were recorded at different inclinations of the sample ψ from the symmetrical position ($\psi = 0$). To obtain the parameters of the line positions, measured data were fitted by pseudo-Voigt function by means of the least-squares refinement. A linear background was supposed. Estimated errors in the line position determination were used as weight factors for subsequent data reduction. For the GAXRD and texture measurements the CuKα radiation (λ = 1.5418 Å) was used, whereas for the strain measurements, the CoKα radiation (λ = 1.7903 Å) was used.

3. Results and discussion

On the basis of the GAXRD measurements performed at several different angles of incidence of the primary beam (i.e. different information depth of diffracted intensity), we can conclude that no depth gradients were observed in our coatings, as can be seen from the nearly constant slopes of the $\sin^2\psi$ dependences constructed from these measurements (figure 1a). Rather high errors of the slopes determined from the linear function fit come from the scatter of measured lattice parameters caused by the pronounced crystal anisotropy present in our coatings, whereas the biggest error calculated for measurement at the lowest angle of incidence ($\gamma = 1°$) is caused solely by the quality of measured diffraction pattern. Equation 1 describes the relation between the measured lattice deformation and the stress tensor. Lattice deformations measured at three different angles ϕ and at several tilts ψ allow us, due to equation 1, to determine all stress tensor components [4]. As it is clearly visible from figure 1b, the coatings are under the state of biaxial, (rotationally symmetrical - $\sigma_{11} = \sigma_{22} = \sigma$) stress, because data measured for different angles ϕ are identical. No shear stresses ($\sigma_{13} = \sigma_{23} = 0$) are present in our samples which can be demonstrated by similar behaviour of lattice parameters measured for positive and negative ψ tilts i.e. no ψ splitting is observable – see figure 1c.

$$\varepsilon_{\phi,\psi}^{hkl} = \frac{1}{2}s_2^{hkl}\left[\left(\sigma_{11}\cos^2\phi + \sigma_{12}\sin 2\phi + \sigma_{22}\sin^2\phi - \sigma_{33}\right)\sin^2\psi + \sigma_{33}\right] + \\ + \frac{1}{2}s_2^{hkl}\left(\sigma_{13}\cos\phi + \sigma_{23}\sin\phi\right)\sin 2\psi + s_1^{hkl}\left(\sigma_{11} + \sigma_{22} + \sigma_{33}\right) \tag{1}$$

Under the assumptions of the biaxial, rotationally symmetrical state of stress, without shear stresses and normal stress component (σ_{33}) equal to zero, equation 1 reduces to a formula, which predicts the linear dependence of ε^{hkl} vs. $\sin^2\psi$ – equation 2.

$$\varepsilon_{\phi,\psi}^{hkl} = \left(2s_1^{hkl} + \frac{1}{2}s_2^{hkl}\sin^2\psi\right)\sigma = 2s_1^{hkl}\sigma + \frac{1}{2}s_2^{hkl}\sigma\sin^2\psi \tag{2}$$

From data measured using the classical $\sin^2\psi$ method, the elastic anisotropy factor A can be derived – equation 3 [5]. Elastic anisotropy factor is given by the ratio between the slopes (ζ) of the dependences ε^{hkl} versus $\sin^2\psi$ measured on (00h) and (hhh) lattice planes.

$$A = \frac{\zeta^{00h}}{\zeta^{hhh}} = \frac{1/2s_2^{00h}\sigma}{1/2s_2^{hhh}\sigma} = \frac{s_2^{00h}}{s_2^{hhh}} \tag{3}$$

The measured dependences of lattice deformations vs. $\sin^2\psi$ were approximated with a linear function (equation 2). We have to mention that there existed deviations from the expected linear behaviour even for *00h* and *hhh* reflections (see figure 2b at low ψ tilts). We suppose that this effect comes from the plastic deformation of highly stressed lattice planes parallel to sample surface. The plastic deformation was recently demonstrated by the presence of screw dislocations in our coatings [6, 7].

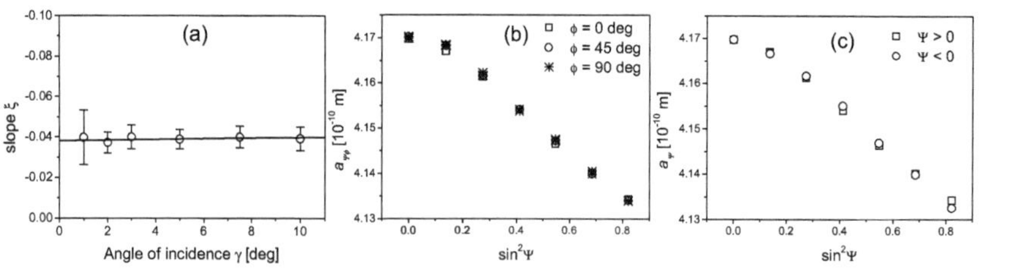

Figures 1a, 1b and 1c. Slopes of the $\sin^2\psi$ plots calculated from GAXRD measurements obtained for different angles of incidence γ for sample $Cr_{0.54}Al_{0.46}N$ (a). Plot of the lattice parameters obtained from the interplanar spacing of the lattice planes (111) versus $\sin^2\psi$ measured for three different ϕ angles (b), measured for positive and negative ψ tilts (c) for sample $Cr_{0.91}Al_{0.08}Si_{0.01}N$.

For highly textured materials, the Crystallite Group Method (CGM) [4] is usually applied. The benefit of this method is, the possibility of getting information on specially chosen crystallite sets, taking into account the angular relations between individual crystallite planes. Using the CGM method the residual stresses for crystallite families having orientation widely incorporated in our samples, i.e. for the crystallites corresponding to the strongest texture component present in the coating, were derived.

3.1 Cr-Al-N and Cr-Al-Si-N system

The XRD phase analysis has show the presence of a single face centered cubic (fcc) phase up to chemical composition of $Cr_{0.54}Al_{0.46}N$ and $Cr_{0.52}Al_{0.43}Si_{0.05}N$. For higher Al contents in the samples, second – wurtzitic AlN phase was identified in coatings. The crystallite size decreases monotonously with increasing Al content from approximately 10 nm for the sample richest in chromium to 4.5 nm for sample #5 in the Cr-Al-Si-N coatings and from approx. 11 nm (sample #1) to 6.5 nm (sample #5) in the Cr-Al-N system. A detailed description of the evolution of microstructural parameters as a function of sample chemical composition are given in [2, 7]. All Cr-Al-N coatings under study were preferentially oriented with the $\langle 111\rangle$ direction perpendicular to the sample surface. Only a slight inclination, less than 5°, of the $\langle 111\rangle$ direction from the sample normal direction was observed. Moreover, all samples exhibit a strong in-plane texture. Similarly, the (111) texture was detected in the Cr-Al-Si-N samples. However, we observed a rapid inclination of the $\langle 111\rangle$ direction from the sample surface perpendicular direction with increasing Al and Si content [8]. The "strength" of the in-plane texture gets weaker with increasing Al and Si concentration.

The anisotropy factor A was smaller than 1 in all cases, i.e. $s_2^{hhh} > s_2^{h00}$. The dependence of the anisotropy factor on aluminium content was much weaker than the dependence of the anisotropy factor on silicon content (figure 2a). The strongest anisotropy (A = 0.4) was observed in the sample $Cr_{0.69}Al_{0.28}Si_{0.03}N$. The slopes of the dependences of ε versus $\sin^2\psi$, which were used for calculation of the anisotropy factors, were obtained with a very high accuracy unlike the residual stresses, which would be affected by the difficulties and uncertainties in the determination of the elastic constants. In figures 2b, 2c and 2d, the slopes ζ^{111} and ζ^{200} are plotted as functions of the chemical compositions of the Cr-Al-(Si-)N coatings. The slope of the ε versus $\sin^2\psi$ dependences is negative for all investigated samples, whereas it is slowly decreasing with increasing Al concentration in Cr-Al-N system and increasing with Al and Si concentration for Cr-Al-Si-N coatings. The magnitude of the slope obtained using CGM lies between the boundary values calculated using the classical method for (111) and (200) lattice planes (compare figures 2b and 2c, 2d). Slight differences between the slopes obtained from the classical method and from CGM occur because these two methods apply different types of averaging. Data obtained from the classical method are averaged over crystallites having different orientation with respect to the sample surface measured on certain lattice planes, whereas CGM results are averaged over different (hkl), but the results correspond to the crystallites having distinct orientation (in our case crystallites corresponding to the major texture component).

Adopting elastic constants given in [9] (ν = 0.214 and E = 520 GPa) and taking into account relations between the Young's modulus and Poisson's ratio and XECs $s_1 = -\nu / E$ and $\frac{1}{2} s_2 = (\nu + 1) / E$ [4], one can calculate the values of the residual stress, but these results can be understood as a very rough approximation as the possible dependence of the elastic constants on the chemical composition was neglected. Residual stress determined using CGM in both types of coatings is compressive, whereas its magnitude is significantly higher for samples without Si (approximately two times for the first four samples), and it is decreasing in Si free samples and increasing in coatings containing Si with increase of [Al] (or [Al]+[Si]) concentration. For the fifth samples - $Cr_{0.46}Al_{0.54}N$ and $Cr_{0.40}Al_{0.52}Si_{0.08}N$, the residual stress magnitude is similar, around -6 GPa for both coating series. Residual stresses determined using the classical $\sin^2\psi$ method exhibit the same trend as described for CGM results, moreover the dependences of residual stress on the composition are slightly smoother than in the CGM results.

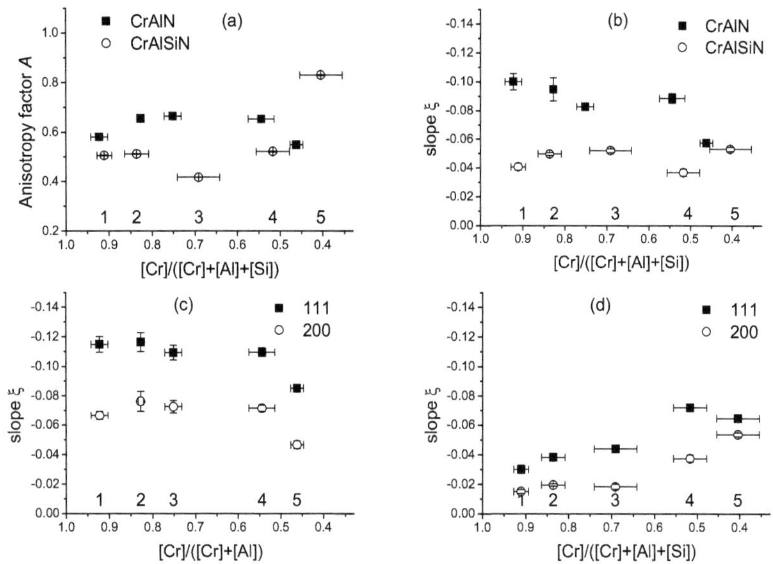

Figure 2a, 2b, 2c and 2d. Plot of the anisotropy factor A (a), slopes of the ε vs. $\sin^2\psi$ dependences measured on crystallite family corresponding to strongest texture component (CGM) (b), and slopes of the ε vs. $\sin^2\psi$ dependences measured on (111) and (200) lattice planes (c- for Cr-Al-N , d- for Cr-Al-Si-N), versus sample composition in chromium based coatings.

3.2 Ti-Al-N and Ti-Al-Si-N system

In Ti-Al-N and Ti-Al-Si-N coatings, the single fcc phase exists up to the chemical composition of $Ti_{0.72}Al_{0.28}N$ and $Ti_{0.71}Al_{0.26}Si_{0.03}N$. Then wurtzitic AlN phase starts to grow and for samples with higher Al and Si content (([Ti]/[Ti]+[Al]+[Si])<0.4) it becomes a dominant phase. The crystallite size decreases from 8 nm for sample #1 to approximately 4.2 nm for sample #4 in Ti-Al-N, and from 7 nm for sample #1 to roughly 2.7 nm for sample #4 in the Ti-Al-Si-N coatings. Crystallites with the highest Ti contents in Ti-Al-(Si-)N coatings are oriented with the $\langle 111\rangle$ direction perpendicular to the sample surface. With increasing amount of Al the $\langle 111\rangle$ crystallographic direction inclines towards the sample surface. In Si

free samples the crystallites are, additionally preferentially oriented in the plane of coatings. The crystallites in the Ti-Al-Si-N do not exhibit such in-plane texture [10, 11].
The anisotropy factor *A* is increasing with increasing Al and Si content and it can be well described by functions $y = 0.79(7)*c_{[Al]}+0.58(3)$, $y = 0.99(9)*c_{([Al]+[Si])}+0.56(9)$ for Ti-Al-N and Ti-Al-Si-N coatings respectively (figure 3a), where $c_{[Al]}$ or $c_{([Al]+[Si])}$ represent Al, or Al+Si concentration. The anisotropy relaxation is faster in the samples containing Si.
The slopes of the ε versus $\sin^2\psi$ dependences obtained using the classical method are negative for both Ti-Al-N and Ti-Al-Si-N as well, and increase linearly with increasing Al and Si concentration up to the [Ti]/([Ti]+[Al]+[Si]) ~ 0.7. The dependences show saturation at approximately -0.07 and -0.055 for Ti-Al-N and Ti-Al-Si-N, respectively – figures 3c and 3d, whereas the values are, in the bounds of errors, similar for (111) and (200) lattice planes (corresponding to the anisotropy relaxation for these samples).

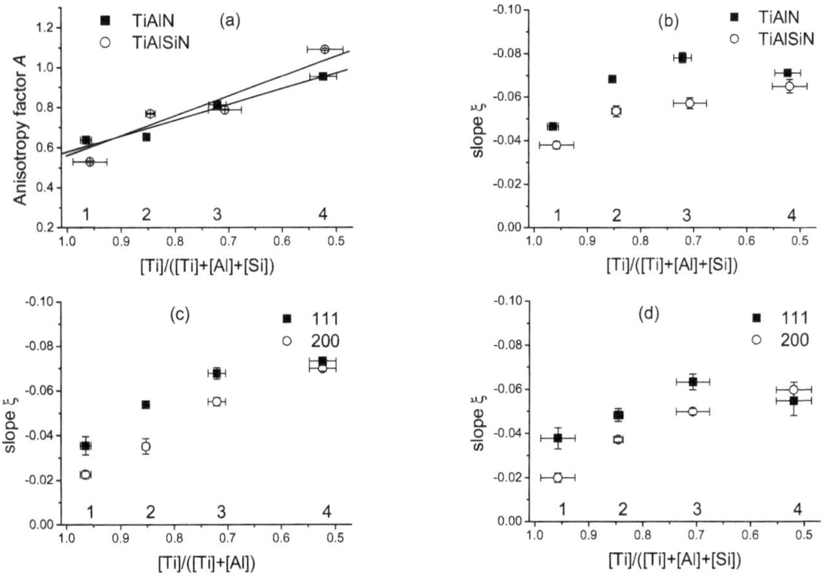

Figure 3a, 3b, 3c and 3d. Plot of the anisotropy factors A, lines correspond to the linear fit (a) Slopes of the ε vs. $\sin^2\psi$ dependences measured on crystallite family corresponding to strongest texture component (CGM) (b), and slopes of the ε vs. $\sin^2\psi$ dependences measured on (111) and (200) lattice planes (c- for Ti-Al-N , d- for Ti-Al-Si-N), versus sample composition in chromium based coatings.

Using the elastic constants $\nu = 0.295$ and E = 640 GPa from [5], the residual stress can be derived from the ε versus $\sin^2\psi$ dependence slopes. Because of the uncertainty in the dependence of elastic constants on composition, the elaborated values of residual stress can again be understood as a rough approximation. Residual stress determined using CGM shows an increase from approximately -5.5 GPa or -4.3 GPa to -8.5 GPa or -7.3 GPa with increasing Al and Si concentration for Ti-Al-N or Ti-Al-Si-N samples respectively. Stress determined for samples without Si is higher than in samples containing silicon, and reach its maximum at -9GPa for the third sample. Comparison of data obtained from CGM and the classical $\sin^2\psi$ method shows that the residual stress determined from the classical $\sin^2\psi$ method (i.e. averaged over all crystallites orientation) is slightly smaller than its value from CGM, which means that preferentially oriented crystallites are in a state of higher compressive residual stress.

4. Conclusions

The Cr-Al-N, Cr-Al-Si-N, Ti-Al-N and Ti-Al-Si-N coatings deposited by CAE have been studied by X-ray diffraction. In all samples no residual stress or lattice parameter gradients as well no shear stresses and only biaxial rotationally symmetrical compressive state of stress were observed. Different types and degrees of preferred orientation of crystallites were found in the investigated samples. Applying the Crystallite Group Method, residual stresses for those families of crystallites having the majority orientation in our coatings were calculated. The anisotropy of the X-ray elastic constants decreased generally with increasing Al and Si concentration for all sample systems, except the case of Cr-Al-N coatings. In these samples rapid inclination of the $\langle 111 \rangle$ direction towards the sample surface were not observed. Anisotropy factors, slopes of ε versus $\sin^2\psi$ dependences and finally residual stresses as a function of sample composition were determined.

References

1. Sugishima, A., Kajioka, H. & Makino, Y., 1997, *Surf. Coat. Technol.*, **97**, 590.
2. Rafaja, D., Dopita, M., Růžička, M., Klemm, V., Heger, D., Schreiber, G. & Šíma, M., 2006, *Surf. Coat. Technol.*, **201**, 2835.
3. Holubář, P., Jílek, M. & Šíma, M., 2000, *Surf. Coat. Technol.*, **133/134**, 145.
4. Hauk, V., 1997, *Structural and Residual Stress Analysis by Nondestructive Methods* (Amsterdam: Elsevier).
5. Perry, A.J., 1990, *Thin Solid Films*, **193/194**, 463.
6. Wüstefeld, Ch., Rafaja, D., Dopita, M. & Klemm, V., *Z. Kristallogr.*, this issue.
7. Rafaja, D., Wüstefeld, Ch., Dopita, M., Růžička, M., Klemm, V., Schreiber, G., Heger, D. & Šíma, M., 2007, *Surf. Coat. Technol.*, **201**, 9476.
8. Dopita, M., Rafaja, D., Wüstefeld, Ch., Růžička, M., Klemm, V., Heger, D., Schreiber, G. & Šíma, M., 2008, *Surf. Coat. Technol.*, **202**, 3199.
9. Attar, F. & Johannesson, T., 1995, *Thin Solid Films*, **258**, 205.
10. Rafaja, D., Poklad, A., Klemm, V., Schreiber, G., Heger, D., Šíma, M. & Dopita, M., 2006, *Thin Solid Films*, **514**, 240.
11. Rafaja, D., Poklad, A., Schreiber, G., Klemm, V., Heger, D. & Šíma, M., 2005, *Z. Metallkd.*, **96**, 736.

Acknowledgements. The authors thank the German Scientific Council (DFG) for financial support of the project # RA-1050/9-1. The work is a part of the research program MSM 0021620834 financed by the Ministry of Education of the Czech Republic.

Z. Kristallogr. Suppl. **27** (2008) 253-260 / **DOI** 10.1524/zksu.2008.0032

In-situ X-ray diffraction investigations of thin films: Determination of thermoelastic constants

Y. Kuru[1], M. Wohlschlögel[1], U. Welzel[1,*], E. J. Mittemeijer[1]

[1] Max Planck Institute for Metals Research, Heisenbergstr. 3, D-70569 Stuttgart, Germany
[*] u.welzel@mf.mpg.de

Keywords: residual stress, diffusion, metal thin films

Abstract. Recipes for in-situ, non-ambient X-ray diffraction stress measurements of thin films have been given in this paper. To this end detailed in-situ X-ray diffraction analyses of stress in Cu and Ni thin films have been performed in the temperature interval between -100°C and 250°C. Apart from focusing on an accurate determination of the coefficients of thermal expansion of film and/or substrate and in-situ calibration of the specimen temperature, attention is paid in particular to the use of measurements below ambient temperature. Thereby the thermoelastic behaviour in the absence of thermally activated processes (i.e. grain growth and defect annihilation) can be studied. This last possibility has rarely been exploited until now.

1. Introduction

X-ray diffraction (XRD) is one of the most frequently used methods for (residual) stress measurement. Besides being non-destructive and highly accurate (stress (variation) of some MPa can be detected), the stress states of all crystalline phases can be obtained separately [1]. Moreover, additional microstructural information, as the crystallographic texture, the density of crystalline defects, such as dislocations, and the crystal size can be acquired from the collected XRD data.

Usually XRD stress analysis is performed at room temperature, but, in principle, in-situ, e.g. during thermal treatments, XRD stress analysis is possible as well [2,3]. As compared to room temperature measurements, the interpretation of XRD data recorded at non-ambient temperatures is more demanding for the following reasons:

(i) The temperature dependence of physical properties of films and the substrate has to be known for data analysis and interpretation: for example, the temperature dependence of the coefficients of linear thermal expansion (CTE) of film and substrate and the temperature dependence of the elastic constants. Such data may be unavailable or ambiguous. Note, for example, that the data for the coefficient of thermal expansion of silicon reported in the lit-

erature exhibit a spread of more than 20 % at a temperature of 25°C (although silicon is an extensively studied material in the semiconductor industry). Moreover, it was found that the CTE of thin films and nanocrystalline materials can differ considerably from the (bulk) literature values [4].
(ii) Precise knowledge of the specimen temperature may lack. It has been shown that the reading of the temperature gauge of the temperature controller of the non-ambient chamber is usually unacceptable as correct indication of the actual specimen temperature [2,5]. Dedicated calibration procedures should be performed.
(iii) Thermally activated processes can influence the stress state of thin films, as stress relaxation or oxidation may occur already at only slightly elevated temperatures (if not already at room temperature). A separation of the various contributions to the state of stress can then be difficult.
The use of in-situ XRD measurements above and, in particular, below ambient temperature for the following investigations is discussed in detail:
(i) Measurement of the coefficients of linear thermal expansion (CTE) of thin films (see also Ref. [4]).
(ii) Calibration of the specimen temperature employing a thin film under investigation (consisting of a specific material with a known CTE) as a reference specimen.
(iii) Determination of the amount of (tensile) stresses created by the thermally activated relaxation processes and observing the accompanying microstructural changes such as variation of grain size and defect concentration.
To this end, detailed in-situ X-ray diffraction analyses of stress in Cu and Ni thin films have been performed in the temperature interval between -100°C and 250°C.

2. Experimental and theoretical background

2.1 Experimental and data evaluation

Cu and Ni thin films were deposited at room temperature onto Si (100) wafers, covered with thin amorphous SiO_2 and Si_3N_4 diffusion barrier layers (each 50 nm thick), by direct current magnetron sputtering under ultra-high vacuum conditions (base pressure about 10^{-10} mbar). Each of the thin films had a thickness of 50 nm. Prior to the deposition, the substrates were sputter-cleaned for about 1 minute with Ar ions of kinetic energy 100 eV. During the film deposition, the substrate holder was rotating and the chamber pressure was maintained at 2×10^{-4} mbar employing Ar as a sputter gas. The Cu and Ni sputter sources were operated at 200 W. After deposition, one Cu thin film was annealed at 500°C for 3 hours in the chamber without breaking the vacuum to induce pronounced grain growth.
The experimental setup employed is of cardinal importance for accurate stress measurements. High brilliance X-ray sources are necessary to investigate variation of residual stresses in thin films (50 nm thick films were used in this study) in-situ. To this end, X-ray diffraction measurements were performed employing a Bruker D8 Discover diffractometer equipped with an Eulerian cradle and Cu-Kα radiation emerging from the focus of a rotating anode microfocus X-ray source (Bruker TXS) operating at 50 kV and 20 mA; the radiation has been converted into a monochromated and quasi-parallel beam by a collimating single reflection X-ray mirror (Xenocs) [6]. The diffracted beam passed a parallel plate collimator

(acceptance angle 0.23°) before being detected by a scintillation counter. For in-situ, temperature-dependent, experiments a heating/cooling chamber DCS350 (Anton Paar) was mounted on the Eulerian cradle (see also [2]). It is possible with this equipment to perform in-situ experiments within the temperature range -100°C to 350°C and at pressures of the order of 10^{-3} mbar.

The specimens were first cooled from room temperature (25°C) to -100°C, then heated back to room temperature. Afterwards the Ni specimen was heated from room temperature to 250°C and finally cooled back to room temperature. The increments in temperature were taken as 25°C throughout the entire heat treatment. The temperature cycle between room temperature and 250°C was repeated four times for the Ni thin film. A delay time of 5 minutes was scheduled for each temperature step in order to guarantee temperature homogenization. The 2θ ranges employed to record the 111 reflections of the layers were 42°-45° for Cu and 41.5°-46.5° for Ni. For every temperature step, 2θ scans were performed at two tilt angles, $\psi = 0°$ and $\psi = 70.54°$, employing the side-inclination method [1]. This resulted in a duration of 22 minutes for one stress measurement.

Calibration of the heating/cooling chamber was performed by a resistance thermometry method (the details of this method have been explained in Ref. [2]) independently of the calibration procedure proposed in this paper (i.e. employing the Cu thin films as reference samples; see Section 2.2). In order to determine the instrumental broadening of the diffractometer as function of diffraction angle 2θ, 13 reflections of a LaB_6 (NIST SRM660a) standard powder sample were measured.

Pearson VII functions were fitted to each reflection by using the software Profit 1.0 from Philips in order to determine peak parameters such as peak-maximum position, shape parameter, integral breadth, Voigt parameter and peak-asymmetry parameter [7].

Residual stresses (i.e., the stress σ, biaxial, parallel to the specimen surface and rotationally symmetric) in the specimens were determined by analyzing the X-ray diffraction data according to the $\sin^2\psi$ method on the basis of the crystallite group method (CGM). This approach can be used for specimens with very sharp and strong texture components. The texture may then be described by single-crystal like ideal orientations. In this limit, single crystal elastic compliances [8] can be used instead of the diffraction stress factors (for details, see [1]; for limitations of the CGM in the case of fibre texture, see Ref. [9]). For the case of the CGM, the direction, ψ^*, of zero lattice strain is given by

$$\sin^2\psi^* = \left(-2s_{12} - \frac{2}{3}\left(s_{11} - s_{12} - {}^{s_{44}}\!/_{2}\right)\right) \Big/ \left({}^{1}\!/_{2}\, s_{44}\right). \tag{1}$$

where s_{11}, s_{12} and s_{44} are the single-crystal compliances in the crystal frame of reference.

Changes of crystallite size and microstrain during the thermal cycles were monitored by single-line broadening analysis with the diffraction vector oriented perpendicular to the specimen surface [10,11]. The single-line method can be employed relatively easily and rapidly (as only one particular diffraction line has to be recorded) to obtain estimates for the size of the coherently diffracting domains D and the microstrain ε [10,11] and, in particular, to observe the evolution of these parameters in-situ as function of temperature. More sophisticated methods involving the measurement of multiple diffraction lines for the analysis of the diffraction line broadening exist. However, these methods are not generally applicable to in-situ measurements because of the long measurement times required.

2.2 Determination of thermal expansion coefficient via the strain-free lattice constant and in-situ temperature calibration

For simplicity, cubic materials, which have isotropic CTEs, are considered for the description of the methodological possibilities. If the specimen temperature has been accurately calibrated, a determination of the temperature-dependent coefficient of thermal expansion of film (and/or substrate) is possible according to:

$$\alpha_F(T) \equiv \frac{1}{a_0(T)} \frac{\partial a_0(T)}{\partial T}, \tag{2}$$

where $\alpha_F(T)$ and $a_0(T)$ denote the CTE and the strain-free lattice constant of the film, respectively.

An in-situ temperature calibration is possible via determination of the strain-free lattice constants provided that an accurate temperature-dependent coefficient of (linear) thermal expansion of the film, $\alpha_F(T)$ (and/or the substrate, $\alpha_S(T)$), is known. In the following, the treatment will be formulated for the case that a film of specific material is employed for the temperature calibration. It holds:

$$\varepsilon_{\text{th}} = \frac{a_0(T_f) - a_0(T_a)}{a_0(T_a)} = \int_{T_a}^{T_f} \alpha_F(T')\,dT', \tag{3}$$

where ε_{th} is the thermal lattice strain, T_a and T_f are the ambient temperature and the final temperature, respectively. Generally, Eq. (3) has to be solved for T_f numerically ($\alpha_F(T')$ is known). Only if $\alpha_F(T')$ is constant or if it can be approximated by a simple analytical expression, Eq. (3) can be solved analytically for T_f.

The practical applicability of Eq. (3) for temperature calibration may be affected by in particular:

(i) Scatter of the literature CTE data which can occur for even widely studied materials (as Si; cf. Ref. [2] and Section 1). This scatter may partly be the result of inaccurate measurements of CTEs, but differences in specimen composition and microstructure may also contribute to it. It has, for example, been found that the CTEs of nanocrystalline materials may differ considerably from the CTEs of corresponding coarse-grained materials [4]. It should be recognized that CTEs measured by dilatometric and diffraction methods are intrinsically different. Whereas dilatometric measurements sense length changes resulting from the lattice expansion and the increase of the vacancy concentration upon increasing the temperature (i.e., dilatometry records the macroscopic, mechanical length change), diffraction methods record the lattice expansion only (see, for example Ref. [12]). Thus, upon approaching the melting temperature (where a considerable vacancy concentration occurs), CTEs determined by diffraction are always smaller than CTEs determined employing dilatometry.

(ii) Erroneous values for the elastic constants used for deriving the (unstrained) lattice constant from the diffraction measurement (cf. Eq. (1)). $\sin^2\psi^*$ is given by a ratio of elastic constants and therefore the sensitivity to errors in the elastic constants may be relatively moderate, e.g. if errors in the different elastic constants, relative to their real values, are about equal.

3. Results and discussion

3.1 Determination of coefficient of thermal expansion and in-situ temperature calibration

In the following sections, two different temperatures are used. The first one is the specimen (Cu thin film) temperature, T, as obtained by direct measurement (resistance thermometry; see Ref. [2]). The other one is the calculated temperature, T_{calc}, derived from the strain-free lattice constants employing Eq. (2) using the recommended thermal expansion function (for Cu [13]). The temperature mentioned throughout the article is the specimen temperature, T, unless otherwise stated.

The as-deposited and annealed (cf. Section 2) Cu specimens have been thermally cycled between room temperature (25°C) and -100°C. The 111 reflection has been measured and the strain-free lattice constants have been obtained by interpolation at $\sin^2\psi^*$ in the d-$\sin^2\psi$ plots, where d is the interplaner spacing of {111} planes, (cf. Section 2.1 and see Fig. 1(a); see also Refs. [14,15]) employing the crystallite group method. The plots of the (thermal) lattice strain ($\varepsilon_{th}=\Delta a_0/a_0^{25°C}$; cf. Eq. (3)) versus temperature are shown in Fig. 1(b) for both Cu layers.

From the linear fit to the corresponding data presented in Fig. 1(b) the average CTE in the investigated temperature interval has been calculated both for the as-deposited specimen and the annealed specimen. The CTE value obtained for the as-deposited Cu thin film is distinctly larger than the literature value whereas for the annealed Cu layer the measured and the literature CTEs agree within the experimental accuracy. Note that ignoring the presence of stress in the Cu layers by deriving CTEs from lattice strains measured at $\psi = 0°$ (diffraction vector oriented perpendicular to the specimen surface normal) would lead to overestimation of the CTE (cf. Fig. 1(a)).

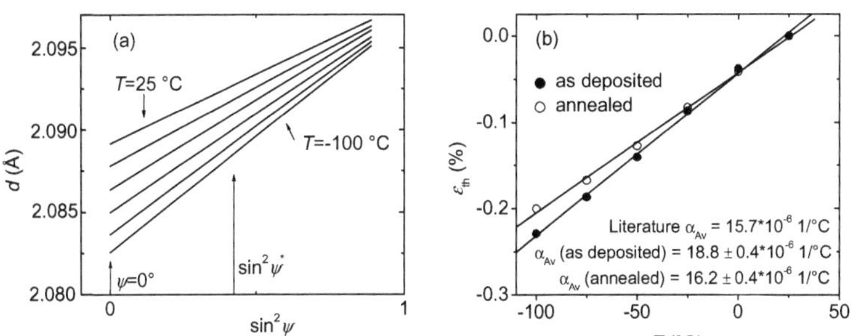

Figure 1. (a) The d-$\sin^2\psi$ plots for the 111 reflection of the annealed Cu layer at different temperatures between 25°C and -100°C. (b)The (thermal) lattice strain, ε_{th}, versus temperature, T.

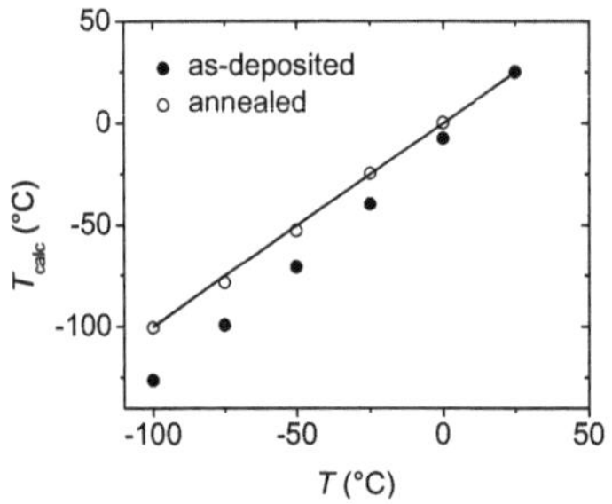

Figure 2. The variation of the temperature, T_{calc}, calculated from the measured a_0 values with the specimen temperature, T, for the as-deposited and annealed Cu layers.

The specimen temperature, T, and the calculated temperature, T_{calc} (see begin of this section), have been plotted along the abscissa and ordinate in Fig. 2, respectively. The line drawn in Fig. 2 pertains to (possible) equality of T and T_{calc}. It follows that, parallel to the above discussion about the measured and literature CTEs, the temperature obtained by the in-situ temperature calibration, T_{calc}, deviates significantly from the specimen temperature as measured directly, T, for the as-deposited Cu thin film whereas a satisfactory agreement occurs for the annealed Cu layer.

The CTE consistency obtained for the annealed Cu layer is a consequence of the appropriateness of the literature value of CTE of (bulk) Cu employed for the calculation of the ordinate values in Fig. 2. After annealing the Cu layer at 500°C for 3 h the lateral crystallite size is about 200 nm (cf. Section 2.1; see Ref. [16]). The considerable deviation of T_{calc} from T is observed for the as-deposited Cu thin film due to its nanocrystalline structure (grain size ~ 25 nm) which results in a larger CTE [16].

3.2 Stress measurements below ambient temperature for characterizing the elastic behaviour of thin films

A Ni thin film has been thermally cycled between room temperature and -100°C. Afterwards, it has been cycled between room temperature and 250°C four times. At each temperature diffraction stress measurements have been performed on the basis of the CGM employing the 111 reflection (cf. Section 2.1). The results are shown in Fig. 3(a).

The linear extrapolation of the thermoelastic behaviour observed during cooling, below room temperature, to temperatures above ambient temperature (Fig. 3(a)) demonstrates unambiguously that already upon heating to temperatures as low as 25°C-50°C, the induced stress change cannot be attributed entirely to the thermal misfit of the film and the substrate (see Fig. 3(a)). Evidently, thermally activated relaxation of the compressive thermal stress in the film, originating from the increasing temperature and the higher CTE of the film than the CTE of the substrate, occurs immediately above ambient temperature or, alternatively, a thermally activated process becomes active that results in the generation of a tensile stress contribution. The magnitude of the amount of compressive stress relaxation above room temperature, or the amount of tensile stress generated, can be estimated from the difference between the observed stress, σ^{obs}, and the expected stress according to the linear extrapolation of the stress evolution below room temperature, σ^{extp}. The difference $\Delta\sigma$ (=σ^{obs}-σ^{extp}), plotted in Fig. 3(b), is substantial upon heating above ambient temperature and reaches a plateau at about 225°C.

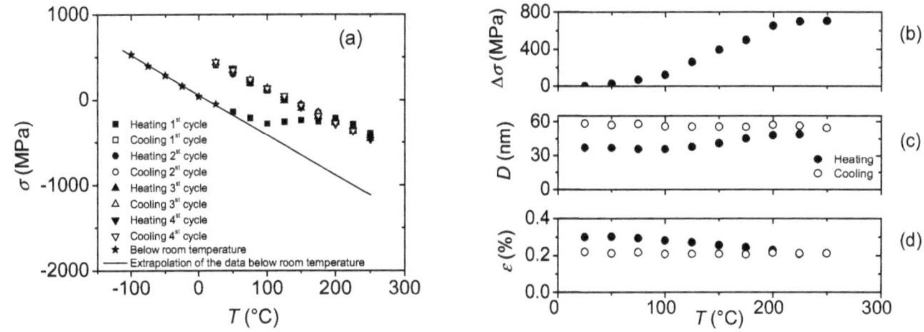

Figure 3. (a) The evolution of residual stress, σ, in the Ni layer with T, (b) the amount of compressive stress relaxed/tensile stress created, $\Delta\sigma\,(=\sigma^{obs}-\sigma^{extp})$, and the variation in (c) crystallite size, D, and (d) microstrain, ε, with T in the first heating cycle.

The corresponding changes of crystallite size, D, and microstrain, ε, in the first heating cycle have also been presented in Fig.3(c) and (d), recognizing that $\Delta\sigma$ can be associated with grain growth and/or the annihilation of crystalline defects (see further).

As is evident from Figs.3 (b)-(d), the heating cycle can be subdivided into two segments. Whereas the annihilation of crystal defects, exhibited by a decrease of the microstrain, proceeds essentially from the beginning of the heating above room temperature (Fig.3(d)), growth of the coherently diffracting domains, in the direction perpendicular to the film surface, starts at a temperature of about 100°C (Fig.3(c)). It appears that in particular upon onset of grain growth a drastic increase of $\Delta\sigma$ occurs (cf. Fig.3(b)).

Adopting a simple treatment proposed by Doerner and Nix [17] (extending the treatment by Chaudhari [18]), that correlates the generation of tensile stress to the elimination of excess volume in the grain boundaries during grain growth, it has been shown here that the amount of tensile stress, $\Delta\sigma$ (~ 700 MPa) arising during the first heating cycle in the Ni layer is well compatible with a tensile stress contribution due to the increase of crystallite size as determined from the diffraction-line broadening analysis, if the excess volume per unit area of a grain boundary is taken as half of an atomic diameter.

After heating to 250°C, a stable microstructure has developed (as evidenced by the observed constancy of diffraction-line position and shape) and thus the first heating segment is followed by a cooling segment and heating and cooling cycles exhibiting fully linear behaviour consistent with the (bulk values of the) CTEs of the Ni layer and the substrate (cf. Fig. 3(a)). The measured thermoelastic slope for the fourth heating cycle is (-4.07 ± 0.1) MPa/°C, which matches the theoretically predicted thermoelastic slope for perfectly {111} fibre-textured bulk Ni (-4.08 MPa/°C) taken as a model for the Ni layer with its sharp and strong {111} fibre texture.

4. Concluding remarks

Strain-free lattice constants extracted from (X-ray) diffraction residual stress measurements can be successfully used to measure average CTEs (if accurate specimen temperature calibration was performed) or, reversely, to calibrate the heating/cooling chamber (if the CTE of the specimen or the substrate is known accurately).

Experimental distinction of the thermal stresses, generated upon heating/cooling due to a difference between the CTEs of film and substrate, and relaxation of stress due to thermally activated processes and build up of (tensile) stress due to as annihilation of crystal defects and grain growth, both *above* ambient temperature, can be realized by investigating the thermoelastic stress-temperature relationship at temperatures *below* ambient temperature, where the thermally activated processes indicated are not operative normally.

References

1. Welzel, U., Ligot, J., Lamparter, P., Vermeulen, A.C. & Mittemeijer, E.J., 2005, *J. Appl. Cryst.*, **38**, 1.
2. Wohlschlögel, M.., Welzel, U., Maier, G. & Mittemeijer, E.J., 2006, *J. Appl. Cryst.*, **39**, 194.
3. Keckes, J., 2005, *J. Appl. Cryst.*, **38**, 311.
4. Kuru, Y., Wohlschlögel, M., Welzel, U. & Mittemeijer, E.J., 2007, *Appl. Phys. Lett.*, **90**, 243113.
5. Beck, M. & Mittemeijer, E.J., 2002, *J. Appl. Cryst.* **35**, 103.
6. Wohlschlögel, M., Schülli, T.U., Lantz, B. & Welzel, U., 2008, *J. Appl. Cryst.*, **41**, 124.
7. *Profit User's Guide*, 1996, (Delft: Philips Analytical X-ray).
8. Gale, W.F. & Totemeier, T.C., 2004, *Smithells Metals Reference Book* (Amsterdam: Elsevier).
9. Welzel, U. & Mittemeijer, E.J., 2004, *Mater. Sci. Forum*, **443**, 131.
10. de Keijser, Th.H., Langford, J.I., Mittemeijer, E.J. & Vogels, A.B.P., 1982, *J. Appl. Cryst.*, **15**, 308.
11. Delhez, R., de Keijser, Th.H. & Mittemeijer, E.J., 1982, *Fresen. Z. Anal. Chem.*, **312**, 1.
12. Simmons, R.O. & Balluffi, R.W., 1960, *Phys. Rev.*, **119**, 600.
13. Touloukian, Y.S., 1975, *Thermal Expansion: Metallic Elements and Alloys* (New York: IFI/Plenum).
14. Hauk, V., 1997, *Structural and Residual Stress Analysis by Nondestructive Methods Evaluation, Application, Assessment* (Amsterdam: Elsevier).
15. Noyan, I.C. & Cohen, J.B., 1987, *Residual Stress* (Berlin: Springer-Verlag).
16. Kuru, Y., Wohlschlögel, M., Welzel, U. & Mittemeijer, E.J., 2008, *J. Appl. Cryst.*, **41**, 428.
17. Doerner, M.F. & Nix, W.D., 1988, *Crit. Rev. Solid State*, **14**, 225.
18. Chaudhari, P., *Vac. Sci. Technol.*, **9**, 520.

Acknowledgements. We thank Mr. F. Thiele and Dr. G. Richter for specimen production.

THIN FILMS

Profiling of fibre texture gradients in thin films by anomalous X-ray diffraction

M. Birkholz[1,*], N. Darowski[2], I. Zizak[3]

[1] IHP, Im Technologiepark 25, 15236 Frankfurt (Oder), Germany
[2] Hahn-Meitner-Institut Berlin, Glienicker Str. 100, 14109 Berlin, Germany
[3] Bessy, Albert-Einstein-Str. 15, 12489 Berlin, Germany
[*] birkholz@ihp-microelectronics.com

Keywords: X-ray diffraction, texture, thin films, texture gradient

Abstract. Anomalous X-ray diffraction is applied to the quantitative determination of fiber texture gradients. Use is made of a theoretical model, in which crystallographic texture is described by an order parameter n and the depth dependent $n(z)$ is assumed to follow a linear behavior (M. Birkholz, 2007, *J. Appl. Cryst.*, **40**, 735). The presented method is based on the extension of kinematic diffraction under inclusion of an inhomogeneous texture and relies on the profiling of different layer depths by tuning the wavelength in the neighborhood of an elemental absorption edge. As an example, thin ZnO:Al layers are investigated and large fiber texture gradients dn/dz between 0.03 and 0.3 m.r.d. nm^{-1} (multiples of a random distribution per nanometer) are revealed to occur in these samples. The approach is concluded to be well suited for the study of microstructure evolution as often observed during the growth of thin films.

1. Introduction

Polycrystalline thin films with thicknesses between some nm and some 100 μm play an important role in modern physics and technology. Many of those layer systems are deposited in vacuum-based processes at growth rates between 10^{-1} to 10^{2} nm s^{-1} and, thereby, develop characteristic microstructures that are accounted for by the ratio of deposition over melting temperature as described in structure zone models [1-3]. Next to their technological significance polycrystalline layers are of fundamental interest to condensed matter physics, since they exhibit remarkable growth phenomena. For example, various microstructure parameters often show pronounced gradients such that the crystallite size, residual stress or the crystallographic orientation may increase with increasing thickness. Particularly the distribution of crystallographic orientation or texture represents an interesting order phenomenon; with the preferred crystallographic direction $\langle HKL \rangle$ typically being rotationally symmetric around the substrate normal and thus being denoted as fiber texture or layer texture.

Since many material properties decisively depend on the orientation distribution function of the polycrystalline grain ensemble [4,5], texture and texture gradients are a highly relevant issue for structure-property relations of advanced materials [6,7] and, in particular, of thin films [8]. For instance, in wurtzite-structured ZnO or AlN layers the crystallites have to align their crystallographic c axis along the substrate normal to enable the intended application in

surface acoustic wave devices. Various texture investigations have been performed for elemental metals [9-11], silicon [12-15], TiN [16-26], ZnO [27-30], high-T_c superconducting and dielectric oxides [31, 32] to mention only a small selection of thin film studies. Most of the experiments were performed by X-ray diffraction (XRD) procedures. Although the usage of X-rays advantageously is a non-destructive technique, the experimental determination of texture gradients from an XRD pattern is not straightforward. In most cases a sample series of varying thickness has to be prepared [10, 12, 19, 26, 29, 30, 32] or investigations have to be performed *in-situ* at synchrotron radiation beam lines [25]. Owing to these limitations, only few texture gradients have been determined in a full quantitative manner so far, which contrasts remarkably to their significance for thin film growth.
Besides the lack of experimental data various theoretical models have been developed. The surface energy minimization approach that explains an evolving $\langle HKL \rangle$ fiber by the lowest energy of (HKL) surfaces has found broad dissemination [18, 27, 28]. In an alternative approach van der Drift argued that the difference in crystallographic growth velocities v_{hkl} causes the selection of the fastest growing direction [33]. Other models addressed the competition between surface energy and strain energy elastically stored in the film volume [16, 17] or the effects of impinging low-energy ion currents [11, 20, 24, 31, 34, 35].
From an experimental point of view interesting results on texture developments were obtained recently by Schell et al., who showed that the $\langle 111 \rangle$ fiber texture in magnetron-sputtered TiN layers evolves through a crossover from a $\langle 200 \rangle$ texture in a post-deposition process [25]. Accordingly, effects of dynamic recrystallization play a more important role at low homologous temperatures ($T_h = T_{dep}/T_{melt}$) than has been assumed so far. Also van der Drift's model received experimental confirmation when it was shown for the first time that a direct consequence of the model, which is the simultaneous increase of fiber texture and deposition rate with deposition time, is in fact observed in Al-doped ZnO layers [30]. Nevertheless, more quantitative results on fiber texture developments are required to improve our understanding of the processes that occur during film growth.
In this work, an X-ray diffraction technique will be introduced by which fiber texture gradients of polycrystalline layers can be determined via probing the sample at different depths. Use will be made of the strong variation of X-ray absorption close to the absorption edge of one of the constituting elements and thereby from the variation of average information depth. Samples of ZnO:Al layers with a pronounced $\langle 00L \rangle$ fiber texture will serve as example systems for demonstrating the technique. X-ray energies were accordingly chosen to lie below and above the Zn K edge of 9.67 keV.

2. Profiling by anomalous diffraction

X-ray diffraction experiments in reflection geometry yield structural information from scattering volumes that are restricted by the penetration depth of the radiation. The latter scales with the inverse linear attenuation coefficient $1/\mu$ that depends on the material exposed and on the wavelength λ of the radiation. Due to the exponential damping of the incoming beam different depths contribute to the diffraction pattern with different intensities. Except for grazing-incidence configurations, the illuminated sample volume in thin film XRD typically encompasses the full depth t of the layer [8, 21]. Fiber textures with the fiber axis parallel to the substrate normal $\boldsymbol{s}_3$ may be determined by measuring Bragg peak intensities I_{HKL} or I_H in a symmetrical configuration at scattering angles 2θ under a set of tilt angles ψ that account

for the inclination of the scattering vector $\boldsymbol{Q}$ with respect to $\boldsymbol{s}_3$, see figure1(a). The measurement yields a set of integrated intensities $I_H(\psi)$, which – in the homogeneous case – contains the texture distribution $T_H(\psi)$. The $T_H(\psi)$ distribution accounts for the volume density of lattice planes (*HKL*) at inclination angle ψ and has to be used in a normalized form in order to cover all crystallites in the sample. Describing for instance the texture distribution by the shape of the $\cos^n\psi$ function, a normalization factor of $(n+1)/2\pi$ has to be chosen. The distribution is characterized by the convenient property that the concentration of lattice planes in the fiber pole – usually given in (intensity) multiples of a random distribution or m.r.d. – directly scales with the order parameter n.
The combined effect of attenuation of the incoming X-ray and tilting of the sample is described by the configuration factor $k = 2/(\sin\theta\cos\psi)$, the product of which with μ has the meaning of an inverse penetration depth. In our approach, distinct $I_H(\psi)$ distributions are measured for different wavelengths λ. The basic principle is sketched in figure 1(b), where the damping of intensity $I/I_0 = \exp(-\mu tk)$ is shown for the ZnO 002 Bragg peak at $\psi = 0°$ and 60° for two different beam energies of 8 and 10 keV. The abscissa is given in units of μt for a thin ZnO:Al film having $\mu t = 0.0113$ and 0.0449 at both energies. It is the difference between the branches of this plot, from which the information on texture gradation is deduced or any other microstructure gradient may be derived.

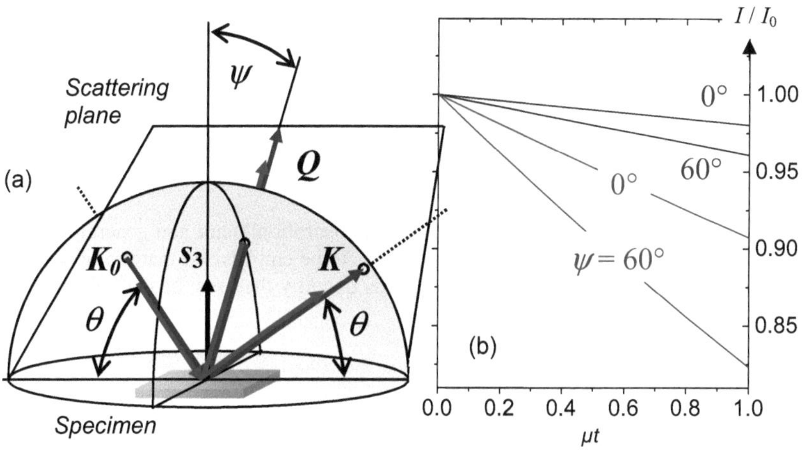

Figure 1. (a) Experimental set-up for θ-2θ measurements under various tilt angles and for different wavelengths, with the latter being symbolized by incoming and exiting wave vectors $\boldsymbol{K}_0$ and $\boldsymbol{K}$ of different color and magnitude. For non zero tilts, $\psi \neq 0$, the scattering vector $\boldsymbol{Q} = \boldsymbol{K} - \boldsymbol{K}_0$ deviates from the direction of the substrate normal $\boldsymbol{s}_3$. (b) Damping of intensity I/I_0 in fiber texture measurements of 002 Bragg reflection from thin ZnO:Al films with two different X-ray energies of 8 and 10 keV (solid and dashed line). I/I_0 values are plotted versus μt and for two tilt angle positions.

Techniques for analyzing non-homogeneous textures in the framework of the kinematic theory were recently put forward [7, 36]. If graded fiber textures are considered, a depth-dependent distribution function $T_H(\psi, z)$ has to be introduced. It has been shown that the kinematical intensity of the fiber peak can be described by the finite Laplace transform [37] of the texture distribution $\mathcal{L}^t(T_H)$ [36], where the transformation is from z space to the penetration depth coordinate $(\mu k)^{-1}$. The general expression for the Bragg peak intensity becomes

$$I_H(\psi, t) = SCF\,\lambda^3\,Lp\left|F_H\right|^2 m_H \frac{k}{2}\mathcal{L}^t(T_H). \tag{1}$$

In this expression SCF, Lp, F_H and m_H stand for an instrumental scaling factor, Lorentz-polarization and structure factor and the multiplicity of the reflection.

For $T_H = ((n+1)/2\pi)\cos^n\psi$ model functions the depth dependence might be included by interpreting the texture order n as a z-dependent polynomial. Analytical solutions of (1) are possible by restricting the polynomial degree to the most simple case, i.e. by setting $n = n_0 + n_1 z$. In passing we note that the fiber texture gradient n_1 is of dimensionality m.r.d. per unit depth. Interpreting n_0 and n_1 in terms of figure 1(b) it means that the single parameter n_0 of the ungraded cases could be deduced from either the difference between the two upper or two lower curves. Intending, moreover, to determine both n_0 and n_1, at least two $I_H(\psi)$ distributions have to be measured at different wavelengths. In case of a linear z dependence of n the finite Laplace transform is solved to yield

$$\frac{k}{2}\mathcal{L}^t(T_H) = \widetilde{N}\cos^{n_0}\psi\frac{k}{2M}\widetilde{A}(t), \tag{2}$$

where $M = \mu k - n_1 \ln(\cos\psi)$ is valid, with Mt now replacing the μtk product, the latter being significant only in the gradient-free texture case. In addition, the abbreviations

$$\widetilde{A}(t) = 1 - \exp(-Mt) = 1 - (\cos^{n_1 t}\psi)\exp(-\mu tk), \tag{3}$$

$$\widetilde{N} = \frac{1}{2\pi}\left[n_0 + n_1 t\left(\frac{1}{Mt} + \frac{1}{1 - e^{Mt}}\right) + 1\right] \tag{4}$$

are used in equation (2) that stand for the generalized absorption factor and generalized normalization factor in the graded texture case [36]. It has to be emphasized that the fiber texture gradient n_1 appears as the product $n_1 t$ and it is this quantity that may be derived from the analysis.

3. Experimental

Layers of ZnO:Al were prepared by reactive magnetron sputtering from a Zn-2wt%Al alloy target, for further details see Refs. [29, 30]. Highly accurate μt products of these samples were derived from atomic area densities Nt determined by Rutherford backscattering [30], applying the connection of both quantities through the molecular weight M_W and mass attenuation coefficient μ_m via $\mu t = \mu_m M_W Nt$. This calculation also allowed for the determination the projected thickness $t_{pro} = \mu t/(\rho\mu_m)$ that is understood as the thickness the layer would have in case its density would compare to the bulk value ρ [8] (in general, the effective thickness of a thin film exceeds the projected one due to inclusion of voids and other excess volume). t_{pro} values of the investigated samples amounted to 118 nm, 246 nm and 413 nm.

Diffraction experiments were performed at BESSY II bending magnet beam line KMC-2 operating with a graded-SiGe/SiGe double-crystal monochromator to achieve an energy resolution $\Delta E/E$ of about 10^{-4} [38]. A six-circle goniometer and a scintillation counter were used for the measurement with beam energies set to 8.048 and 10 keV (abbreviated by 8 and 10 keV in the following). This choice of X-ray energies was motivated, firstly, to allow for comparison of one set of results with those from laboratory experiments by employing a conventional copper tube. Secondly, the higher beam energy is sufficiently above the Zn *K* edge to ensure that EXAFS oscillation in the mass attenuation coefficient μ_m are damped out [39] and to validate the use of the free atom value. Since highly precise intensity values were required for the modeling of equation (1) the experiments were set out to measure only few reflections with a high peak-to-background ratio.

Intensity $\theta/2\theta$ scans of 002 Bragg peaks were collected from each sample for both energies, while inclinations ψ were adjusted to 0, 10, 20, 30, 45 and 60°. Measured intensity data were corrected for peak asymmetry introduced by the Lorentz-polarization factor, which was applied in the appropriate form for in-plane polarized synchrotron radiation, $Lp = (\sin\theta \sin 2\theta)^{-1}$. Lp-corrected Bragg peaks were fitted by pseudo Voigt function profiles yielding the desired integral intensity distributions $I_H(\psi)/Lp$ with estimated standard deviations on the order of 1..2% for maximum intensities at $\psi = 0$.

The results are displayed in figure 2 for the thickest sample with error bars comparable to the size of the dots. Those values were subsequently divided by $\lambda^3 |F_H|^2 m_H$ and subjected to a non-linear regression in terms of equation (1) and (2). Effects due to the polar space group of ZnO were estimated to be of minor significance and were thus neglected by usage of an averaged structure factor F_H. Normal and anomalous scattering factors f, f', f'' as well as mass attenuation coefficients μ_m for both X-ray wavelengths were consistently taken from the Berkeley X-ray server [40], see table 1. Unit cell parameters a, c and oxygen positional parameter u were from Abrahams and Bernstein [41].

Table 1. Normal and anomalous scattering factors f, f' and f'' and mass attenuation coefficients μ_m of the three elements contained in the samples; f values are specified for the 002 peak position at $\sin\theta/\lambda = 1.92$ nm^{-1}. All data according to [40].

		8 keV			10 keV		
	f	f'	f''	μ_m (cm^2 g^{-1})	f'	f''	μ_m (cm^2 g^{-1})
Zn	24.56	-1.56	0.7035	56.35	-2.69	3.694	237.9
O	5.768	0.0522	0.0337	11.13	0.036	0.021	5.673
Al				46.88			24.76

4. Results and discussion

The regression of corrected intensities $I_H(\psi)/(\lambda^3 Lp|F_H|^2 m_H)$ was performed for each sample to both energies simultaneously by assuming n_0 and $n_1 t$ to be independent of energy. Only the scaling factors decomposed into distinct values SCF(E_1) and SCF(E_2). The resulting fitting curve is also presented in figure 2. It is realized from this plot that an excellent agreement of the measured distributions was achieved by modeling the fiber texture gradient with equations (1) to (4).

The values obtained by the regression for n_0 and n_1t are given in figure 3 as a function of sample thickness. It is realized from this figure that n_1t shows a falling tendency, while the opposite trend is observed for n_0. This reflects an increase of preferred orientation with increasing thickness of ZnO:Al layers, albeit the grading of texture is diminishing. Both results are fully in accordance with previous studies [29] justifying the reliability of the approach. The inset of the figure depicts the course of n_1, the values of which were derived from the n_1t/t_{pro} ratio. For the thinnest sample, having t_{pro} = 118 nm, the texture gradient n_1 approaches a value of 0.3 m.r.d. nm^{-1} averaged over the full depth of the layer, which means that every three nanometers the texture order n increases by one m.r.d. unit.

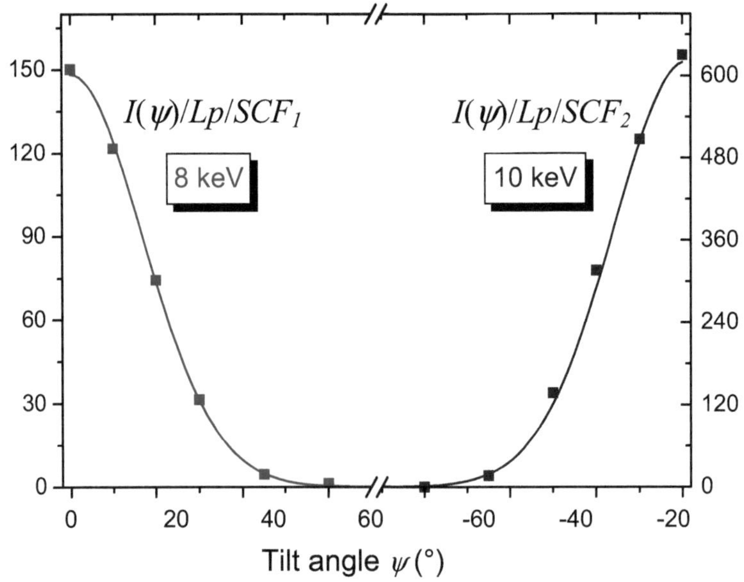

Figure 2. Experimental results of measured $I_H(\psi)/Lp$ values (black dots) and model curve (continuous line) with n_0 and n_1t values given in figure 3. Experimental data and model curve were both divided by the instrumental scaling factor SCF derived from the fit.

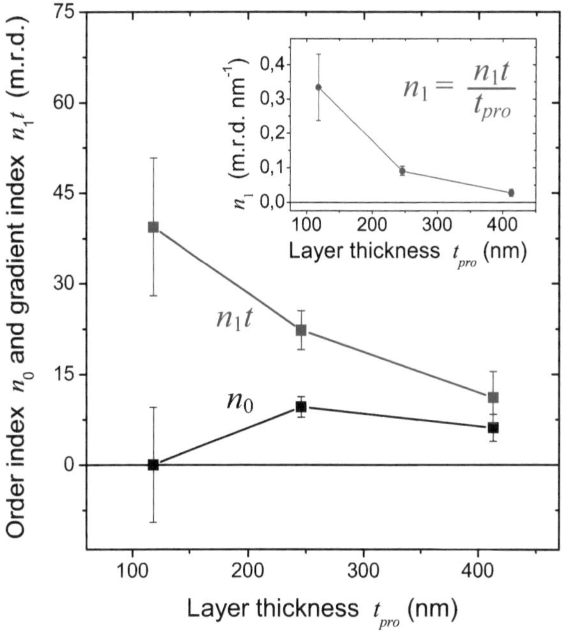

Figure 3. Quantitative results and estimated standard deviations for the initial texture index n_0 and the product of thickness and texture gradient n_1t for the investigated samples. The inset shows the pure texture gradient n_1 in units of m.r.d. nm^{-1}.

This unexpected and remarkably large value underlines the strong increase of fiber texture with increasing thickness in these samples. It indicates strong re-ordering processes to occur on an atomic scale within already deposited layers that might either be due to bulk diffusion or plastic deformation processes.

The results also show that the strongest texture gradient is observed close to the layer-substrate interface. This contrasts to the gradients typically encountered for residual stress, where the strongest variations are mostly observed in the vicinity of the layer surface. An ongoing discussion is concerned with the relation between fiber texture and residual stress evolution and whether one could be the cause of the other microstructural property [11, 16, 17, 20, 22-24, 26].

Apparently, this discussion will benefit from the availability of quantitative results of fiber texture gradients as they may be determined by the formalism presented in this work. It may be expected that future investigations of combined texture, residual stress and other microstructure gradients will yield a more comprehensive picture of individual thin film growth processes. With the introduction of the technique for evaluating fiber texture gradients as presented in this work a further tool for this long-term task has been provided.

5. Conclusions

The quantitative determination of fiber texture gradients has successfully been performed for the example of thin polycrystalline films. The presented technique may be applied to a single thin film specimen and lifts the burden of preparing complete sample series of different thickness or to rely on *in-situ* XRD techniques. To the best of our knowledge this is the first work, where a fiber texture gradient has been determined quantitatively in the appropriate dimensions of multiples of a random distribution per unit depth. Large texture gradients on the order of 0.03 to 0.3 m.r.d. nm^{-1} have been uncovered for thin ZnO:Al films, with the strongest texture increase occurring close to the interface to the substrate. The presented concept is completely general and may equally be applied to other material systems and other orientation distributions than of the $\cos^n\psi$ type. It can be expected that the quantitative determination of texture gradients will deliver valuable experimental input to extend the spare data base of microstructure gradients as required for a perspective full theoretical modeling of thin film growth.

References

1. Movchan, B.A. & Demchishin, A.V., 1969, *Phys. Met. Metallogr.*, **28**, 83.
2. Thornton, J.A., 1974, *J. Vac. Sci. Tech.*, **11**, 666.
3. Machlin, E.S., 1995. *The Relationship between Thin Film Processing and Structure.* Materials Science in Microelectronics (Croton-on-Hudson: Giro Press).
4. Bunge, H.-J., 1982. *Texture Analysis in Materials Science* (London: Butterworth).
5. Wenk, H.-R. & van Houtte, P., 2004, *Rep. Prog. Phys.*, **67**, 1367.
6. Bonarski, J.T., Bunge, H.J., Wcislak, L. & Pawlik, K., 1998, *Textures Microstruc.*, **31**, 21.
7. Bonarski, J.T., 2006, *Prog. Mat. Sc.*, **51**, 61.
8. Birkholz, M., 2006. *Thin Film Analysis by X-ray Scattering* (Weinheim: Wiley-VCH), chapter 5 and pp 64-66.
9. Harper, J.M.E., Cabral, C., Andricacos, P.C., Gignac, L., Noyan, I.C., Rodbell, K.P. & Hu, C.K., 1999, *J. Appl. Phys.*, **86**, 2516.
10. Okolo, B., Lamparter, P., Welzel, U. & Mittemeijer, E.J., 2004, *J. Appl. Phys.*, **95**, 466.
11. Birkholz, M., Genzel, C. & Jung, T., 2004, *J. Appl. Phys.*, **96**, 7202.
12. Birkholz, M., Selle, B., Conrad, E., Lips, K. & Fuhs, W., 2000, *J. Appl. Phys.*, **88**, 4376.
13. Birkholz, M., Conrad, E. & Fuhs, W., 2001, *Jpn. J. Appl. Phys.*, **40**, 4176
14. Reinig, P., Fenske, F., Fuhs, W., Alex, V. & Birkholz, M., 2002, *J. Vac. Sc. Technol. A*, **20**, 2004.
15. Morales, M., Leconte, Y., Rizk, R. & Chateigner, D., 2004, *J. Appl. Phys.*, **97**, 034307.
16. Pelleg, J., Zevin, L.Z., Lungo, S. & Croitoru, N., 1991, *Thin Solid Films*, **197**, 117.
17. Oh, U.C. & Je, J.H., 1993, *J. Appl. Phys.*, **74**, 1692.

18. Knuyt, G., Quaeyhaegens, C., D'Haen, J. & Stals, L.M., 1995, *Thin Solid Films*, **258**, 159.
19. Quaeyhaegens, C., Knuyt, G., D'Haen, J. & Stals, L.M., 1995, *Thin Solid Films*, **258**, 170.
20. Greene, J.E., Sundgren, J.-E., Hultman, L., Petrov, I. & Bergstrom, D.B., 1995, *Appl. Phys. Lett.*, **67**, 2928.
21. Sloof, W.G., Kooi, B.J., Delhez, R., de Keijser, T.H. & Mittemeijer, E.J., 1996, *J. Mater. Res.*, **11**, 1440.
22. Genzel, C. & Reimers, W., 1998, *phys. stat. sol. (a)*, **166**, 751.
23. Leoni, M., Scardi, P., Rossi, S., Fedrizzi, L. & Massiani, Y., 1999, *Thin Sol. Films*, **345**, 263.
24. Rauschenbach, B. & Gerlach, J.W., 2000, *Cryst. Res. Technol.*, **35**, 675.
25. Schell, N., Bottiger, J., Matz, W. & Chevallier, J., 2003, *Nucl. Instr. Meth.*, **B 199**, 133.
26. Abadias, G. & Tse, Y.Y., 2004, *J. Appl. Phys.*, **95**, 2414.
27. Fujimura, N., Nishihara, T., Goto, S., Xu, J. & Ito, T., 1993, *J. Cryst. Growth*, **130**, 269.
28. Jiang, X., Jia, C.L. & Szyska, B., 2002, *Appl. Phys. Lett.*, **80**, 3090.
29. Birkholz, M., Selle, B., Fenske, F. & Fuhs, W., 2003, *Phys. Rev. B*, **68**, 205414.
30. Fenske, F., Selle, B. & Birkholz, M., 2005, *Jpn. J. Appl. Phys.*, **44**, L662.
31. Berdahl, P., Reade, R.P., Liu, J., Russo, R.E., Fritzemeier, L., Buczek, D. & Schopp, U., 2003, *Appl. Phys. Lett.*, **82**, 343.
32. Ricote, J., Poyato, R., Alguero, M., Pardo, L., Calzada, M.L. & Chateigner, D., 2003, *J. Am. Ceram. Soc.*, **86**, 1571.
33. van der Drift, A., 1967, *Philips Res. Rep.*, **22**, 267.
34. Bradley, R.M., Harper, J.M.E. & Smith, D.A., 1986, *J. Appl. Phys.*, **60**, 4160.
35. Rauschenbach, B. & Helming, K., 1989, *Nucl. Instr. Meth.*, **B42**, 216.
36. Birkholz, M., 2007, *J. Appl. Cryst.*, **40**, 735.
37. Doetsch, G., 1970. *Einführung in die Theorie und Anwendung der Laplace-Transformation* (Basel: Birkenhäuser).
38. Erko, A., Packe, I., Gudat, W., Abrosimov, N. & Firsov, A., 2000, *A Graded Crystal Monochromator at BESSY II*, in *SPIE conf. proc.* **4145**, edited by A.K. Freund et al., p. 122.
39. Decremps, F., Datchi, F., Saitta, A.M., Polian, A., Pascarelli, S., Di Cicco, A., Itié, J.P. & Baudelet, F., 2003, *Phys. Rev. B*, **68**, 104101.
40. Henke, B.L., Gullikson, E.M. & Davis, J.C., 1993, *Atomic Data Nucl. Data Tab.*, **54**, 181.
41. Abrahams, S.C. & Bernstein, J.L., 1969, *Acta Cryst.*, **B25**, 1233.

Acknowledgements. We thank Daniel Chateigner, Christoph Genzel, Maarten Heyn, Peter Zaumseil and Harald Beyer for helpful discussion and Frank Fenske for leaving these otherwise already fruitful samples for this investigation.

Z. Kristallogr. Suppl. **27** (2008) 273-285 / **DOI** 10.1524/zksu.2008.0034

X-ray residual stress analysis in CVD multilayer systems: Influence of steep gradients on the line profile shape and -symmetry

M. Klaus[1], Ch. Genzel[2,*], H. Holzschuh[3]

[1] Technical University Berlin, Institute of Materials Science and Technologies, Ernst-Reuter-Platz 1, D-10587 Berlin, Germany
[2] Hahn-Meitner-Institute Berlin (c/o BESSY), Department of Structural Research, Albert-Einstein-Str. 15, D-12489 Berlin, Germany
[3] Walter AG, Derendinger Str. 53, D-72072 Tübingen, Germany
[*] genzel@hmi.de

Keywords: X-ray stress analysis, multilayer systems, residual stress gradients, diffraction line profile

Abstract. Residual stresses in multilayer systems are usually not uniform but occur in form of gradients which are balanced either within the individual layers or between different layers and the substrate. Starting from a formalism which extends the method of X-ray residual stress depth profiling in monolayer thin films to the multilayer case, as-grown and grit-blasted Al_2O_3/TiCN hard coating systems are analyzed with respect to the in-plane residual stress state. For the Al_2O_3-layer on top the blasting process results in a steep residual stress gradient leading to a nonlinear $d_\psi^{hkl} - \sin^2\psi$ -distribution as well as asymmetric diffraction line profiles, the shape and integral breadth of which vary during strain depth profiling.

1. Introduction

For the evaluation of residual stresses in polycrystalline materials numerous diffraction methods have been developed in the past decades [1-3]. Concerning the investigation of thin films the situation is more complicated compared to bulk materials, because many of the problems scientists are faced with in X-ray stress analysis (XSA) are 'concentrated' within a thickness of a few microns or less. The microstructure is usually not uniform over the film thickness but occurs in form of gradients of the chemical composition [4], the crystallographic texture [5] and/or the residual stresses [6], which superimpose each other [7]. For a comprehensive survey of problems related to thin film X-ray analysis the reader is referred to [8,9].

The present paper addresses a special problem arising in the XSA of residual stress gradients generated by mechanical surface processing in multilayer systems. Depending on the steepness of the gradient and the experimental method used for depth profiling, the measured profiles becomes more or less asymmetric [10]. From the literature of diffraction analysis it is well-known that asymmetric diffraction line broadening may have various reasons such as faulting in cold-worked metals [11] or compositional variations in chemically treated surface layers [12]. A systematic study regarding the influence of steep residual stress gradients on the line profile shape, however, is still missing.

Starting from an extension of the basic equation of X-ray stress depth-profiling to the case of multilayer systems, which allows to take into account buried layers, the formalism is applied to the study of cemented carbide (WC/Co) cutting tools, which were coated by chemical vapour deposition (CVD) with a combination of Al_2O_3 and TiCN layers. To tailor the residual stress state and to improve the toughness of the tools, grit blasting was used to introduce compressive stresses in form of steep intra-layer gradients into the Al_2O_3-layer on top. The experimental results are compared with simulations of the XSA measurements showing that the line asymmetry observed in the diffraction patterns is caused by the steep in-plane residual stress gradient.

2. Depth-resolved X-ray stress analysis in multilayer systems

2.1 Extension of the basic equation of XSA to multilayer systems

The analysis of residual stress fields in the near surface region of quasi-isotropic polycrystalline bulk materials and thin films is based on the measurement of reflections *hkl* and the evaluation of lattice strains $\varepsilon_{\phi\psi}^{hkl} = \left[d_{\phi\psi}^{hkl} - d_0^{hkl} \right] / d_0^{hkl}$ (d_0^{hkl} - stress free lattice spacing) for different azimuth and inclination angle sets (ϕ, ψ) of the diffraction vector $\mathbf{g}^{hkl}$ with respect to the sample reference system. Assuming the residual stress state to be biaxial and of rotational symmetry, i.e. $\sigma_{11} = \sigma_{22} = \sigma_{\parallel}$, $\sigma_{i3} \equiv 0 \, (i = 1,2,3)$, which applies in most cases for thin films, the basic equation of XSA in its depth dependent form becomes

$$\varepsilon_{\psi}^{hkl}(\tau) = \left[1/2\, s_2^{hkl} \sin^2 \psi + 2 s_1^{hkl} \right] \sigma_{\parallel}(\tau) \tag{1}$$

where $1/2\, s_2^{hkl}$ and s_1^{hkl} are the diffraction elastic constants (DEC) and τ is the 1/e penetration depth of the X-rays in the material [2]. Within the kinematical diffraction theory a general formulation for τ is given by

$$\tau = \frac{1}{\mu k} = \frac{\sin^2 \theta - \sin^2 \psi + \cos^2 \theta \sin^2 \psi \sin^2 \eta}{2 \mu \sin \theta \cos \psi} \tag{2}$$

where θ and μ are the Bragg angle and the linear absorption coefficient, respectively, and η describes the rotation of the sample around the diffraction vector $\mathbf{g}^{hkl}$ [13]. The angle-dependent terms in equation (2) are usually summed up in the geometry term k which depends on the method used for depth profiling. The correlation between the exponentially (i.e. absorption) weighted 'Laplace space' depth profiles $\varepsilon_{\psi}^{hkl}(\tau)$ and $\sigma_{\parallel}(\tau)$ and the actual 'real space' depth profiles $\varepsilon_{\psi}^{hkl}(z)$ and $\sigma_{\parallel}(z)$ is given by

$$\varepsilon_{\psi}^{hkl}(\tau) = \frac{\int_0^D \varepsilon_{\psi}^{hkl}(z) e^{-z/\tau} dz}{\int_0^D e^{-z/\tau} dz}, \quad \sigma_{\parallel}(\tau) = \frac{\int_0^D \sigma_{\parallel}(z) e^{-z/\tau} dz}{\int_0^D e^{-z/\tau} dz} \tag{3}$$

[14], where D is the layer thickness. Hence, for monolayer systems equation (3) allows for the evaluation of the actual stress depth profiles $\sigma_{\|}(z)$ from the Laplace depth profiles $\sigma_{\|}(\tau)$, which are obtained directly by means of equation (1) from the experimentally measured lattice strain distributions $\varepsilon_{\psi}^{hkl}(\tau)$.

In multilayer systems the situation is more complex and requires a modified treatment. We consider a multilayer system consisting of an alternating sequence of layers D_i with thickness t_i^D of identical crystallographic structure and composition, which contribute to the diffraction signal, and layers A_i of thickness t_i^A, which do not fulfil the diffraction condition for the same Bragg angle θ as the layers D_i, but attenuate the X-ray beam by absorption (figure 1). In this case the contribution of each diffracting layer D_i has to be weighted by the absorption contribution of the layer A_i above. For the Laplace stresses we obtain with $\tau = 1/\mu k$:

$$\sigma_{\|}(\tau) = \frac{e^{-\mu k t_1^{A,eq}} \int_{z_1}^{z_2} \sigma_{\|}(z) e^{-\mu k z} dz + e^{-\mu k t_2^{A,eq}} \int_{z_3}^{z_4} \sigma_{\|}(z) e^{-\mu k z} dz + \ldots}{e^{-\mu k t_1^{A,eq}} \int_{z_1}^{z_2} e^{-\mu k z} dz + e^{-\mu k t_2^{A,eq}} \int_{z_3}^{z_4} e^{-\mu k z} dz + \ldots}. \tag{4}$$

In the above equation, the thickness of the diffracting layers is $t_i^D = z_{2i} - z_{2i-1}$.

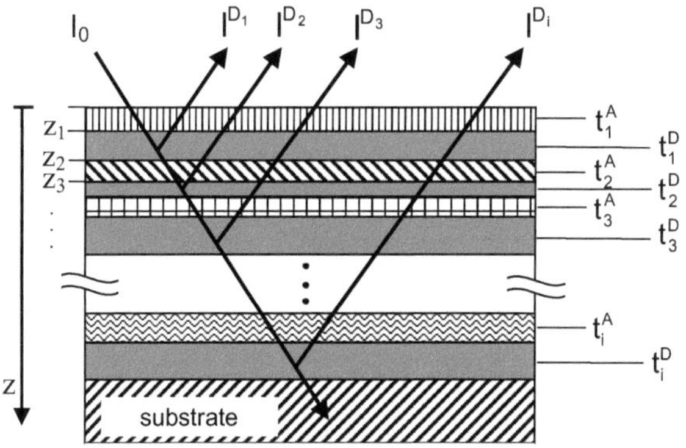

Figure 1. Schematic illustration of X-ray diffraction at a multilayer system. See text for details.

The actual thickness t_i^A of the absorbing layers A_i was replaced by an ‚equivalence' thickness $t_i^{A,eq}$ which is defined by the condition that it results in the same attenuation as the layer A_i of thickness t_i^A, but with its linear absorption coefficient μ^* replaced by that of the layer sort D_i, given by μ:

$$\mu t_i^{A,eq} = \mu^* t_i^A \tag{5}$$

[12]. The replacement (5) enables to use $\tau = 1/\mu k$ as consistent variable for stress depth profile evaluation by describing $\sigma_{\parallel}(z)$ in the diffracting layers D_i by appropriate functions and making use of equation (4) to calculate the respective Laplace stress function $\sigma_{\parallel}(\tau)$. Hence, equation (4) represents a generalized formulation of the $\sigma_{\parallel}(\tau) - \sigma_{\parallel}(z)$ - transformation problem (3) being valid for any multilayer system.

2.2 Simulation of depth-resolved XSA experiments

We assume the sample to be composed of infinitesimal small sublayers δz parallel to the surface, each of them giving rise to a diffraction signal $I^{hkl}(2\theta, z)$ (see figure 1). The total signal $I^{hkl}_{total}(2\theta)$ is obtained by absorption weighted integration over all sublayers δz. Within the kinematical theory the intensity diffracted by a sublayer δz in a depth z is given by

$$I^{hkl}(2\theta, z) = I_0 \frac{r_e^2 \cdot \lambda^3 \cdot \left|F_{hkl}\right|^2 \cdot H \cdot e^{-M}}{R^2 \cdot V_{EZ}^2} P(\theta) L(\theta) G(\theta) A(\theta, z)\, LSF\left[2\theta, 2\theta^{hkl}_{\psi,\max}(z)\right] dz \qquad (6)$$

(I_0 - primary beam intensity, r_e - electron radius, F_{hkl} - structure factor of the reflection hkl, H -multiplicity factor, M – Debye-Waller factor, R - distance between sample and detector, V_{EZ} - volume of the unit cell). *LSF* stands for the *L*ine *S*hape *F*unction which will be discussed later on. $P(\theta)$, $L(\theta)$ and $G(\theta)$ are the polarization-, Lorentz- and geometry factor, respectively:

$$P(\theta) L(\theta) G(\theta) = \frac{1 + \cos^2 2\theta}{2} \frac{1}{\sin 2\theta} \frac{1}{2 \sin \theta} \qquad (7)$$

The absorption coefficient $A(\theta, z)$ for each sublayer δz at depth z is given by

$$A(\theta, z) = \frac{S}{\sin \alpha(\theta)} \exp\left\{ -\mu \left[\frac{1}{\sin \alpha(\theta)} + \frac{1}{\sin \beta(\theta)} \right] z \right\} \qquad (8)$$

(S - primary beam cross section), where $\alpha(\theta)$ and $\beta(\theta)$ are the angles which the primary and the diffracted beam, respectively, include with the sample surface (see [15]). For the line shape function $LSF[2\theta, 2\theta^{hkl}_{\psi,\max}(z)]$, which describes the intensity distribution diffracted by each sublayer, we use the Pseudo-Voigt (*PV*) function:

$$LSF\left[2\theta, 2\theta^{hkl}_{\psi,\max}(z)\right] = w(z) \cdot \exp\left\{ -\frac{\lambda^2}{8\left[\delta_d^{hkl}(z)\right]^2} \left[\frac{1}{\sin 1/2\, 2\theta} - \frac{1}{\sin 1/2\, 2\theta^{hkl}_{\psi,\max}(z)} \right]^2 \right\}$$

$$+ \left[1 - w(z)\right] \cdot \left\{ 1 + \frac{\lambda^2}{\left[\delta_d^{hkl}(z)\right]^2} \left[\frac{1}{\sin 1/2\, 2\theta} - \frac{1}{\sin 1/2\, 2\theta^{hkl}_{\psi,\max}(z)} \right]^2 \right\}^{-1} \qquad (9)$$

Equation (9) is based on the following assumptions: 1. The intensity distribution $I^{hkl}(2\theta, z)$ diffracted by an infinitesimal sublayer δz is supposed to be symmetrical. 2. The intrinsic broadening of each sublayer profile being defined on the lattice spacing scale by the dispersion parameter $\delta_d^{hkl}(z)$ does not depend on the orientation (ϕ,ψ). 3. The line shape specified by the weight factor $w(z)$ is allowed to vary with depth between a pure Gaussian (*G*) ($w = 1$) and a pure Lorentzian (*L*) shape ($w = 0$). Therefore, both $\delta_d^{hkl}(z)$ and $w(z)$ provide the basis for introducing depth dependent microstructure models into the formalism. 4. The maximum position $\theta_{\psi,\max}^{hkl}(z)$ of the individual sublayer profiles $I^{hkl}(2\theta,z)$ is determined by the residual stress depth distribution $\sigma_{ij}(z)$ and thus dependent on the measuring direction ψ. In the presence of a composition gradient within a diffracting layer D_i, the local maximum $\theta_{\psi,\max}^{hkl}(z)$ may become a function of the stress free lattice parameter $\theta_0^{hkl}(z)$:

$$\theta_{\psi,\max}^{hkl}(z) = \theta_0^{hkl}(z) - \tan\theta_0^{hkl}(z)\,\varepsilon_\psi^{hkl}(z)\frac{180}{\pi} \tag{10}$$

with $\theta_0^{hkl}(z)$ being related to $d_0^{hkl}(z)$ by Bragg's law. The strain gradient $\varepsilon_\psi^{hkl}(z)$ is related to the stress gradients $\sigma_{\|}(z)$ by the fundamental equation (1), if the τ-profiles are replaced by the respective *z*-profiles defined within each diffracting layer D_i.

The integral breadth of each sublayer profile $I^{hkl}(2\theta,z)$ includes the information on the film microstructure (domain size, lattice defects, micro-strain [11]) and is obtained by integrating equation (9):

$$\beta_{intr.}^{hkl}(z) = w(z)\cdot\beta_{intr.}^{hkl,(G)}(z) + [1-w(z)]\cdot\beta_{intr.}^{hkl,(L)} = w(z)\cdot\sqrt{2\pi}\,\delta_{2\theta}^{hkl}(z) + [1-w(z)]\,1/2\,\pi\,\delta_{2\theta}^{hkl}(z) \tag{11}$$

where $\beta_{intr.}^{hkl,(G)}(z)$ and $\beta_{intr.}^{hkl,(L)}(z)$ are the integral breadths of the Gaussian and the Lorentzian profile, respectively. The subscript *intr.* is to indicate, that the integral breadth in the above expression refers to the *intrinsic* line width of the individual sublayer profiles $I^{hkl}(2\theta,z)$, which is in general different from the integral breadth β_ψ^{hkl} of the measured or simulated line profiles $I_{total}^{hkl}(2\theta)$ (see below).

The relation between the dispersion parameters δ_d^{hkl} and $\delta_{2\theta}^{hkl}$ on the lattice spacing and the 2θ- scale used in the equations (9) and (11), respectively, is given by

$$\delta_d^{hkl}(z) = 1/2\cot\theta\, d^{hkl}\,\delta_{2\theta}^{hkl}(z)\frac{\pi}{180}\,. \tag{12}$$

The total line profile diffracted by a multilayer system as shown in figure 1, is finally obtained by integrating over all diffracting layers D_i the system consists of:

$$I_{total}^{hkl}(2\theta) = \frac{1}{A^\infty(\theta)}\left[e^{-\mu k t_1^{A,eq}}\int_{z_1}^{z_2} I^{hkl(D_1)}(2\theta,z)\,d + e^{-\mu k t_2^{A,eq}}\int_{z_3}^{z_4} I^{hkl(D2)}(2\theta,z)\,dz + \ldots\right] \tag{13}$$

where $A^\infty(\theta) = (1+\tan\psi\cot\theta\cos\eta/2\mu)$ is the normalization factor [15].

For the evaluation of the measured and simulated diffraction line profiles, respectively, we used an *A*symmetric *P*seudo-*V*oigt function (*APV*) defined by

$$APV(2\theta) = a_1 \left\{ a_4 \exp\left[\frac{(2\theta - a_2)^2}{2(a_3^2 + a_5^2)} \right] + (1 - a_4) \frac{1}{1 + 4\dfrac{(2\theta - a_2)^2}{a_3^2 + a_5^2}} \right\} \left\{ 1 + Erf\left[\frac{a_5(2\theta - a_2)}{\sqrt[\sqrt{2}a_3]{a_3^2 + a_5^2}} \right] \right\} \quad (14)$$

where the parameter a_5 quantifies the line profile asymmetry. The macroscopic line shape parameter a_4 corresponds to $w(z)$ in the equations (9) and (11), which defines the profile shape on the microscopic scale.

3. Experimental

3.1 Samples

For the investigations commercial cutting inserts for turning steel were used from Walter AG (WPP10 CNMA120412). The 20μm multilayer coatings were grown by chemical vapour deposition on cemented carbide inserts. One coating system was analyzed in the as-grown state, whereas the top TiN layer of the second sample was removed by blasting the rake face of the insert. Figure 2a shows the cross section of a typical as-grown sample. In the example considered in the following, the thickness ratio of Al_2O_3 (top layer) to TiCN was 14μm/4μm. Because the investigated system consists of various crystalline phases with different crystal structure, at least two of them (Al_2O_3, WC) being of non-cubic symmetry, the diffraction patterns contain a multitude of diffraction lines (figure 2b). The reflections for the XSA experiments were chosen by the condition that they do not overlap with neighbouring diffraction lines. A texture analysis revealed less pronounced fibre textures with $\langle 001 \rangle$ and $\langle 211 \rangle$ being the main components in the topmost Al_2O_3-layer and in the buried TiCN layer, respectively.

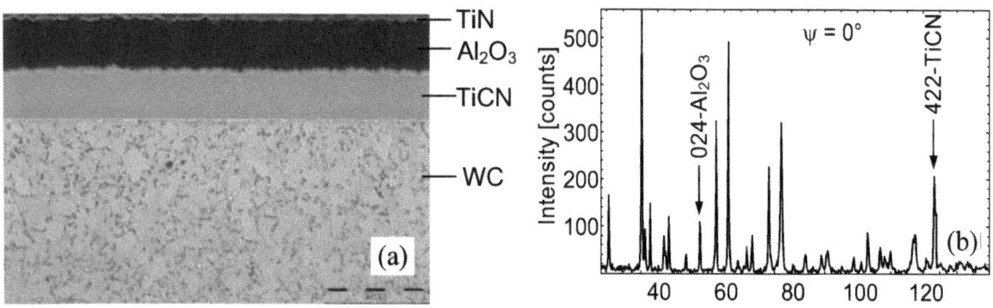

Figure 2. (a) Cross section micrograph of an as-grown 20 μm multilayer coating. (b) X-ray diffraction pattern of the investigated multilayer system (as-grown state). The reflections hkl used for residual stress analysis are marked by arrows.

3.2 X-ray diffraction

The X-ray experiments were carried out on the 5-circle diffractometer ETA using a thin film attachment for grazing incidence diffraction [6] (table 1).

Table 1. Parameters used for the X-ray measurements and residual stress evaluation.

Radiation	CuKα (0.154056 nm), long fine focus
Operating conditions	32 kV / 50 mA, no K_β-filter
X-ray optics	polycapillary semilens (primary beam) soller slit (0.4°) + flat 001 LiF monochromator (diffracted beam)
Reflections *hkl* / 2θ-range	Al_2O_3: 024 (2θ = 52.6°) / [51.0° ... 55.0°] TiCN: 422 (2θ = 123.3°) / [120° ... 127°]
Step width / counting time	Δ2θ = 0.05° / t = 30s
ψ-range / step width (for the Al_2O_3 layer on top)	[0° ... 89.5°], $\Delta \sin^2\psi = 0.05$ ($\psi \leq 80°$) / $\Delta\ \psi = 0.5°$ ($81° \leq \psi \leq 89.5°$)
Diffraction elastic constants	Al_2O_3: $s_1^{024} = -0.55 \cdot 10^{-6}$ MPa^{-1} $\frac{1}{2}s_2^{024} = 2.96 \cdot 10^{-6}$ MPa^{-1} TiCN: $s_1^{422} = -0.47 \cdot 10^{-6}$ MPa^{-1} $\frac{1}{2}s_2^{422} = 2.83 \cdot 10^{-6}$ MPa^{-1}

Both, the topmost Al_2O_3-layer (after blasting) and the TiCN-layer beyond were analyzed with respect to the in-plane residual stress state. For the evaluation use was made of the DEC calculated from the single crystal elastic constants for Al_2O_3 [16] and TiN [17], respectively, by the Eshelby/Kröner model. The application of the DEC being valid for quasi-isotropic materials can be justified by the fact, that both phases exhibit only a small single crystal elastic anisotropy.

4. Results

4.1 Residual stress analysis

In a preliminary study performed under different azimuths ϕ the residual stress state in all investigated coatings was found to be of rotational symmetry, regardless of the coating design and state of surface processing. Figure 3 shows the $\sin^2\psi$-distributions of the lattice spacings d_ψ^{hkl} and the integral breadth β_ψ^{hkl} for both, the as-grown and the blasted state. Due to the absorption by the layers above, the accessible ψ-range for the TiCN-layer was restricted to 45° $(\sin^2\psi = 0.5)$. For the as-grown state, in both layers almost linear $d_\psi^{hkl} - \sin^2\psi$-distributions with positive slopes were obtained, which allow for residual stress evaluation by means of the conventional $\sin^2\psi$-method.

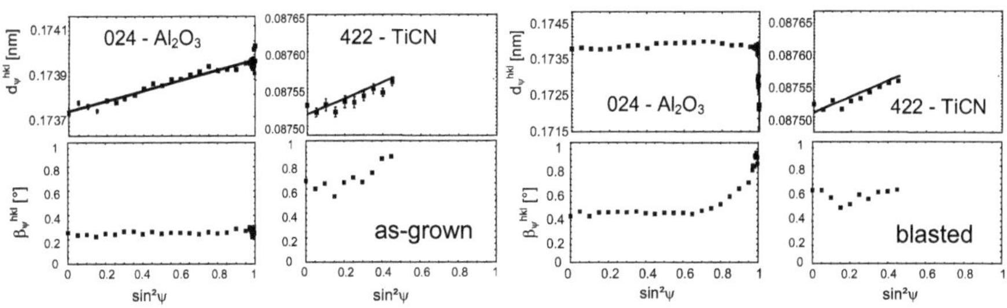

Figure 3. $\sin^2\psi$ *-distributions of the lattice spacings* d_ψ^{hkl} *(top) and the integral breadth* β_ψ^{hkl} *(bottom) obtained for the* Al_2O_3 *- and the TiCN - layer, as-grown state (left), blasted state (right).*

The average in-plane residual stresses $<\sigma_{\|}>$ were found to be +439 ± 20 MPa (Al_2O_3) and +443 ± 50 MPa (TiCN). The integral breadth β_ψ^{hkl} observed for the Al_2O_3-layer, which is the more interesting one concerning the subsequent surface processing step, is small compared to the blasted case shown below and remains constant within the investigated ψ -range.

For the blasted state no significant changes were observed for the buried TiCN layer. The $d_\psi^{422} - \sin^2\psi$ -distribution remains linear with a positive slope resulting in an average tensile in-plane residual stress of $<\sigma_{\|}> = +492 \pm 43$ MPa. A completely different behaviour was found for the topmost Al_2O_3 layer. The $d_\psi^{024} - \sin^2\psi$ -distribution remains nearly horizontal up to $\psi \approx 80°$. For larger ψ , however, the d-values drop down drastically by an amount of $\Delta d \approx 2 \cdot 10^{-3}$ nm. Furthermore, compared to the as-grown state the integral breadth β_ψ^{024} is enlarged by a factor of about 1.5 and increases significantly with increasing ψ , i.e. towards the blasted surface.

The residual stress distribution in the Al_2O_3 layer after blasting was evaluated from the β_ψ^{024} -data by means of the formalism introduced in subsection 2.1. We described the stress depth profile in the z-space by an exponentially damped first order polynomial function $\sigma_{\|}(z) = (b_0 + b_1 z)e^{b_2 z}$. From figure 4a it can be seen that the maximum compressive stress at the surface takes a value of about -8 GPa and drops down to zero within a depth range of about 2μm. The high surface compressive stresses are compensated within the layer by tensile stresses which have a maximum of about +2.5 GPa in a depth of about 3μm. The recalculated $d_\psi^{024} - \sin^2\psi$ -curve fits well into the experimental data.

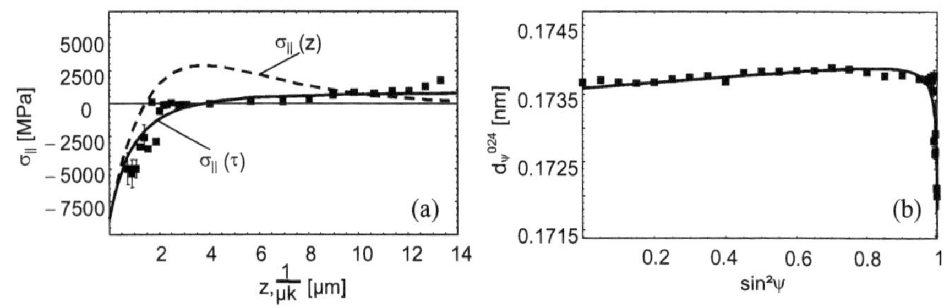

Figure 4. (a) Residual stress depth profiles $\sigma_{\|}(\tau)$ and $\sigma_{\|}(z)$ in Laplace- and real space, respectively, evaluated for the Al_2O_3- top layer, blasted state. (b) Recalculated $d_\psi^{024} - \sin^2\psi$ - curve.

4.2 Line profile evaluation

A striking effect concerning the line profile shape is shown in figure 5, where the 024 line profiles of the Al_2O_3 layer measured for different orientations ψ are compared for the as-grown and the blasted state.

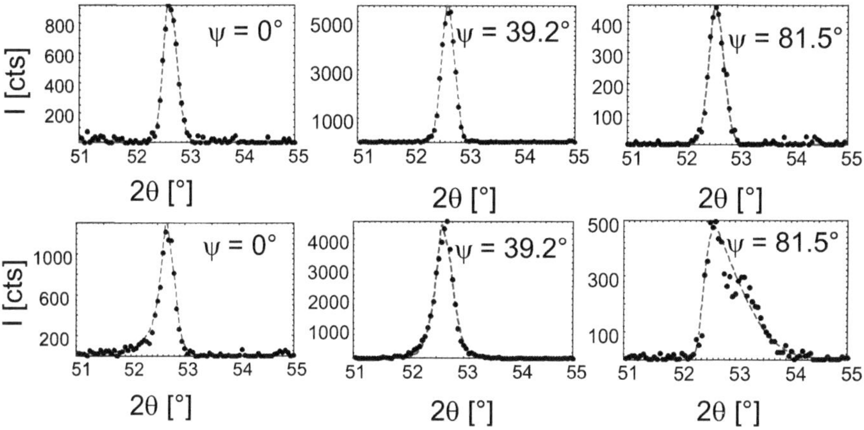

Figure 5. 024-diffraction line profiles of the Al_2O_3-layer, as-grown state (upper diagrams) and blasted state (lower diagrams). ψ = 39.2° is close to the strain-free direction of the biaxial stress state, ψ^*_{024} = 37.3°. The profiles were fitted using the asymmetric Pseudo-Voigt function defined by equation (14).

Two significant differences become obvious: 1. As mentioned before, the blasting process obviously leads to considerable line broadening, which becomes especially apparent for large ψ, i.e. small penetration depths (cf. figure 3). 2. After blasting the profiles are strongly asymmetric with longer tails on the small (large) 2θ angle side for $\psi < \psi^*$ $(\psi > \psi^*)$, where $\psi^*_{hkl} = \arcsin\sqrt{-2s_1^{hkl}/\tfrac{1}{2}s_2^{hkl}}$ is the strain-free direction of a biaxial stress state of rotational symmetry.

4.3 Simulation of the XSA experiment

To verify the experimental results in the previous section and to elucidate the reason for the observed variation of the diffraction line profile shape and –width we applied the formalism outlined in section 2 to simulate the $\sin^2\psi$ -measurement performed on the blasted Al_2O_3-layer. The input parameters needed for the calculation of the line profiles $I^{024}(2\theta, z)$ diffracted by each sublayer δz in a depth z (see equations (6) and (9)) were obtained from the experimental data. The $\theta^{024}_{\psi,\max}(z)$ -depth profile was evaluated by means of equation (10) with the in-plane residual stress gradient $\sigma_{\|}(z)$ from figure 4. For the determination of the line shape parameter depth profile we described $w(z)$ by an empirical function $w(z) = c_0 + c_1 \ln(z + c_2)$ and calculated the parameters c_i by fitting the respective Laplace transform $w[\tau(\sin^2\psi)]$ to the experimental $a_4 - \sin^2\psi$ -distribution (figure 6a).

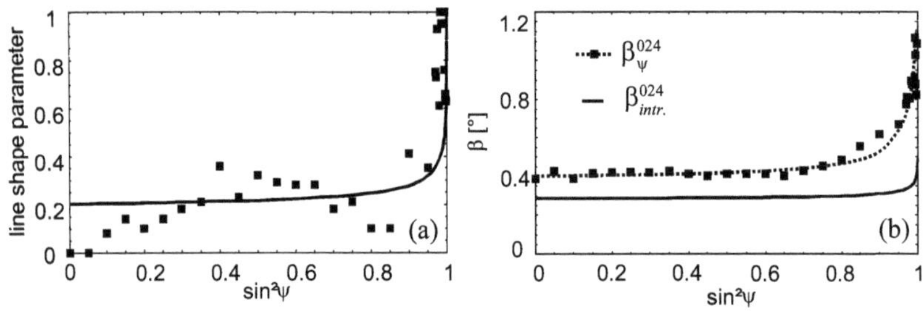

Figure 6. Calculated input parameters for the XSA simulation as function of $\sin^2\psi$ *,* Al_2O_3*-layer, blasted state. (a) Line shape parameter* a_4 *obtained from measured diffraction line profiles (dots) and shape parameter w used for the calculation of the sublayer profiles (solid line). (b) Intrinsic line width* $\beta^{024}_{\text{int}\,r.}$ *used for describing the intrinsic broadening of the sublayer profiles and integral breadth* β^{024}_{ψ} *of the measured line profiles (cf. figure 3).*

The most sophisticated step was the evaluation of the depth profile $\delta^{024}_{2\theta}(z)$, which defines by equation (11) the intrinsic broadening $\beta^{024}_{\text{int}\,r.}(z)$ of the in individual sublayer intensity profiles $I^{024}(2\theta,z)$. It cannot be calculated directly from the $\beta^{024}_{\psi}-\sin^2\psi$ -plot shown in figure 3, because β^{024}_{ψ} is the sum of micro- and macro-strain broadening, the latter being due to the stress gradient $\sigma_{\parallel}(z)$. We used an empirical approach $\delta^{024}_{2\theta}(z)=e_0+e_1\ln(z+e_2)$ and determined the coefficients e_i by simultaneous matching the simulated and measured intensity profiles, $I^{024,sim.}_{total}(2\theta)$ and $I^{024,\exp.}_{total}(2\theta)$, respectively, for the investigated ψ -angles. From figure 6b it can be seen, that the resulting $\beta^{024}_{\text{int}\,r.}-\sin^2\psi$ -curve shows nearly horizontal progression up to large ψ and is very close to the $\beta^{024}_{\psi}-\sin^2\psi$ -distribution obtained experimentally for the as-grown Al_2O_3-layer (cf. figure 3). The results of the line profile simulation for selected inclination angles ψ are shown in figure 7.

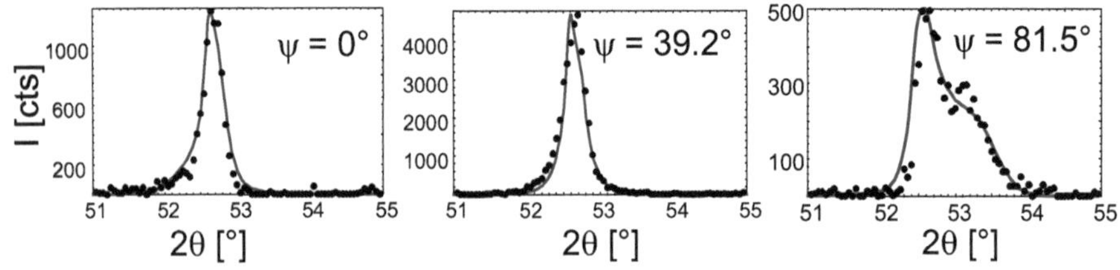

Figure 7. Diffraction line profiles (solid curves) simulated for the same orientations ψ *as shown in figure 6. The profiles were scaled up to the absolute intensities of the corresponding experimental profiles (dotted curves).*

In figure 8 the results obtained by experiment and simulation, respectively, are compared for the $\sin^2\psi$-plots of the three parameters integral breadth, line shape and -asymmetry, which can be used to assess the accuracy of the applied model. We found the experimental and the simulated distributions to agree well for all parameters, which also applies to the measured and recalculated $d_\psi^{024}-\sin^2\psi$-plots and stress depth profiles, respectively, which are shown in figure 9.

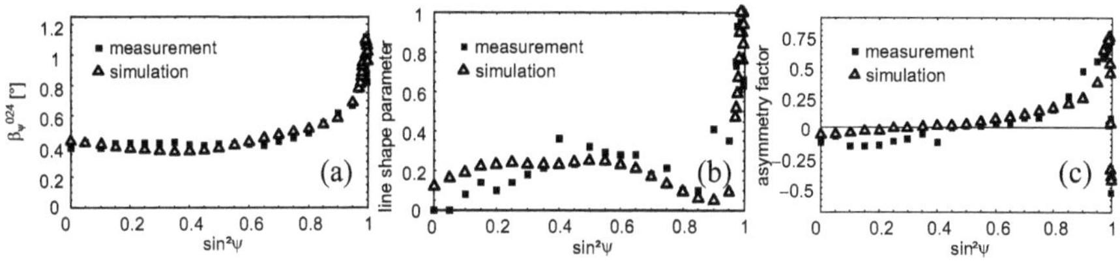

Figure 8. Measured and simulated $\sin^2\psi$ *-distributions of the integral breadth (a), the line shape parameter (b) and of the asymmetry factor (c), respectively.*

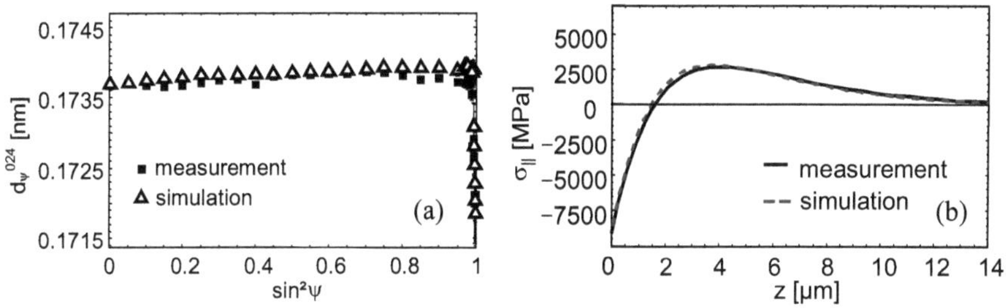

Figure 9. Comparison of measured and recalculated XSA results. (a) $d_\psi^{024}-\sin^2\psi$ *-distribution and (b) residual stress depth profile* $\sigma_{||}(z)$.

5. Discussion

The XSA results show that grit blasting is an appropriate tool for tailoring the residual stress state in CVD hard coating systems. The average in-plane tensile residual stresses obtained for the as-grown state (figure 3, left side) correspond to the idea of extrinsic thermal (growth) stresses due to the mismatch in the coefficients of thermal expansion. The depth distribution of the compressive stresses introduced by blasting depends strongly on both, the coating design and the process parameters. For thin CVD Al_2O_3/TiCN bilayer systems with a total thickness of 5 µm it was shown in [18] that it is possible to generate compressive residual stresses in form of long-range 'interlayer-gradients' [19] into the total coating system as well as in the near interface substrate zone and therefore, to shift the tensile residual stresses into deeper substrate regions. In case of thicker coating systems as considered here, however, the compressive residual stresses are already balanced in form of a steep 'intralayer gradient' [19] within the topmost Al_2O_3 layer (thickness 14 µm), whereas the TiCN layer below is less affected.

Due to the steep residual stress gradient in the Al_2O_3 layer the corresponding diffraction line profiles become strongly asymmetric. Here the perhaps most striking finding concerns the run of the asymmetry factor (figure 8c). Both, experiment and simulation reveal a twofold change of the line profile asymmetry from the small to the large angle side of the profiles and vice versa. In a previous work [10] we showed this effect, which occurs in the presence of residual stress gradients, to be caused by the variation of the penetration depth τ during a $\theta-\theta$-scan. The first change being observed in the strain-free direction at $\sin^2\psi^* \approx 0.37$ is accompanied by a (slight) minimum of the integral breath (figure 8a) thus indicating that the stress-induced macro-strain has a minimum is this direction. The second zero-crossing at very large ψ (i.e. very small τ) can be explained by the steepness of the in-plane stress gradient. So the change from the compressive- into the tensile state of residual stress occurs within a very thin surface layer of about 2µm, thus leading to the observed symmetric line broadening on a high macro-strain level as indicated by the large integral breadth in figure 8a. Therefore, the conclusion can be drawn, that the observed line broadening in the present case is mainly caused by the macro-stress gradient $\sigma_{\parallel}(z)$.

The model introduced here for the evaluation of steep residual stress gradients in thin film structures with complex geometry and for the analysis of asymmetric line broadening due to residual stress- and microstructure gradients can be extended and applied to various other cases concerning X-ray depth profiling. Gradients of the chemical composition within individual layers, for example, may be implemented by $\theta_0^{hkl}(z)$ in equation (10). By simulating complete XSA measurements for different approaches based, for example, on the $\sin^2\psi$-method (this paper) or on the scattering vector method [13], the model can be used as a tool for 'feasibility' studies in order to assess experiments with regard to accuracy to be expected.

6. Summary

Based on the Laplace method for the X-ray analysis of residual stress gradients in the near surface region of bulk materials and thin films, a formalism was introduced which allows for the evaluation of non-uniform residual stress distributions in multilayer systems. The application of the method was demonstrated by the example of CVD hard coatings consisting of Al_2O_3- and TiCN-layers, which were grit-blasted after deposition in order to generate beneficial compressive stresses within the system. The blasting process was shown to result in steep nonlinear in-plane residual stresses gradients in the topmost Al_2O_3-layer. The high surface compressive stresses are balanced within the Al_2O_3-layer, whereas the deeper coating zones are less affected.

The measured diffraction line profiles of the blasted state were found to be strongly asymmetric with a characteristic variation of their shape during depth profiling. By means of simulations on the basis of the kinematical diffraction theory it was shown that the observed line broadening is mainly due to the macro-strain which is induced by the extremely steep residual stress gradient.

References

1. Noyan, I. C., & Cohen, J. B., 1987, *Residual Stress Measurement by Diffraction and Interpretation* (New York: Springer).
2. Hauk, V., 1997, *Structural and Residual Stress Analysis by Nondestructive Methods* (Amsterdam: Elsevier).
3. Spieß, L., Schwarzer, R., Behnken, H. & Teichert, G., 2005, *Moderne Röntgenbeugung* (Wiesbaden: Teubner).
4. Kötschau, I. M. & Schock, H. W., 2006, *J. Appl. Cryst.*, **39**, 683.
5. Birkholz, M., 2007, *J. Appl. Cryst.*, **39**, 735.
6. Genzel, Ch., 2001, *Adv. X-Ray Anal.*, **44**, 247.
7. Somers, M. A. J. & Mittemeijer, E. J., 1990, *Metall. Trans. A*, **21**, 901.
8. Birkholz, M., 2006, *Thin Film Analysis by X-ray Scattering* (Weinheim: Wiley-VCH).
9. Welzel, U., Ligot, J., Lamparter, P., Vermeulen, A. C. & Mittemeijer, E. J., 2005, *J. Appl. Cryst.* **38**, 1.
10. Genzel, Ch., 2005, *Mat. Sci. Technol.*, **21**, 10.
11. Warren, B. E., 1959, *Progress in Metal Physics*, **8**, 147.
12. Dehlez, R., de Keijser, Th. H. & Mittemeijer, E. J., 1987 *Surf. Eng.*, **3**, 331.
13. Genzel, Ch., 1994, *Phys. Status Solidi (A)*, **146**, 629.
14. Dölle, H. & Hauk, V., 1979, *Härterei Tech. Mitt.*, **34**, 272.
15. Genzel, Ch., 1997, *Phys. Status Solidi (A)*, **159**, 283.
16. Landoldt-Börnstein, 1979, *New Series*, Group III, Vol. 11, (Berlin: Springer).
17. Kress, W., Roedhammer, P., Bilz, H., Teuchert, W. & Christensen, A. N., 1978, *Phys. Rev. B*, **17**, 111.
18. Klaus, M., Genzel, Ch. & Holzschuh, H., *Thin Solid Films*, in print.
19. Noyan, I. C., Huang, T. C. & York, B. R., 1995, *Crit. Rev. Solid State Mater. Sci.*, **20**, 125.

Magnetron deposited TiO_2 thin films - crystallization and temperature dependence of microstructure and phase composition

R. Kužel[1,*], L. Nichtová[1], Z. Matěj[1], J. Šícha[2], J. Musil[2]

[1] Department of Condensed Matter Physics, Faculty of Mathematics and Physics, Charles University in Prague, Ke Karlovu 5, 121 16 Praha 2, Czech Republic
[2] Department of Physics, Faculty of Applied Sciences, University of West Bohemia in Pilsen, Czech Republic
[*] kuzel@karlov.mff.cuni.cz

Keywords: powder diffraction, titanium dioxide, crystallization, nanocrystalline thin films

Abstract. Temperature dependence of microstructure and phase composition of two sets of nanocrystalline TiO_2 thin films magnetron deposited on glass and silicon substrates have been studied by X-ray scattering. X-ray powder diffraction patterns were measured in parallel beam optics and evaluated by the total pattern fitting. After annealing, the amorphous films are crystallized, the nanocrystalline films consisting of anatase and rutile are partially crystallized while pure nanocrystalline anatase films remain nearly unchanged. With increasing temperature above 400 °C significant increase of crystallite size and transformation of anatase into rutile after 700 °C can be observed.

1. Introduction

Titanium dioxide films are widely used because of their excellent properties as chemical stability, mechanical hardness and optical transmittance with high refractive index. They are used as ultra thin film coatings, nanostructured membranes etc., in electrochemistry and electrocatalysis, in microoptics and electrooptics and in photocatalysis. These applications particularly depend on the layers properties (stability, transparency, photoconductivity, photoactivity, etc.) and their microstructure (surface area, pore size distribution, particle size, crystallographic structure). The photocatalytic activity of TiO_2 [1-3] can result in the decomposition of organic compounds on the TiO_2 surface or the reduction of the contact angle between water and the TiO_2 surface under ultraviolet irradiation (excellent hydrophilicity) [see 4 for more references]. The films can be prepared by several techniques but the magnetron deposition is favourable from the point of view of mechanical durability required for practical applications. Depending on the deposition conditions, the films can be prepared as amorphous or nanocrystalline.

The amorphous TiO_2 films do not have good hydrophilicity. Therefore, crystalline form of the oxide is required. This can be obtained either by annealing of amorphous films [4] or by setting of deposition parameters in such a way that crystalline or nanocrystalline films are obtained immediately. Studies of the former case were described in [5]. It was found that the amorphous TiO_2 films can be crystallized at approximately 250 °C and their crystallite size was then immediately above 100 nm. The films with the thickness below 400 nm crystallized at higher temperatures (350 °C), diffraction patterns did not change significantly after annealing up to 600 °C. Tensile residual stresses have been found in crystallized films and relaxed after annealing at 500 °C. Fibre texture was usually not very strong and of the (100) type for thicker films (above 1500 nm) and (101) for thinner films.

Studies of the deposited nanocrystalline films with different thickness and films prepared at different oxygen pressure were briefly described in [6]. It was found that rutile grows on the interface with the substrate. It is replaced by anatase with increasing distance from the substrate. Therefore thinner films contained mainly rutile while in thicker films anatase was dominating. The stresses in as-deposited nanocrystalline films were small (< 100 MPa) but complicated (triaxial, possible gradients) and the texture not always fibre and inclined with respect to the surface.

In the present work the evolution of the microstructure and phase composition of these films was studied by XRD.

2. Experimental

TiO_2 films were sputtered by dual magnetron equipped with two Ti(99.5) targets of 50 mm in diameter and supplied by a dc–pulsed Advanced Energy Pinnacle Plus+ 5kW power supply unit operating in bipolar asymmetric mode at repetition frequency f_r = 100kHz and duty cycle $\tau/T = 0.5$; here τ and T are the length of pulse and the period of pulses. Films were deposited on unheated microscope glass slides (26 × 26 × 1 mm^3) and silicon (100) substrates (26 × 26 × 1 mm^3) at substrate to target distance $d_{s\text{-}t}$ = 100 mm, total working pressure p_T = 0.9 Pa, average pulse discharge current $I_{da1,2}$ = 3 A and average pulse power density $W_{da} \approx 55$ Wcm^{-2}. Maximum substrate surface temperature during the deposition was lower than $T_{surf} \leq 160$ °C. Further details on dual magnetron system are given elsewhere [6].

Two sets of samples were studied. Nanocrystalline TiO_2 films with different thickness h = 100, 220, 515, 935, 2000 nm were deposited on the glass and silicon substrates in the oxide mode of sputtering at oxygen partial pressure p_{O2} = 0.15 Pa and deposition rate a_D= 9 nm/min. The other samples with the thickness of h ≈ 1000 nm were deposited at different oxygen partial pressure p_{O2} = 0.05 Pa, 0.085 Pa and 0.15 Pa with deposition rates a_D = 45 nm/min, 20 nm/min and 9 nm/min. The samples were annealed in the air in the temperature range 200-1000 °C with the step of 100 °C for 30 minutes at each temperature.

The measurements were performed mainly on Philips (Panalytical) X'Pert MRD in parallel beam setup, 2θ scans with the angles of incidence 0.5-1.5° with parallel plate collimator placed in the diffracted beam and the Goebel mirror inserted in the primary beam. The annealed samples were also studied with the Panalytical system MPD in double-mirror setup (mirror also in the diffracted beam) because this arrangement has about three times better resolution, of course on the account of intensity. In these measurements, phase analysis and X-ray line profile analysis were mainly performed. The measurements of residual stresses were carried out by using a polycapillary and Eulerian cradle by the $\sin^2\psi$ method for several different peaks. In addition, X-ray reflectivity curves were measured for all the samples.

3. Methods of evaluation

Two methods of evaluation of XRD profiles were applied – *fitting of peak clusters* and *total pattern fitting*. In the former method the diffraction peaks were fitted with the Pearson VII or pseudo-Voigt functions. The correction for instrumental broadening was performed with the aid of a NIST LaB_6 standard and the Voigt function method. The modified Williamson-Hall plots (integral breadth β (in 1/d) vs. sin θ) were used for estimation of crystallite size and microstrain. However, this method has severe limitation because of strong and not completely correct preassumptions on shape of distribution of both effects. It can be used better if one of them is clearly dominating. The method is also difficult to apply in the case of nanocrystalline films when all the profiles are strongly overlapped. In this case fitting of profiles must be performed very carefully with fixing of some parameters (e.g. peak positions).

Therefore, the diffraction patterns must be evaluated by whole pattern fitting. There are different programs available nowadays. The first group consists of famous Rietveld type programs like Fullprof or GSAS. They are written mainly for structure refinement but include also phenomenological description of line broadening (strain and size) even in with possibility of line broadening anisotropy. The second group of programs [7-8] is concentrated on physically correct description of real structure parameters (crystallite size distribution, dislocations). However, none of these programs include correct evaluation of 2θ scans taken for thin films at low angles of incidence. This should consider appropriate absorption correction, treatment of residual stress and texture. For this case, only the program Maud [9, 10] seems to be available. Nevertheless, in order to have everything under own control, we have decided to write a new program and used for this very flexible modular system of Crystal Objects and Fox program [11, 12] developed by Vincent Favre-Nicolin. Currently, the system is still under development and it was modified primarily for analysis of Ti-O system.

Applied model for this system include the following features. In most case two phases are considered – rutile and anatase. Peak position are determined by variable lattice parameter a, c and zero shift error. Specimen displacement error is not considered for parallel beam geometry. The effect of a possible residual stress in the films to diffraction peaks positions has not been implemented in our model yet. The main reason is a strong anisotropy in residual stress found in the samples [5] and missing elastic constants of anatase. On the other hand, natural physical broadening of diffraction lines in the measured patterns is so large that a possible small shift of such broad peaks is difficult to observe. Peak intensities are calculated by the ObjCryst library from a known atomic structure model so that appropriate intensity values are included. In addition of instrumental corrections also absorption in the thin film and a texture correction can be included. The model of absorption correction for our TiO_2 samples consist of a simple anatase and rutile layered structure on the Si or glass substrate. Individual layers are characterized with their thickness and linear absorption coefficient. Texture correction can either be obtained from known model ODF after numerical integration over all crystallites with diffracting {hkl} planes perpendicular to direction of the measured diffraction vector, or alternatively, the peak intensities can be varied independently, if the texture is unknown or difficult to describe. Pole figures (PFs) of samples, where both rutile and anatase phases are still nanocrystalline are hard to measure, hence it is difficult to find out an appropriate model ODF function. A few measured PFs indicate a complex (not fibre) texture related to the deposition geometry (dual magnetron). Therefore, for the fitting of diffraction patterns of these samples, the peak intensity of each (hkl) diffraction determined by the structure factor can be multiplied by a free factor. Of course, this leads to more correlation in the fitting, it introduces a need for more interactive work of the user and should be used only if inevitable.

Peak profiles are given by the numerical convolution of the known instrumental function and physical profiles including several free parameters. The instrumental function is described by the pseudo-Voigt (pV) function the half width *FWHM* and shape parameter k of which are described by the commonly used relations for angular dependence in 2θ (Caglioti polynomial) with free parameters determined by the fitting of standard diffraction pattern (LaB_6) measured on the same instrumental arrangement as the one used for measurement of the samples investigated. Hence, the description of the instrumental function is exactly the same as the verified one used in [7-10]. The size broadening effect is described by the model function for log-normally distributed spherical crystallites like in [7, 9, 10]. Since the type of crystal defects in TiO_2 crystalline phases that are related to the microstrain observed in x-ray measurements could not be simply described or observed, the phenomenological microstrain broadening function was used. The peak broadening dependent on the diffraction vector magnitude is described by the pV function. Its *FWHM* dependence with diffraction angle 2θ is given by the Caglioti polynomial with only the quadratic term nonzero. This means that the *FWHM* in the reciprocal space units is linearly increasing with the diffraction vector magnitude. The shape factor of the pV-function common for all the $2\theta_{hkl}$ diffraction peaks of a particular phase can also be refined. The relation for *FWHM* can be converted to the integral breadth β and by using the formula from [13] one parameter - microstrain e is determined:

$$FWHM(2\theta[rad])^2 = U * \tan(\theta);\ \beta(2\theta[rad])^2 = 4e * \tan(\theta);\ 4e = \left(\frac{1-k}{\varphi_G} + \frac{k}{\varphi_C}\right)\sqrt{U},$$

where $\varphi_G = 2(\ln(2)/\pi)^{1/2}$, $\varphi_C = 2/\pi$ are the well known Gaussian and Lorentzian shape parameters.

4. Results

Unlike previous work [5, 6] all the diffraction patterns were evaluated by total pattern fitting. However, in most cases the values of mean crystallite size that is the dominating effect on line profile broadening were quite close to the values obtained from the Williamson-Hall plots. Larger differences were found for microstrain. Diffraction patterns of the nanocrystalline films with different thickness were essentially distinct. The patterns for thin films (thickness 100 and 220 nm) showed dominating rutile phase with a small indication of anatase while in thicker films (above 500 nm) mainly anatase was observed [5, 6]. With annealing there were no significant changes neither in phase composition nor microstructure of the films up to the temperatures of 600-700 °C. Then the transformation of anatase into rutile begins. The corresponding fits are shown on figures 1 and 2.

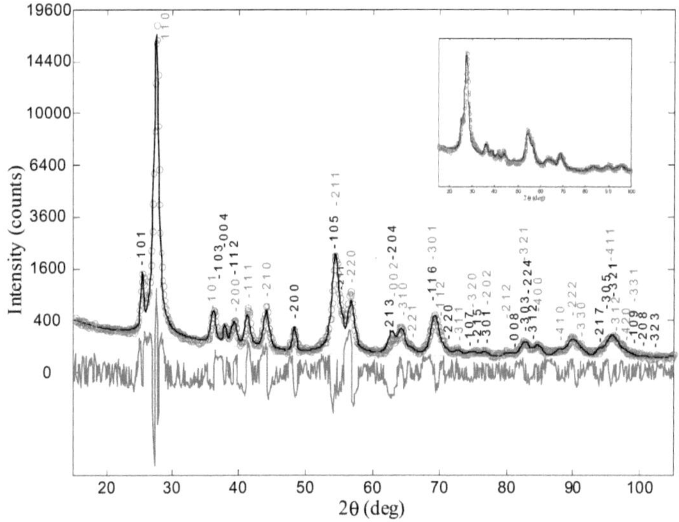

Figure 1. Fitting of the diffraction pattern for the thin film (220 nm) after annealing at 700 °C, (for as-deposited film in the inset). Thicker and thin labels correspond to anatase and rutile, respectively.

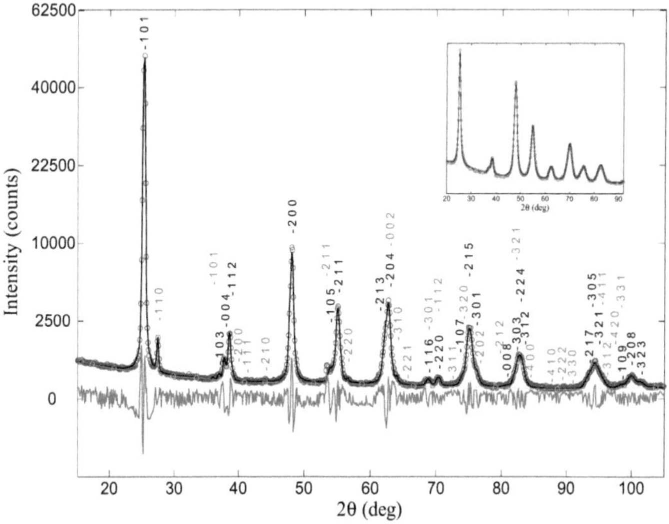

Figure 2. Fitting of the diffraction pattern for the thick film (2000 nm) after annealing at 700 °C, (for as-deposited film in the inset). Thicker and thin labels correspond to anatase and rutile, respectively.

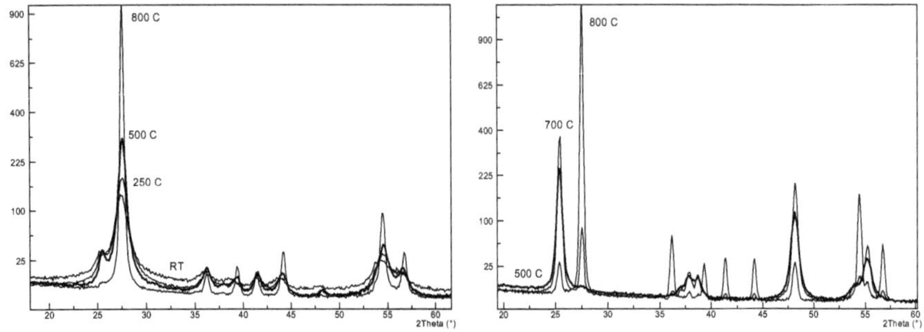

Figure 3. Diffraction patterns of TiO_2 films annealed at different temperatures. Strongest peaks of anatase and rutile are 101 at 25° and 110 at 27°, respectively – a) thinner film (220 nm) with dominating rutile and gradually increasing crystallite size with annealing (500 °C, thick line), b) thicker film (935 nm) with dominating anatase after deposition and nearly unchanged crystallite size up to 500 °C (thick line).

For thinner films where anatase is minor phase its reduction could be observed after annealing at 700 °C (figure 3a) and completely disappeared at 1000 °C. For thicker films rutile peaks clearly appeared after annealing at 700 °C (figure 3b). At 800 °C both phases coexisted, for 2000 nm thick sample in about equal fraction but for 935nm-film only less than 10 % of anatase was observed. The anatase peaks were negligible for 900 °C and entirely missing after annealing at 1000 °C.
Microstructural evolution can be seen on figure 4. In two-phase films the rutile often forms smaller crystallites than anatase. The mean crystallite size in the as-deposited films is 5-7 nm with exception of the thickest film (14 nm). Only slow increase can be observed with annealing temperature up to 400 °C. Then significant changes start and the crystallite size after annealing at 1000 °C reaches the limit of XRD sensitivity (several hundreds of nm). In thinner two-phase films the crystallites of anatase are a little larger in the whole temperature range. Microstrain in rutile in these films is approximately constant (about 1 %) up to 800 °C.
The diffraction patterns for the second set of samples prepared with different partial oxygen pressure are shown on figure 5. For low oxygen pressure (0.05 Pa) as-deposited amorphous films are well crystallized after annealing at 400 °C (crystallite size 150 nm, microstrain 0.2 % (figure 5a). With medium pressures (0.0875 Pa, figure 5b) mixture of anatase and rutile was obtained. After annealing, the rutile crystallites did not changed (5 nm) and anatase crystallites slightly grew (from 21 to 33 nm), the corresponding microstrain dropped from 0.8 to 0.1 %. Rutile has significantly smaller crystallites. This seems to be typical for a mixture of both phases. The higher pressure (0.15 Pa) leads to dominating anatase in the film. The annealing of this sample at 400 °C did not have significant effect on microstructural parameter (crystallites of about 8 nm, microstrain 0.7 % (figure 5c).

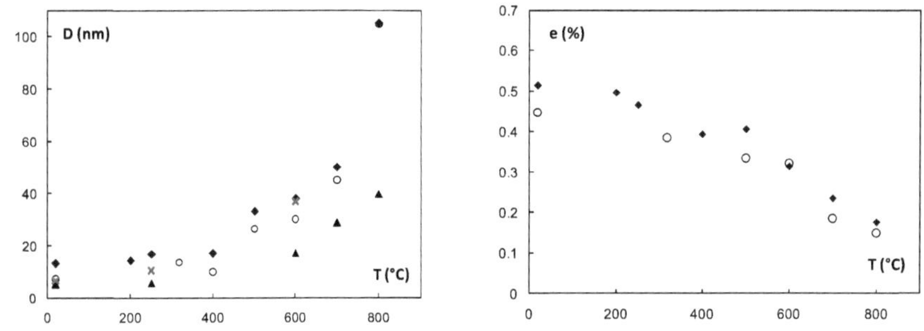

a) b)

Figure 4. Dependence of crystallite size (a) and microstrain (b) on annealing temperature; for anatase - 2000 nm thick film (♦), 935 nm (◦), 220 nm (×) and rutile 220 nm (▲).

a)

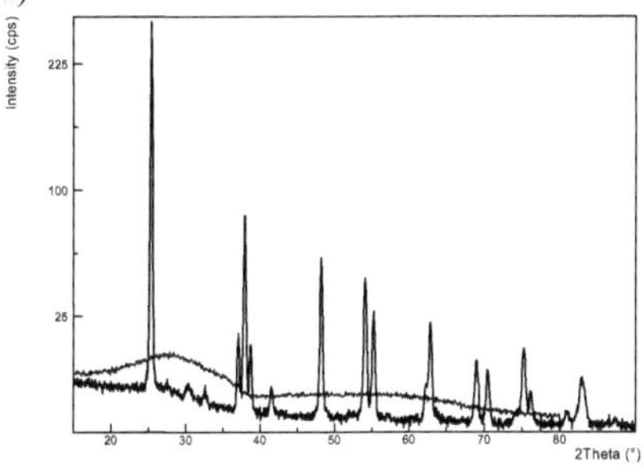

c)

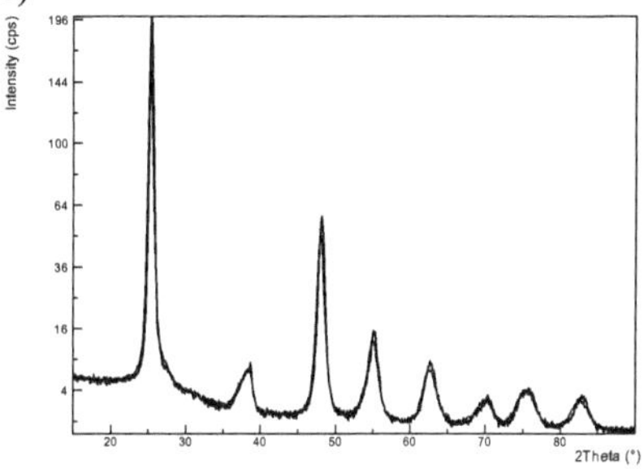

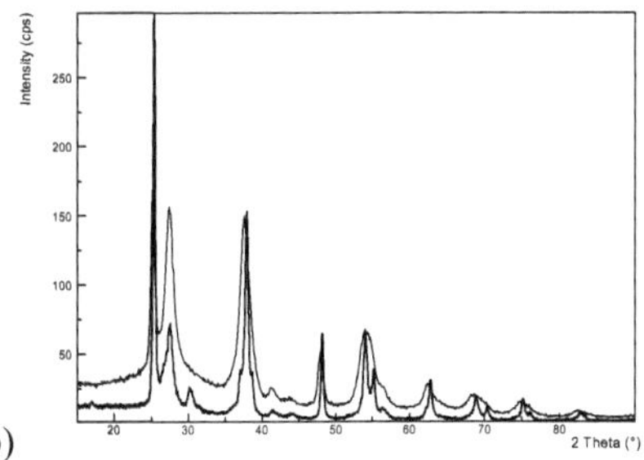

b)

Figure 5. Diffraction patterns of three samples deposited at different partial oxygen pressures (0.05 Pa – a, 0.0875 Pa – b, 0.15 Pa - c), as-deposited (thinner lines, upper curves) and after annealing at 400 °C (thicker curves).

Simple (uni- or biaxial) tensile stress and only a weak texture were found for the amorphous films after crystallization. Only for the thinnest films (~ 100 nm), the 101 texture (anatase) was found. In case of nanocrystalline films the stress was low but not uniaxial. This is related to significantly stronger and more complicated texture due to the dual magnetron geometry.

Reflectivity measurements revealed increasing surface roughness with both increasing thickness and annealing temperatures.

5. Conclusions

Temperature dependence of microstructure and phase composition of two sets of nanocrystalline TiO_2 thin films magnetron deposited on glass and silicon substrates have been studied. Significant differences were found for the films with different initial microstructure and phase composition. After annealing at temperatures higher than 250 °C, the amorphous films are crystallized with crystallite size larger than 100 nm, the films consisting of anatase and rutile are partially crystallized while pure nanocrystalline anatase films remain nearly unchanged. With increasing temperature above 400 °C significant increase of crystallite size and transformation of anatase into rutile after 700 °C can be observed.

References

1. Sopyan, N., Watanabe, N., Murasawa, S., Hashimoto, K. & Fujishima, A., 1996, *J. Photochem. Photobiol.* **A 98**, 79.
2. Byun, I., Jin, Y., Kim, B., Lee, J.K. & Park, D., 2000, *J. Haz. Mat.*, **B73**, 199.
3. J. Negishi, M. Takeuchi, and T. Ibusuki, 1998, *J. Mater. Sci.,* **33**, 5789.
4. Kužel, Jr., R., Černý, R., Valvoda, V., Blomberg, M. & Merisalo, M., 1994, *Thin Solid Films*, **247**, 64.
5. Kužel, R., Nichtová, L., Matěj, Z., Heřman, D., Šícha, J. & Musil, J., 2008, *Z. Kristallogr.* Suppl. **26**, 247-252.
6. Kužel, R., Nichtová, L., Matěj, Z., Heřman, D., Šícha, J. & Musil, J., 2008, *Z. Kristallogr.* Suppl. **26**, 241-246
7. Scardi, P. & Leoni, M., 2002, *Acta Cryst. A*, **58**, 190.
8. Ribárik, G., Ungár,T. & Gubicza, J., 2001, *J. Appl. Cryst.*, **34**, 669.
9. Lutterotti, L., Chateigner, D., Ferrari, S. & Ricote, J., 2004, *Thin Solid Films*, **450**, 34.
10. Lutterotti, L., http://www.ing.unitn.it/~maud.
11. Favre-Nicolin, V. & Černý, R., 2002, *J. Appl. Cryst.*, **35**, 734.
12. Favre-Nicolin, V., Free Objects for Crystallography, http://objcryst.sourceforge.net/Fox
13. Wilson, A.J.C., 1949, *X-Ray Optics* (London: Methuen), p. 5.

Acknowledgements. The work is supported by the Grant Agency of the Czech Republic under number 106/06/0327 and also as a part of the research plans MSM 0021620834 and MSM 4977751302 financed by the Ministry of Education of the Czech Republic.

Z. Kristallogr. Suppl. **27** (2008) 295-302 / **DOI** 10.1524/zksu.2008.0036

Influence of reactive plasmas on thin nickel films

M. Quaas[1,*], O. Ivanova[2], C. A. Helm[2], H. Wulff[1]

[1] University of Greifswald, Institute of Biochemistry, Felix-Hausdorff-Str. 4, 17487 Greifswald, Germany
[2] University of Greifswald, Institute of Physics, Felix-Hausdorff-Str. 6, 17487 Greifswald, Germany
[*] quaas@chemie.uni-greifswald.de

Keywords: reactive plasmas, thin films, grazing incidence x-ray diffraction

Abstract. 20 nm thick Nickel films were treated in a microwave plasma (SLAN, 2.45 GHz) with Argon as working gas. Oxygen and hydrogen containing plasma were subsequently applied to achieve oxidation and reduction of the metallic nickel. The plasma-treated samples were investigated by complementary techniques: grazing incidence x-ray diffractometry (GIXD), x-ray reflectometry (XR) and atomic force microscopy (AFM). The combination of these techniques allows a detailed analysis of the occurring plasma-chemical processes.
Oxygen plasma treatment results in the formation of NiO on the metallic Ni grains. The domain size of the formed NiO is appreciable smaller than that of the Ni domains.
The kinetics of the oxidation process can be described using the "shrinking core" model. The oxide formation rate depends on the sample history, i.e. on the number of oxidation/reduction cycles carried out. The first oxidation process is controlled by oxygen diffusion into the film, the kinetics of the subsequent oxidation cycles is more reaction controlled.
Hydrogen plasma treatment causes a fast reduction up to the level of metallic Ni.

1. Introduction

The effect of low energy plasma species onto solid surface layers is one focus of plasma research. A specific challenge in plasma science is the investigation of chemical reactions in solid surface layers as response to external plasma parameters. Investigations of energy and particle influx onto the substrate have to be combined with measurements describing the plasma-wall interaction. Thus the investigation of the structural properties of the plasma-treated surface is crucial.
So far there are only few experimental data showing the effect of low energy plasmas onto solid walls. Structural investigations of surface layers after exposure to plasma convincingly describe surface layers with different physical and structural properties [1-4].
Two main problems are in the centre of interest:

(i) Which influence do energy and particle flux exert onto the wall? How do crystal and/or defect structure of the surface layer change?

(ii) Are chemical reactions within solid walls solely caused by reactive plasma species in front of the wall? Are chemical reactions only limited by diffusion of reactive species within in the surface layer?

Oxidation of bulk samples of many metals has been studied widely for more than a century, but comparatively little attention has been directed to the oxidation of thin films or small particles. Many studies on Ni nanoparticles are qualitative observations of the oxidation, or structure determination of the Ni-oxide. A detailed study of oxidation of nickel nanoparticles at different temperatures was recently performed by Niklasson and Karmhag [5, 6].
The present study focuses on the microwave enhanced oxidation and reduction of thin polycrystalline Ni-films. The experiments were performed on Ni-films of 20 nm thickness alternately exposed to oxygen and hydrogen microwave plasma.
To describe the kinetics of a chemical reaction the degree of conversion has to be determined depending on time. X-ray reflectometry (XR) as well as grazing incidence x-ray diffractometry (GIXD) are well suited tools for the investigation of thin crystalline films. Using standard x-ray tubes these techniques allow a non-destructive determination of reaction parameters by evaluating the integral intensities of diffraction profiles (GIXD) or film thickness (XR). In addition information on density as well as surface and interface roughness can be obtained by XR. The surface topology and grain size at the wall can be investigated with Atomic Force Microscopy (AFM). Thus, complementary information about porosity or density and surface roughness is obtained. The combination of these techniques enables a detailed investigation of plasma induced chemical processes on solid surface layers.

2. Experimental

Thin pure Ni coatings (20 nm thickness) deposited on Si substrates with 800 nm SiO_2 and 1 nm Cr buffer layers were treated in a 2.45 GHz microwave plasma source (SLAN) [7]. No external substrate heating was applied. Plasma power was kept constant at 700 W. The experiments were carried out at 50 Pa. Successive oxidation and reduction cycles were performed. Gas mixtures of 10 sccm Ar / 10 sccm H_2 for reduction and 10 sccm Ar / 10 sccm O_2 for oxidation experiments were used, respectively. After every 30 minutes the samples were taken out of the plasma chamber and characterized by GIXD, XR and AFM. Then the plasma treatment was continued.
With a reference sample thermal annealing experiments were performed at 600°C for 2 h. Thus conceivable thermally induced processes could be monitored.
For GIXD and XR measurements of a Siemens D 5000 θ-2θ diffractometer equipped with a dedicated reflectometry sample stage was used. Cu-K_α radiation was used for measurements. For GIXD a special parallel beam attachment (Soller collimator, secondary monochromator) was applied.
GIXD measurements (asymmetric Bragg case) at an incidence angle $\omega = 0.5°$ were analyzed regarding position, profile and intensities of observed Bragg reflections. A pseudo-Voigt fit was applied to the measured profile and the domain sizes were calculated using Warren-Averbach plot. A detailed description of the method is given in [8].
Diffrac^{plus} Leptos 3.03 program package by Bruker AXS was used to determine thickness, density and roughness of the samples from the XR measurements.

In addition the sample topology was imaged by tapping mode AFM in air (Nanoscope IIIa, Digital instruments) using Si tips with 10 nm radius.

3. Results and discussion

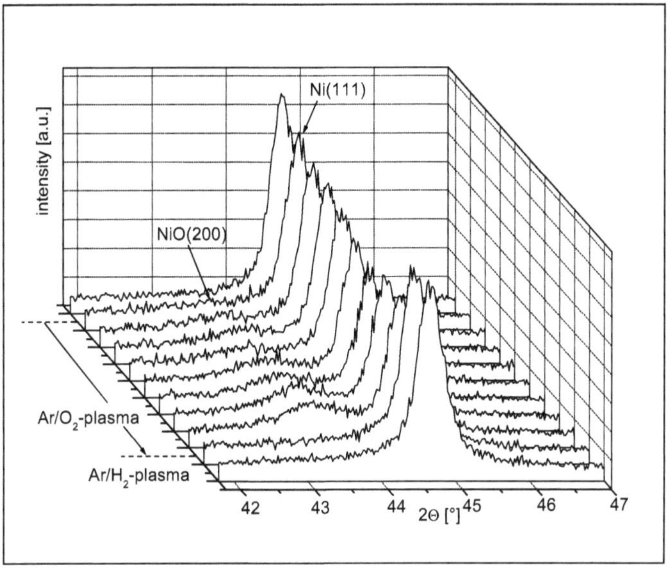

Figure 1. GIXD characterization of a 20nm thick Ni film before and after different plasma treatments. The 2nd, 3rd , .. 8th measurement were taken after 30, 60, 90, ..., 240 min exposure to the oxygen plasma. The two last measurements were taken after 30 and 60 min exposure to hydrogen plasma.

No thermal induced chemical reactions of the Ni samples occur, as could been shown by the annealing experiments of a reference sample. However, an oxygen plasma treatment results in the formation of nickel oxide. Reactive species in the plasma like O^*, O^+, O_2^+ or O^- seem to be necessary to induce a chemical reaction of the metallic nickel. Figure 1 shows GIXD measurements of a sample as deposited and after different plasma treatments. The diffraction profile of Ni(111) is observed, and after 30 min of exposure to Ar/O_2 plasma, a peak corresponding to the NiO (200) signal appears, whose intensity grows if the exposure to oxygen plasma is prolonged. After about 6 hours exposure to oxygen plasma, it has reached saturation. The NiO formation stops at a mass ratio of Ni to NiO of about 3 to 2 as deduced by comparison of the integrated x-ray reflection intensities. Further oxygen plasma treatment does not cause any additional changes in phase composition. A subsequent hydrogen plasma treatment restores metallic Ni. The formation of nickel hydride (Ni_2H or NiH) could only be observed when a negative substrate voltage was applied. The corresponding data will be published later.

As shown in figure 1, the width of the x-ray diffraction profile of both phases remains nearly constant during plasma treatment. However, the NiO x-ray diffraction profiles are strongly broadened in comparison with the Ni profiles. Domain sizes determined from the diffraction profiles [8] are 18 nm for metallic Ni and 6 nm for nickel oxide, respectively.

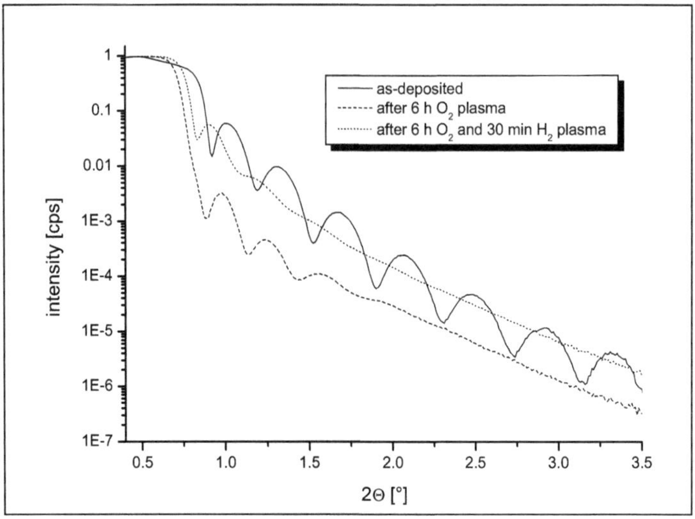

Figure 2. Reflectometry measurements of a 20 nm thick Ni film as deposited, after 6 hours oxidation and after subsequent reduction.

Figure 2 shows XR measurements of the film as deposited and after exposure to oxygen and then to hydrogen plasma. For the original film an interference pattern is observed which is typical for a homogeneous film. The slow decay of the oscillations observed for the as deposited film indicates a small film/air roughness. After oxygen plasma treatment, the frequency of the oscillations increases, indicating a thicker film. Also, the interference pattern is damped much faster, indicating an increased film/air roughness. Furthermore, the critical angle is shifted to much lower values, which is due to a decreased electron density of the film. A subsequent hydrogen plasma treatment results in a small increase of film density and decreasing thickness, yet the increased roughness observed after oxygen plasma treatment remains. As shown in figure 3 the total film thickness increases from 20 nm to 26 nm due to oxygen plasma treatment. Then, 30 min exposure to hydrogen plasma causes thinning to about 19 nm. This value is smaller than the one obtained for the as deposited sample. The difference can be explained by a weak sputter process due to the energy and particle influx to the sample. For the subsequent oxidation/reduction cycles similar changes of the film thickness were observed, but the thickness increases much faster, as shown in figure 3.

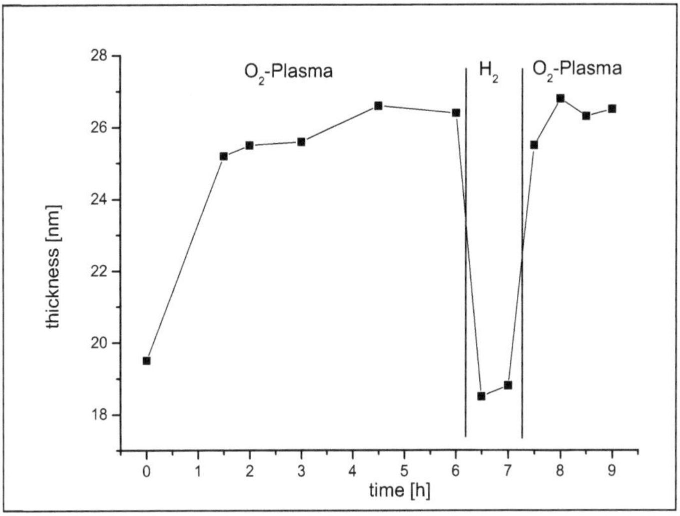

Figure 3. Layer thickness depending on time and type of plasma treatment. The parameters were obtained from the data shown in figure 2 and additional XR measurements.

An increased thickness is expected for oxidation because of the larger molar volume of NiO compared to Ni. Considering the ratio $V_m^{NiO} / V_m^{Ni} = 1.7$ and the 40% conversion of the 20 nm thick Ni film, a total film thickness of 23 nm should be obtained. As figure 3 shows, the experimentally observed thickness is larger. As a consequence the simple model of a compact Ni layer covered by an oxide layer is not appropriate for our experiments.

AFM images show the surface topology (figure 4). Nearly spherical grains are observed for the original film (figure 4, left). The particles are covered by much smaller grains after oxygen plasma treatment (figure 4, right), the images are reminiscent of cauliflowers. Both the XR and AFM results suggest nucleation and growth of NiO domains on the surface of comparable large Ni grains, starting at the film/air interface, eventually leading to a porous layer consisting of Ni and NiO. An increase in roughness and decrease in density should be the final result, which is indeed observed with XR measurements.

To describe the kinetics of the plasma oxidation of the metallic Ni a degree of oxidation α is introduced. Evaluating the diffractometry measurements α can be calculated from the integrated intensity of NiO(200) as a function of the plasma treatment time t. To normalize α, it is divided by the integrated intensity at saturation, when the maximum value is reached:

$$\alpha(t) = \frac{I_{NiO}(t)}{I_{NiO}(\max)}. \qquad (1)$$

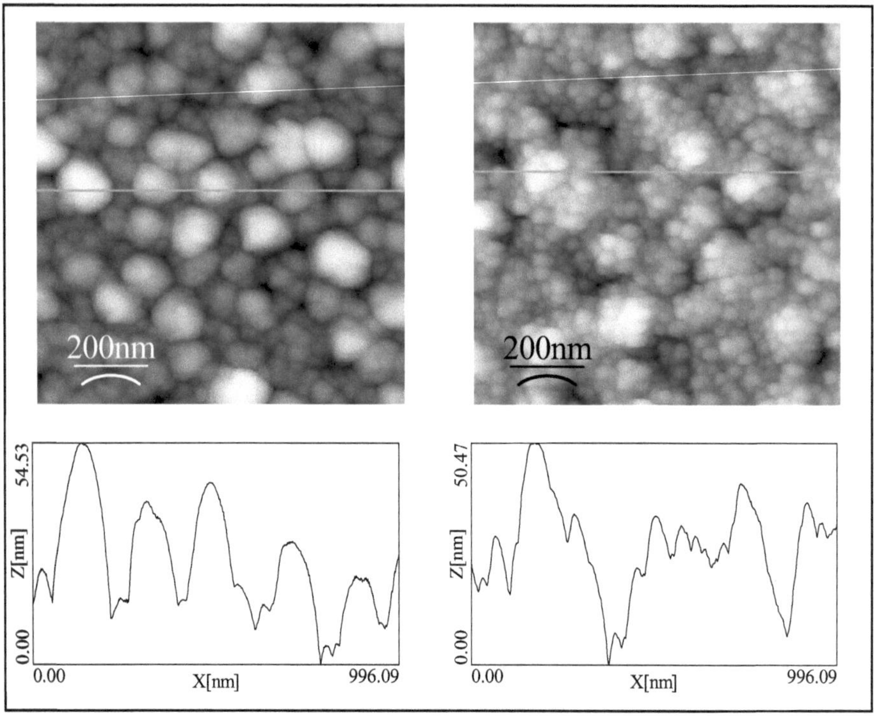

Figure 4. Top: Tapping mode AFM images of a 20 nm thick Ni film, as deposited (left) and after 5 h oxygen plasma treatment (right). For both films height histograms through the indicated horizontal lines are shown.

The AFM images suggest spherical grains. The chemical reaction starts at the surface of a spherical Ni grain, converts the surface layer and thus reduces the radius of the remaining Ni grain, which shows a "shrinking core". Therefore, during the oxidation process, the reaction interface decreases. A flat surface layer exhibits a very different reaction geometry, there the interface has a constant area, which is moving from the film/air towards the film/substrate interface during the reaction. However, the film conversion can be controlled either by diffusion of the reactive species through the film or by the reaction itself. Ginstling and Brounshtein have developed a model for the kinetics of three dimensional diffusion controlled processes [10]:

$$1-\tfrac{2}{3}\cdot\alpha-\sqrt[3]{(1-\alpha)^2}=kt, \tag{2}$$

with k giving the rate constant. Sato et al. give the description of reaction controlled processes for spherical geometry [11]:

$$1-\sqrt[3]{1-\alpha}=kt. \tag{3}$$

In most cases the experimental data cannot be described by only one of these models. Different processes of transport, activation and conversion of reactive species can determine the kinetics of a chemical reaction.
The so called "mixed reaction control" model by Frade and Cable holds if diffusion and reaction limited processes superpose [12]:

$$\left(1-\sqrt[3]{(1-\alpha)}\right)\cdot a-\left(\tfrac{2}{3}\alpha-1+\sqrt[3]{(1-\alpha)^2}\right)\cdot b=kt\,. \qquad (4)$$

The parameters *a* and *b* describe the fraction of the respective processes (with $a+b=1$). In this case the ratio *a/b* is a criteria to find the respective importance of the processes involved. This can be achieved by rephrasing equation 4 and plotting it as shown in figure 5 for different oxidation cycles.

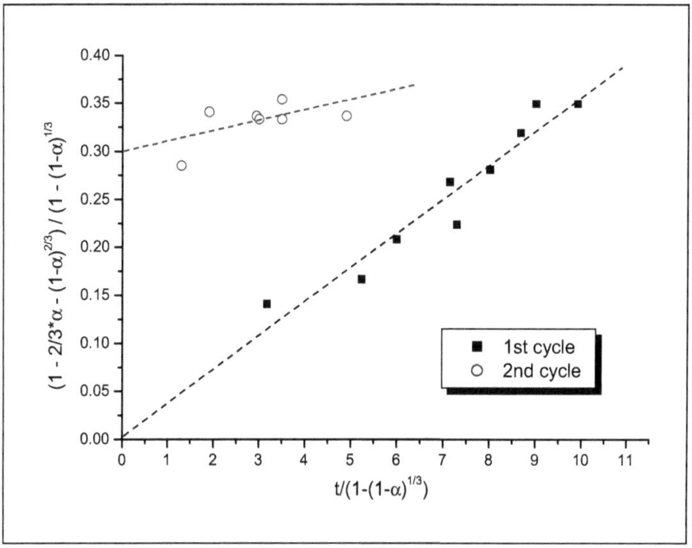

Figure 5. The kinetics of first and second oxidation cycle plotted according to equation 4. The ratio a/b can be determined from the intercept of the linear fit.

Obviously the sample history is important for the kinetics of oxide formation. The first oxidation cycle is clearly diffusion limited with $a/b = 0.005$. The subsequent reduction in hydrogen plasma unfortunately is to fast to observe it ex-situ. However, if the sample is treated in oxygen plasma again, the occurring oxidation seems to be more reaction controlled than in the first oxidation cycle ($a/b = 0.3$). The reason for this different mechanism seems to be the larger reaction surface due to the cauliflower structure of the sample once oxidised.

4. Conclusion

The effect of low energy plasma treatment on 20 nm thick Ni-surface layers was studied. Oxidation and reduction reactions within the solid Ni-wall solely caused by reactive plasma

species have been observed. Oxygen plasma treatment results in the formation of small NiO domains on the Ni grains. A cauliflower like topology can be observed.
The kinetics of the oxidation was described by the „shrinking core“ model. A superposition of diffusion controlled and reaction controlled processes has to be considered. The oxidation rate depends on the sample history. While the first slow oxidation process is clearly diffusion controlled, the influence of the chemical reaction is increasing for the next cycle.

References

1. Jacob, W., 1998, *Thin Solid Films,* **324**, 1.
2. Quade, A., Steffen, H., Hippler, R. & Wulff, H., 2002, *Anal. Bioanal. Chem.,* **374**, 720.
3. Itakura, A.N., Shimoda, M. & Kitajma, M., 2003, *Appl. Surf. Sci.,* **216**, 41.
4. Jiang, J.C., Lian, K. & Meletis, E.L., 2002, *Thin Solid Films,* **411**, 203.
5. Niklasson, G.A. & Karmhag, R., 2003, *Surf. Sci.,* **532-535**, 324.
6. Karmhag, R., Niklasson, G. & Nygren, M., 2001, *J. Appl. Phys.,* **89**, 3012.
7. Werner, F., Kozec, D. & Engemann, J., 1998, *Plasma Sources Sci. Technol.,* **3**, 473.
8. Wulff, H. & Steffen, H., 2001, “Characterization of thin films” in: *Low temperature Plasma Physics,* edited by R. Hippler, S. Pfau, M. Schmidt & K.H. Schoenbach, (WILEY-VCH), pp. 253.
9. Sato, T., Sato, S., Tamura, K. & Okuwaki, A., 1992, *Br. Ceram. Trans. J.,* **91,** 117.
10. Ginstling, A.M. & Brounsthein, B.I., 1950, *J. Appl. Chem. USSR (Consultant Bureau Translation),* **23**, 1327.
11. Sato, T., Murakami, T., Ende, T., Shimada, M., Komuya, K., Komatsu, M. & Kameda, T., 1991, *J. Am. Cer. Soc.,* **26**, 1749.
12. Frade, J.R. & Cable, M.J., 1995, *J. Am. Cer. Soc.,* **78**, 90.

Acknowledgements. Financial support for this work was provided by Deutsche Forschungsgemeinschaft (SFB TR24 “Fundamentals of Complex Plasmas”). We gratefully acknowledge Dr. K. Ellmer, HMI Berlin, for preparation of the Ni-films.

Author Index

Bei Fragen zur Produktsicherheit wenden Sie sich bitte an:
If you have any questions regarding product safety,
please contact:

Walter de Gruyter GmbH
Genthiner Straße 13
10785 Berlin
productsafety@degruyterbrill.com